Ontology and the Ambi

T0177435

Many significant problems in metaphysics are tied to ontological questions, but ontology and its relation to larger questions in metaphysics give rise to a series of puzzles that suggest that we don't fully understand what ontology is supposed to do, nor what ambitions metaphysics can have for finding out about what reality is like. Thomas Hofweber aims to solve these puzzles about ontology and consequently to make progress on four metaphysical problems: the philosophy of arithmetic, the metaphysics of ordinary objects, the problem of universals, and the question of whether reality is independent of us. Crucial parts of the proposed solution involve considerations about quantification and its relationship to ontology, the place of reference in natural languages, the possibility of ineffable facts, the extent of empirical evidence in metaphysics, and whether metaphysics can properly be esoteric.

Overall, Hofweber defends a rationalist account of arithmetic, an empiricist picture in the philosophy of ordinary objects, a restricted form of nominalism, and realism about reality, understood as all there is, but idealism about reality, understood as all that is the case. He defends metaphysics as having some questions of fact that are distinctly its own, with a limited form of autonomy from other parts of inquiry, but rejects several metaphysical projects and approaches as being based on a mistake.

Thomas Hofweber is Professor of Philosophy at the University of North Carolina at Chapel Hill. He works in metaphysics, the philosophy of language, and the philosophy of mathematics.

'rich, skilfully crafted, and mandatory for anyone working in (meta-)ontology or nearby areas in the philosophy of mathematics and language . . . It will repay a careful reader's interest many times over.'

Sam Cowling, *Philosophical Quarterly*

'Thomas Hofweber's *Ontology and the Ambitions of Metaphysics* is ambitious, thought provoking, and a good read.'

Karen Bennett, *Philosophy and Phenomenological Research Book Symposium*

'rich and rewarding . . . Hofweber offers a beautiful picture of natural numbers, properties, and facts.'

Thomas Sattig, *Philosophy and Phenomenological Research Book Symposium*

'Thomas Hofweber has written a very rich book.'

Gabriel Uzquiano, *Analysis Book Symposium*

'Thomas Hofweber's book brings together, inter-weaves, and expands upon the author's seminal work on the ontology of numbers—as well as properties, propositions, and ordinary objects—from the past ten or so years. The book is thematically varied and rich in interesting discussions and insights. The writing is lucid, discussions of rival positions are fair, arguments are developed carefully, and crucial junctures are appropriately sign-posted. Thus, the book is both an enjoyable and an instructive read and can be highly recommended to everyone interested in any of the topics covered.'

Robert Schwartzkopff, *Zeitschrift für philosophische Forschung*

Ontology and the Ambitions of Metaphysics

Thomas Hofweber

OXFORD
UNIVERSITY PRESS

OXFORD
UNIVERSITY PRESS

Great Clarendon Street, Oxford, OX2 6DP,
United Kingdom

Oxford University Press is a department of the University of Oxford.
It furthers the University's objective of excellence in research, scholarship,
and education by publishing worldwide. Oxford is a registered trade mark of
Oxford University Press in the UK and in certain other countries

First published 2016
First published in paperback 2018

Published in the United States of America by Oxford University Press
198 Madison Avenue, NewYork, NY 10016, United States of America

British Library Cataloguing in Publication Data
Data available

Library of Congress Cataloging in Publication Data
Data available

ISBN 978-0-19-876983-5 (Hbk.)
ISBN 978-0-19-880223-5 (Pbk.)

for Rebecca

Contents

Preface

This book hopes to make progress on a number of metaphysical problems that are closely tied to ontological questions. The key to making progress in this area, I will argue, is to have a better understanding of ontology itself: what questions it is supposed to answer, how these questions can be answered, who should answer them, and for what the answer matters. Our main focus will be ontological questions about natural numbers, ordinary objects, properties, and propositions, and the corresponding, larger metaphysical debates about the philosophy of arithmetic, the metaphysics of material objects, the problem of universals, and the picture of reality as the totality of facts. An overriding concern related to solving these problems will be what ambitions metaphysics can have in answering questions about what the world is like. In the end I will defend that metaphysics has more work to do than most philosophers think, but less than most metaphysicians think.

In outline, this book proceeds as follows:

Chapter 1 discusses four examples of metaphysical debates that are closely tied to ontological questions. These four examples will be our main examples throughout: natural numbers, ordinary objects, properties, and propositions. It then raises three puzzles about how we should understand these ontological questions and how metaphysical problems could possibly be so closely tied to them. Our overall goal will be to solve the metaphysical problems by understanding the puzzles that ontology gives rise to.

Chapter 2 starts to solve the first puzzle by looking at the relationship between ordinary non-metaphysical statements and apparently equivalent ones that are about numbers, properties, propositions, truth, and other metaphysically loaded things. These pairs of statements play a crucial role in trivial arguments for the existence of numbers, properties, and propositions, and solving the first puzzle is closely tied to understanding these arguments.

Chapter 3 discusses quantification in natural language and how it is tied to ontological questions. I propose, on the basis of considerations about ordinary communication, that quantifiers are polysemous, with one reading that is tied to ontology and another reading that isn't. This second reading is spelled out in more detail and the idea is generalized to generalized quantifiers in an appendix to the chapter. At the end of this chapter we will be in a position to solve the first puzzle.

Chapter 4 outlines a strategy for answering ontological questions. It distinguishes internalism from externalism about a domain of discourse, and spells out why it is crucial to find out which one is correct. This chapter concludes the first part of this book. After having seen what needs to be done to make progress, it is now time to try to do that for our four cases.

Chapter 5 starts with the first case: natural numbers. It investigates what we do when we talk about natural numbers in ordinary conversations as well as in mathematics. A proposal is made about why number words systematically play a double role in natural language and what their semantic function is. Chapter 6 attempts to solve various problems in the philosophy of arithmetic in light of the account of talk about natural numbers developed in chapter 5.

Chapter 7 focuses on the metaphysics of ordinary objects, our second case. It considers what we do when we talk about objects and defends that we have good empirical reasons for an affirmative answer to the ontological question about ordinary objects.

Chapter 8 investigates talk about properties and propositions, our remaining two cases, as well as some general issues in semantics. Chapter 9 deals with the most important objection to the account proposed in chapter 8. Chapter 10 considers whether there could be ineffable facts and how reality, understood as all that is the case, relates to reality, understood as all there is. Chapter 11 attempts to make progress on the problem of universals and on metaphysical questions about properties more generally. This concludes the second part of the book, which deals with ontological questions and the metaphysical problems they are closely tied to. In this part a solution to the second puzzle about ontology will be proposed.

In the third, and final, part we return to ontology and metaphysics more generally. In chapter 12 I try to find out whether ontological questions should properly be seen as metaphysical ones, and how metaphysical questions relate to questions in other parts of inquiry. Here we will see a solution to our third and final puzzle about ontology. Chapter 13 criticizes approaches to metaphysics that hold that metaphysics has a unified subject matter that is, in one form or another, tied to what is fundamental or some kind of metaphysical priority. A brief concluding chapter 14 sums up and looks ahead.

The position defended in this book, and the solutions advocated for the various metaphysical problems, are based on a number of core ideas and theses, which will be more fully articulated, explained, and defended in the chapters to come. They include:

1. *The polysemy of quantifiers.* Quantifiers in natural language have more than one reading, and we have a need for each of them in ordinary everyday communication.
2. *The non-referentiality of certain singular terms.* Expressions that syntactically are like names can have very different semantic functions from names.
3. *Reference free discourse.* Some domains of discourse systematically do not involve any attempts at reference at all.
4. *Rationalist arithmetic.* A rationalist philosophy of arithmetic based on an empirical defense of natural number talk as a reference free discourse.
5. *A world of only objects.* There are material objects, but no properties, no facts, and no natural numbers.

6. *The effability thesis.* Every truth can be stated by us, in our present language.
7. *Conceptual idealism without ontological idealism.* Reality, as the totality of things, is independent of us, but reality, as the totality of facts, is not.
8. *A domain for metaphysics.* Some questions of fact are properly addressed in metaphysics, and some ontological questions are among them.
9. *Autonomy without freedom.* Ontological questions in the domain of metaphysics all have the same answer.

The view defended here draws upon some ideas put forth in various papers I wrote over the last several years. The present book is an attempt to put these ideas together, to develop them properly, to show how they are connected, to work out a coherent picture of metaphysics and the role of ontology in it, and, in particular, to make progress on a series of metaphysical problems. A few chapters are based on previously published papers, but most are new. In particular, chapter 2 is based on Hofweber (2007a), chapter 5 on Hofweber (2005a), chapter 9 on parts of Hofweber (2006a), and parts of chapter 10 overlap with parts of Hofweber (2016a). In each case the material has been updated and revised. The remaining nine chapters are new, although on a few occasions a smaller section overlaps with a section from a previously published paper, and sometimes ideas defended in the new chapters come up here or there in earlier work. In particular, chapter 13 is a much more detailed development of the criticism of esoteric metaphysics outlined in Hofweber (2009a), and chapter 3 does the same for the view of quantification defended in Hofweber (2000b) and Hofweber (2005b). The overall setup is inspired by Hofweber (2005b). Of course, no knowledge of any of these papers is presupposed in the following.

All the readers of drafts of this book agreed on two things: first, it is too long for anyone to be expected to read it all the way through, and, second, I really need to say more about X (although there was a wide disagreement about what X should be). I agree with this as well. I didn't intend this to be a long book. To the contrary, I planned on a short and snappy one, but it didn't turn out that way. It became clear to me that to do this properly I needed to start from scratch and motivate the problems I hope to solve, to more properly develop the aspects of the overall view tied to considerations about language, to include a discussion of all of the four cases that are now included, to deal with the most important objections, and so on. Once the manuscript got sufficiently long, I had to make it even longer to make the individual chapters reasonably self-contained. This adds up to a lot of pages overall, and to many large-scale philosophical debates engaged in a single book. It is natural to think that I am trying to do too many things in one book, but for the overall position to come into view these topics need to be addressed side by side. It is crucial, it seems to me, to discuss all four of our cases of metaphysical problems tied to ontological ones, to set up the general puzzles about ontology, to properly develop the account of quantifiers, the larger picture of metaphysics, and so on. They all need to be in one book for the story to come together. There is a shockingly long list of topics which I had to cut from being discussed here, several of which were nicely written up, or so I thought, but deleted to save space. What remains is what I feel needs to be there. I know it is too long for anyone to be expected to read it all the way through, but I can't make it shorter without losing the big picture or having big gaps in the argument. In fact, I can assure you that if you read it all the way through you will agree with the others who have done so that (a) it is too long, but (b) I really need to say more about X. And as I said, I agree with this as well.

This book does, and I believe needs to, discuss many large-scale philosophical topics together: the nature of metaphysics, ontology, the semantics of quantifiers, focus and syntax, number words in natural language, the philosophy of arithmetic, perceptual beliefs, the problem of universals, idealism, ineffable facts, and several others. Collectively the literature on these topics is beyond vast and it would be impossible to do justice to all of it. This book doesn't aim to do the impossible. It just can't be done to argue in each case why this alternative view is wrong, or why that objection will go nowhere, or how the present position is different from another one. All these are worthwhile things to do, but it being impossible to do them all should not prohibit us from talking about all these topics together. In my discussion of the literature I focused on the cases that struck me as especially relevant and important.

I submitted what I thought was pretty close to a final version of this book to OUP in 2010. Since then it has gone through several rounds of revisions in response to substantial comments from six referees and many other philosophers. During those revisions I changed the overall setup, dropped some chapters, added some new ones, rewrote various parts, while the larger view remained stubbornly the same. I have tried to keep up with the literature that appeared in the meantime and to incorporate it in the text when possible, but often all I could do is mention a major new book or article in a footnote rather than discuss it in the detail it deserves, and I suspect for many I did not even do that. The emphasis of the cited literature is consequently often less on the very latest, but more on what was most vivid when the basic things were put in place.

Acknowledgments

Over the years I received help from many people with this project. Some helped me to get it started and to get it on the right track, some to wrap it up, some saved me from clear mistakes along the way, some had great objections that led to improvements. Some helped me for a long time, some only in a short email exchange or a single conversation. My sincere thanks to all of you! There surely must be others, but for each of the following I can still pinpoint where or how they helped, although I won't try to spell it out case by case. My thanks to Bob Adams, Jody Azzouni, Kent Bach, Brendan Balcerak Jackson, Dorit Bar-On, Jefferson Barlew, Karen Bennett, Johan van Benthem, Jason Bowers, Otávio Bueno, Shamik Dasgupta, Catharine Diehl, Matti Eklund, Anthony Everett, Sol Feferman, Katharina Felka, Hartry Field, Krasi Filcheva, Kit Fine, Peter Godfrey-Smith, Charles Goodman, Tristan Haze, Randy Hendrick, Eli Hirsch, Andreas Kemmerling, Boris Kment, Joshua Knobe, Toni Koch, Dan Korman, Matt Kotzen, Robert Kraut, Phil Kremer, Wolfgang Künne, Marc Lange, Pat Lewtas, Bill Lycan, Matt McGrath, Friederike Moltmann, Ram Neta, Jill North, Jeff Pelletier, John Perry, Jesse Prinz, Michael Raven, Agustín Rayo, Mike Resnik, Tobias Rosefeldt, Richard Samuels, Jonathan Schaffer, Kevin Scharp, Stephen Schiffer, Benjamin Schnieder, Robert Schwartzkopff, Ted Sider, Allan Silverman, Keith Simmons, Peter Simons, Rob Smithson, Rich Thomason, Amie Thomasson, Martin Thomson-Jones, David Velleman, Lisa Vogt, Warren Whipple, Al Wilson, Julia Zakkou, Ed Zalta, and surely several more.

I owe a special debt to six anonymous referees whose comments together were the length of a decent size book and which led to many substantial changes. My thanks to Peter Momtchiloff for finding these terrific referees, somehow persuading them to take on this task, and his long-standing support of this project.

Some parts of this book are based on material published previously, as detailed in the Preface. I would like to thank Duke University Press and Oxford University Press for permission to reuse the material. Thanks to the *Philosophers' Imprint* for not requiring any permission to reuse one's work. My thanks also to those who volunteered their time and skill to give us LATEX and TexShop, which were used to write this book.

Finally, special thanks to Rebecca Walker for lots and lots of all kinds of things. I dedicate this book to you, with gratitude for our life together.

How to Read This Book

Obviously, the proper answer is: cover to cover. It is written with one main line of argument in mind, starting with the motivation of a problem, and ending with a proposed solution, going over four main cases in the middle, each of which has an important role in the overall story. The next best thing would be to read the early chapters 1–4 and then to read the chapters on just one of our four main cases, before returning to the general discussion of metaphysics in chapter 12. Unless you have a particular interest in one of the four cases, I think it might be best to focus on the case of arithmetic and thus chapters 5 and 6, skipping chapters 7–11. Anyone who is not tempted to understand metaphysics as being concerned with fundamental reality or metaphysical grounding can also skip chapter 13, which criticizes these approaches. So, a second best route to the main ideas of this book would be to read chapters 1–6 and then chapter 12. If that is still too much, I would recommend reading a chapter on the topic that interests you most. I put in an effort to make the discussion of the various separate topics reasonably self-contained, and so I hope individual chapters can be read by themselves with some benefit. Thus anyone who would like to focus simply on quantification (chapter 3), the philosophy of arithmetic (chapter 6), the problem of universals (chapter 11), the ineffable (chapter 10), the critique of esoteric metaphysics (chapter 13), or any other chapter, will hopefully find that to be a possibility. A few chapters have one or more appendixes that deal with further issues from that chapter. They are important, but they can be skipped without loss of the overall argument by anyone satisfied enough with the treatment of the issue in that chapter. But before spending too much time on deciding on which parts to read and which parts to skip, let's just start from the beginning and see where it goes from there.

1
Ontology and Metaphysics

1.1 Introduction

The grandest and most ambitious part of philosophy is metaphysics: the project of finding out, within philosophy, what reality is like in general. A central part of metaphysics is ontology: the project of finding out what kinds of things make up reality, what exists, or what there is. Metaphysics could thus roughly be divided into two parts. The first is ontology, which is supposed to tell us what there is in general, or what kinds of things make up reality. The second is the rest of metaphysics, which is supposed to tell us, among others, what these things are like in various general ways. Many of the great philosophical problems are problems in metaphysics. And many of these are closely tied to problems in ontology. We will see some examples shortly.

There also is a long tradition in philosophy to find metaphysics and ontology suspect. Prominent worries are that metaphysical statements are meaningless, or that metaphysical investigations won't lead to knowledge. But such worries are usually based on a very strict requirement for something to be meaningful or to be knowledge, a requirement so strict that most other things wouldn't be meaningful or knowledge either. I don't find much to worry about in these objections, but there is another worry which I think is terribly serious. It casts doubt on metaphysics in general, or at least suggests we don't quite understand what we are supposed to do when we engage in metaphysics. This worry isn't that the question isn't meaningful or that the answer can't be known, but rather the opposite: the questions are perfectly meaningful, but the answers are already known. There is no work for metaphysics to do, the worry goes, since whatever work there is supposed to be done, it has already been done elsewhere. The allegedly deep questions that metaphysics, in particular ontology, hoped to answer have long been answered and their depth was an illusion. The depth was an illusion not because the questions are somehow defective, or in some way relative, but because they are perfectly non-defective and objective questions that have long been answered rather trivially in other parts of inquiry. A prime example of this situation, to be discussed in detail shortly, is the ontology of mathematical objects. The metaphysician seems to want to know whether there are any mathematical objects: things like numbers. But the mathematician has proven long ago that there are infinitely many prime numbers, and thus that there are numbers.

This worry, once properly spelled out, suggests that the project of metaphysics and ontology isn't as well defined as one should hope. In light of this worry one might be tempted to give up on metaphysics, but this would be premature, since there are two promising ways to defend it. One way to it is to hold that metaphysics has a distinct and unified subject matter that is sufficiently separate from other parts of inquiry. Metaphysics is not supposed to find out what reality is like in general after all, but instead what reality is *ultimately* or *fundamentally* like. And these notions are to be taken in a metaphysical sense, which makes sure that those questions aren't trivially answered in other parts of inquiry. A second way to defend metaphysics is to hold that once we pay close attention to what we try to do when we do ontology and metaphysics, and how we articulate the questions we hope to ask, then we can see that these questions are neither trivial nor already answered. This second approach doesn't require a distinct subject matter for metaphysics, but instead a careful study of the language we employ and the goals we have when we ask the relevant questions. Both approaches defend metaphysics, but they do so in different ways. Both hold that metaphysics can have real ambitions to contribute to inquiry, but their ambitions will be different. Whichever defense one adopts will lead to rather different versions of what the discipline should aim to do, and how its problems are to be solved. We will discuss both approaches in detail, but I will defend the second one in this book.

Thinking about metaphysics as a discipline is often an anti-metaphysical activity. Many philosophers who wondered about metaphysics as a whole did so in order to reject metaphysics, with the logical positivists being the prime example. But this does not have to be so. There are real puzzles about metaphysics and ontology, and their discussion needn't be tied to attempts to reject metaphysics. Those fond of metaphysics often react to questions about metaphysics in general negatively, taking them to be an attempt to demand some outside justification for a practice that is doing well by its own standards. Their motto is: don't ask what metaphysics is supposed to do, or how it is possible to do that, just do metaphysics. But both of those attitudes are mistaken. Thinking about problems for ontology and metaphysics in general is not anti-metaphysical, but instead the key for answering some of the most important metaphysical and ontological questions, or so I hope to argue in this book. The goal in the following will be to show that we can make progress on a number of significant metaphysical problems that are tied to ontology by first thinking about metaphysics and ontology in general. I will argue that a crucial obstacle to finding a solution to these problems has been a mistaken conception about what ontology is supposed to do and how it is supposed to be done. Once we correct this error we will be able to see the answer to several ontological questions, and consequently to a series of larger metaphysical questions.

Throughout this book we will focus on four ontological problems which are closely tied to four large-scale metaphysical debates. These problems are our focus because of their significance within metaphysics and because they bring out a variety of different ways a metaphysical problem can turn out. As we will see, two of them are similar to each other, but the other two are different from all the others. What we will conclude about these four ontological problems in particular can't be expected to carry over to any other ontological problems, but what we will conclude about ontology applies to ontology in general. But before we can get there we have to briefly reintroduce the ontological problems we will focus on, as well as the metaphysical questions they are tied to.

1.2 Four Problems in Ontology

Often, but not always, metaphysical problems are closely tied to problems in ontology. In this section we will consider four cases where this seems to be the case. I will motivate the metaphysical problems and their connection to ontological questions from scratch in a way that is quite common, but that we will need to critically evaluate in the following. Some of this might seem a bit elementary, but bear with me. I will argue below that although some of this is exactly right, some of it is completely wrong.

1.2.1 Numbers

A good example of the connection of large-scale philosophical questions to issues in ontology is the philosophy of mathematics. Mathematics is philosophically puzzling in a number of ways. First of all, it is a discipline that is highly objective. The results it achieves should be accepted by everyone, no matter what one's cultural background, preference, or opinions. Mathematics is not alone in this regard, of course. Other disciplines are equally objective. The natural sciences, which try to find out various things about the natural world, often have that feature as well. And this suggests that mathematics might be objective for a similar reason that the natural sciences are objective. The latter aim to correctly describe a certain aspect of the natural world, which in turn exists independently of us and our opinions. In this sense particle physics aims to correctly describe subatomic particles, biology aims to describe living things, and so on. These are all objective disciplines since the objects they aim to describe, as well as their features, exist independently of us and our opinions. But if mathematical objectivity is understood along these lines, then what part of reality does it aim to describe? It is tempting to think that it aims to describe its own, distinct part of reality: mathematical objects. These objects would likely be neither in space nor time, and quite different from ordinary objects we normally interact with. But if they exist independently of us and mathematics aims to describe them, then no wonder mathematics is so objective.

On the other hand, the method with which mathematics achieves its results seems to be in prima facie tension with it having mathematical objects as its subject matter. Mathematics achieves results paradigmatically with proofs, and these proofs can in general be carried out with one's eyes and ears closed, by just thinking. But if mathematics is about some objects that exist independently of us, how can we find out about them by just thinking? There might thus be a better way to think of the source of mathematical objectivity, one that can live with there being no mathematical objects, and the following is one way this might go. When we carry out the proofs we all rely on certain assumptions, certain principles, or axioms, at least implicitly, that characterize the subject matter about which we hope to carry out proofs. But wouldn't those assumptions alone be enough to account for the objectivity of mathematics? Wouldn't it be enough if we all shared the assumptions that are the starting points of our proofs? If the objectivity can come from our shared assumptions, then the mathematical objects might simply fall out of the picture. Mathematics thus might not aim to describe a certain part of reality after all, but do something else. And whatever else that is more precisely, it might make clear that thinking alone is enough to achieve mathematical results, that mathematics is fully objective, while nonetheless there are no mathematical objects that mathematics aims to describe.

These are two large-scale views in the philosophy of mathematics that are closely tied to the ontological question of whether or not there are any mathematical objects in the first place. If reality contains some special part of mathematical objects, then this is what mathematics will have to capture correctly in order for its results to be true. But if there are no such objects, then this is not how mathematics and the source of its objectivity are to be understood. A crucial dividing line in the philosophy of mathematics is thus what one says in answer to the question

(1)   Are there mathematical objects?

If one says "yes" one will be in one camp, if one says "no" one will be in a different one.

In the following, for reasons that will become clear later, we will in particular focus on the philosophy of arithmetic, the mathematical discipline that deals with the natural numbers. Here the general questions are just the same, but the mathematical objects, if there are any, that arithmetic would be about are special objects: natural numbers. A crucial question for the philosophy of arithmetic is thus the question

(2)   Are there natural numbers?

However one answers this question will put one in one of two camps about how to understand the philosophy of arithmetic. Progress in the philosophy of arithmetic is thus tied to finding the answer to this ontological question.

1.2.2 Ordinary Objects

The objects we interact with most in everyday life are artifacts like tables and chairs, biological organisms like human beings and trees, and inanimate natural objects like rocks. All these are midsize objects, which can in principle be divided up into smaller pieces from which they are made up or composed. Even though we think of the world as containing such objects, there are a number of worries about whether this is indeed correct. One of them is tied to how the midsize objects relate to the things from which they are built up. All these objects are made up from smaller parts like atoms. But do we really have good reason to think that there are these objects besides the atoms, or is all there is just the atoms arranged in certain ways? The main alternative to the standard view that there are ordinary objects is some version of atomism, that is, a view that holds that there are just the simplest things, but nothing that is built up from them, in particular no tables. There are some good arguments that suggest a form of atomism is true: doesn't science support that all there is is the subatomic, physical world? Don't the atoms do all the work when it comes to causing things, including our experiences, and so anything in addition to them, like tables, are just extra baggage? Isn't it somewhat arbitrary and mainly due to us which collections of atoms we call "an object" and thus put together into a unit? Doesn't thinking about the world in terms of ordinary objects reflect more on our thinking than on how the world actually is? Whether the atomist or the ordinary conception of the material world is correct seems to be closely tied to an ontological question, the question

(3)   Are there (ordinary) objects?

If the answer is "yes", then one large-scale picture of the material world is correct; if the answer is "no", then a very different one is correct.

1.2.3 Properties

The world, as we commonly think of it, contains at least ordinary objects. Whether this conception of the world is accurate in the end is up for dispute, as we just saw above, but even if it is correct there is a further question about whether or not that is all there is. Are there just particular things, or is there more? One of the most compelling arguments for there being more is simply the following argument. The particular things are a certain way. Some are red, some are green. Those that are red have something in common: being red. So, there is something all red things have in common. But what is that thing? Not another particular red thing, which has a particular location, but something different from and separate from all the red things. It is a property or universal. It is something that all the red things share, that they partake in, or somehow relate to. Thus reality contains more than just particulars, it also contains universals which those particulars share, or so the argument goes.

But if there really are properties or universals beside individuals, then where are they and how do they relate to the individuals that have them? To say how this relationship is supposed to go has been notoriously difficult for those who believe in such universals. If the universal is where the thing is that has it, then it seems that universals are at more than one place at a time. If the universal is somewhere else, maybe outside of space and time, then it seems to leave the object by itself as a completely featureless thing or bare particular. And it leaves the question open how these bare particulars relate to the universals which are somewhere totally different. Maybe then the world consists just of individuals, and not also properties that they have in common. But does that mean that individuals have nothing in common? Do red roses not all have one thing in common: being red, that is, the property or universal of being red?

This group of questions is one of the oldest and most widely discussed problems of all of metaphysics: *the problem of universals*. There are subtly different versions of what the problem is more precisely, but in a nutshell it is the problem of saying (a) whether there are besides individuals also other things which are the properties that they have, and (b) if so, what these properties are, where they are, and how they relate to the individuals that have them.

The problem of universals is related to another one, which seems even more important. It is the problem of whether there is an objectively best description of the world in the following sense. We describe the world with sentences that are mostly of a subject–predicate form. In the simplest case the subject term picks out an individual that we are making a claim about, and the predicate specifies what we claim of the individual. So, "Fido is a dog" says something of Fido, namely that he is a dog. Predicates group things together. There are many other things besides Fido of which it can be truly said as well that they are a dog. But among all the different ways in which things can be grouped together by the predicates we use in our language, are there some that are objectively better than others, ones that the world itself suggests, not simply that we prefer? This is, of course, a little metaphorical, but one could see how it might be made sense of by talking about properties. If there are properties, if the world contains them beside the individuals that have them, then it might seem that the world recommends our grouping things together with predicates the same way it groups them together with properties. If there is a property of being a dog, one that groups all and only the dogs together, then it might be a good idea to have a predicate in our language that expresses just that property, and that groups things together in the same way. In particular, maybe only some predicates express properties, which would make these predicates special. With predicates we can group things together any way we want, but properties are how the world itself groups things together. But if there are no properties, if the world contains just individuals, then it would seem that it is up to us, and us alone, to group things together. The world remains silent on which things go together, we can take our pick.

Whether the world we live in is a world of only individuals and whether there is a distinguished description of it are questions that are closely related to the question of whether or not there are any properties besides the individuals. One crucial dividing line in understanding the world of individuals is thus what the answer is to this question:

(4)   Are there properties?

This is a question in ontology and one of central importance for metaphysics.

1.2.4 Propositions

It is natural to hold that thoughts involve at least two things. One is the person, who is thinking the thought. The other is the content of the thought, that is, what one is thinking. When I think that Obama is the president then I am the thinker, and the content of the thought is that Obama is the president. These contents are generally also called propositions. Thus when I think that Obama is the president I have a thought that has the proposition that Obama is the president as its content. Contents or propositions give rise to two related philosophical debates.

First there is the question of whether thinking indeed involves relating to some thing which is the content of one's thought, and if so, how we mange to relate to such things as propositions. Propositions presumably are not ordinary objects, so the question remains how I manage to relate to this one when I have one thought and that other one when I have a different thought. More importantly, one might think that this picture of the contents of thoughts as relating to propositions is mistaken. An alternative would be to hold that our thoughts have contents, but we do not relate to things which are contents. So, when I think that Obama is the president I relate only to Obama, and think of him that he is the president, not to some other thing which is a content. On such a picture there would be no such things as contents to which we relate, even though our thoughts have contents in the sense that they are contentful.

A second issue connected to propositions is not directly about the contents of thoughts, but "the propositional" more broadly. The proposition that Obama is the president is closely tied to the truth, or the fact, that Obama is the president. If there are such things as propositions then it is not unreasonable to think that there are such things as truths or facts. The world then would not simply contain objects and possibly also properties, but in addition also facts or truths. And if the world contains such things then this might suggest a certain form of realism: what can in principle be truly said about the world is already contained in the world itself, as one of its building blocks. To say something truly is just to say something which is a truth, and what truths there are is something that is a part of what there is in general. The truths are all there already, so to speak, and we can only hope to capture as many of them as possible. But if the world contains only material objects, say, then there are no such things as propositions, and no such things as facts or truths. We will still be able to think truly about the world, and say true things about it, but the world itself does not contain any truths or facts. And this suggests a rather different picture of what we do when we aim to describe the world. This contrast might be made clear when we think about whether it could in principle be that there are some truths that creatures like us could never say or think. If facts or truths exist as part of the world, then they are just there, waiting to be thought. It might well be that we can get all of them in the end, but since they are just there, as an independent part of the world, it would be reasonable to expect that we can not get them all. Some might well be completely beyond us, that is beyond what creatures like us can think or say. On the other hand, if the world does not contain any such facts or truths or propositions, then this might be very different. What can in principle be thought is not directly tied to what propositions are part of the world, and so it might be tied to us in a way that we can think everything there is to think.

The questions how we manage to have thoughts with contents and whether we can think all there is to think are tied to an ontological question:

(5)   Are there propositions?

Whatever one says in answer to this question will lead to different views of the contents of thought and its limits.

All these large-scale metaphysical issues discussed over the last couple of pages thus seem to be closely tied to ontological questions. To solve these problems seems to require an answer to the ontological questions. We will need to see in the following whether this is correct. That is, we will need to see whether the ontological questions indeed do have a central role in these debates, what this role might be, and, in particular, how we should understand the ontological questions more precisely. As we will see shortly, to give ontological questions a central role in metaphysics is problematic, as are the ontological questions in the first place. To understand the metaphysical problems outlined better we will need to understand ontology better. And once we do that we will be in a position to answer these metaphysical questions, or so I hope to show. But first we need to look at ontology more closely.

1.3   Two Kinds of Ontological Questions

Although the precise statement of the questions that ontology is trying to answer is not clear, and subject to controversy, we can still approach it with a first stab as follows. Ontology first and foremost tries to find out what there is, what the stuff is that reality contains. But things don't end there. Whenever one holds that reality contains certain things, a number of further questions immediately arise. The problem of universals is a good example, but others would do just as well. If one holds that there are universals, then the question suggests itself what they are like, in general. Are they concrete or abstract, are they everywhere their instances are or nowhere at all, and so on. These kinds of questions are generally taken to be part of ontology a bit more broadly conceived, and to respect this it is useful to distinguish two kinds of questions that are commonly asked in ontology. First there is the question about whether there are the relevant things at all, whether reality contains any of them. Then, assuming the answer to this question is affirmative, there is the question of what these things are like in the most general way. Some problems in ontology focus more on the first of these questions, some focus more on the second. We can thus distinguish the *primary ontological question* about some kind of thing F, from the *secondary ontological question*. For both of them there will be a real issue about how they should properly be expressed, something we will discuss shortly. But a natural first try is simply this: the primary ontological question about Fs is the question of whether there are any.

(POQ)   *The primary ontological question* about Fs: are there Fs?

If the answer is "no" then this would seem to settle the issue and no further questions about Fs would follow. But if the answer is "yes" then it seems natural to follow up with a request for more information about what these things are in general like.

(SOQ)   *The secondary ontological question* about Fs: what are Fs in general like?

Some philosophical problems in ontology are primary ontological questions, some are secondary ontological questions. Sometimes the two questions get mixed together. The problem of universals is often stated as the problem of saying whether there are any universals, and if so how they relate to the particulars that have them. Much of the work that is actually carried out is usually in the part that deals with the secondary ontological question, in part to justify the answer one gives about the primary ontological question. So, if one says "yes" to the primary question, then one better have some answer to the secondary ontological question. And even though it is easy to say "yes" to the primary ontological question, it turns out to be hard to answer the secondary ontological question in a satisfactory way. Sometimes the secondary ontological question comes first in the order of inquiry. We might first determine what ordinary objects would have to be like, if there were any, and then use this characterization to argue that there aren't any, since nothing is or can be like that. Secondary ontological questions also overlap metaphysics more broadly understood, in the way that goes beyond ontology. Here there are not clear dividing lines, and it is simply a matter of from which angle one approaches a set of issues. Still, we will in the following at first focus mostly on the primary ontological questions, and only return to the secondary ones a little later. It will turn out that a proper understanding of the primary ontological question is what is central for the answer to both and for metaphysics more broadly.

1.4 Three Puzzles About Ontology

1.4.1 Is Philosophical Ontology an Incoherent Project?

That something is not quite right with ontology as we normally think of it can be seen quite easily. Consider again the question about whether there are any mathematical objects, and its relationship to the philosophy of mathematics. The project of a philosophy of mathematics was intended to be a philosophical investigation of mathematics from the outside. We philosophers want to understand mathematics from a philosophical point of view. We take a look at mathematics with all its great achievements and results, accept them just like any insider would, but wonder about mathematics. We don't want to change it or to challenge its achievements, but to understand it. And when we try to do this we think that our understanding of mathematics is tied to what we should think the answer to certain questions is, including the question of whether or not mathematics aims to describe an independently existing domain of things—the mathematical objects—which are the source of mathematical objectivity, the subject matter of mathematics, and so on. And thus a crucial fork in the road is what one says is the answer to the question of whether or not there are any mathematical objects. And that question is intended to be part of the philosophical project of trying to understand mathematics, to see what it looks like from a philosophical point of view.

But this project seems to be completely incoherent. One can't coherently accept the results of mathematics, but still wonder whether there are mathematical objects. Mathematical objects are paradigmatically things like numbers, functions, or sets. But the results of mathematics themselves imply that there are mathematical objects such as numbers, functions, or sets. Take Euclid's Theorem, which was proven about 2,500 years ago and has been beyond question ever since. It states that

(6)   There are infinitely many prime numbers.

This immediately implies that there are infinitely many numbers, and thus that there are numbers. But how then can we ask whether there are mathematical objects, things like numbers, without thereby also questioning the results of mathematics? It just seems incoherent to have this philosophical stance, and to engage in the project of trying to understand mathematics from the outside by asking such questions.

But on the other hand it seems perfectly coherent, and philosophers have been doing it for quite a while with a strong sense of a meaningful project. They want to know whether mathematics aims to describe a special part of reality: the mathematical objects. And this is tied to the question of whether or not there are such objects in the first place, whether or not reality contains besides the ordinary stuff in space and time also other things that mathematics is about, things like numbers. What should we think about this situation?

The situation is actually even worse for the philosophical project than I just made it out to be. If we wonder whether or not there are any natural numbers, we don't have to rely on mathematics to conclude that there are such numbers. We can conclude this already from the fact that I have two hands, and thus the number of my hands is two, and thus there is at least one number: the number of my hands, i.e. two. Can I really answer the ontological question about mathematical objects by drawing a couple of trivial inferences from the fact that I have two hands? Even worse still, can't I simply conclude that there are numbers by giving examples of them: the number two is one, the number three is another?

Ontology was intended to be part of a metaphysical project that asks a further question, one that goes beyond the questions that are addressed, and often answered, within a discipline or domain of discourse that the metaphysician tries to theorize about. Metaphysics so understood was ambitious in that it had further facts to uncover than those that were already investigated in the discipline it was trying to understand. And these further facts were further in the sense that those from within the discipline did not themselves settle the corresponding questions. There was really more work to do, and ambitious metaphysics set out to do it. But how the ontological questions can be seen as ambitious metaphysics in this sense is just what seems to be incoherent. The question we philosophers want to answer is apparently already answered within mathematics itself. Whether the philosophical project of ontology is coherent, what question it is supposed to answer, and whether these questions can play an important role in the philosophy of mathematics and metaphysics in general is what we hope to make progress on in this book.

There are really several different aspects of this worry. They concern both the philosophical project of asking ontological questions, as well as giving the answer to these questions a significant role in one's philosophical account of a certain discipline or domain of discourse. There are issues about whether we expressed the question we wanted to ask properly, how it could be that this question is answered the way it seems to be answered, what the answer matters for, and why any of this should be seen as answering a philosophical question. It is good to bring out the different aspects of this issue more clearly. I will in the following isolate three puzzles that, as I will argue, are especially important. These puzzles will be with us for most of this book, and to solve them will require quite a bit of work in each case. The three puzzles are in essence three different aspects of the worry outlined above. Not all of them apply equally to all ontological questions, but the fact that each one of them applies to some important ontological questions should give rise to the worry that we don't understand ontology until we understand each of them better.

1.4.2 The First Puzzle: How Hard?

Ontological questions are naturally stated as questions about what stuff reality contains, what exists, or what there is. Among the important ontological questions are those about numbers, properties, and propositions, which are central to a variety of philosophical debates. The first puzzle arises when we ask

(Q1)   How hard is it to answer ontological questions?

in particular the ones considering these three cases. This question has two apparently equally good, but contrary answers.

ANSWER 1: VERY HARD

Ontological questions are questions about what reality contains. There is no easy way to figure this out. In fact, it might seem too hard for philosophy even to attempt. After all, some of these things might be abstract and outside of space and time. How could philosophy hope to establish that there are such things? Maybe one can argue that there are such things if the best theory of the world as a whole requires there to be such things. But figuring that out is at least as hard as figuring out what the best theory of the world as a whole is. And most of that won't be work for philosophy to do. So, answering ontological questions is very hard.

ANSWER 2: TRIVIAL

The questions of whether or not there are numbers, properties, or propositions can all be answered trivially in the affirmative. There are trivial arguments that establish that there are numbers, properties, and propositions, and these arguments use only completely uncontroversial premises, and just a couple of easy steps of inference. Here are some examples:

(7) a. Jupiter has four moons.[1]
     b. Thus: the number of moons of Jupiter is four.
     c. Thus: there is a number which is the number of moons of Jupiter, namely four.
     d. Thus: there are numbers, among them the number four.

(8) a. Fido is a dog.
     b. Thus: it's true that Fido is a dog.
     c. Thus: something is true, namely that Fido is a dog.
     d. Thus: there are propositions, among them the proposition that Fido is a dog.

(9) a. Fido is a dog.
     b. Thus: Fido has the property, or feature, of being a dog.
     c. Thus: there is a property Fido has, namely being a dog.
     d. Thus: there are properties, among them the property of being a dog.

Of course, each of these arguments can be disputed. Each of them involves several steps where a philosopher can, and has, come along and deny that it is a valid inference. But let's be honest: are there really any easier and more compelling arguments anywhere in philosophy? Is there really anything we can establish more easily? The questions of whether there are numbers, properties, or propositions are trivial, the answer is obviously: yes. And that this is so can be seen, among other ways, by the above trivial arguments. Nothing could be easier than to see that there are numbers, properties, and propositions. These trivial arguments don't carry over to all ontological questions. For example, there seems to be no analogous argument for there being ordinary objects, and we should try to understand where this difference comes from. But that there are such trivial arguments for answers to some ontological questions is enough to give rise to a puzzle about ontology.

[1] Jupiter in fact has more than four moons. Nonetheless, I will use this example, as do many others, since it is a classic example of Frege's, from his Frege (1884).

UPSHOT OF THE FIRST PUZZLE

The first puzzle highlights an issue that often arises in ontological debates. On the one hand, ontological debates seem to be deep and substantial debates. Whether or not there are mathematical objects such as numbers seems to be a deep philosophical question connected to the nature of mathematics and the source of its objectivity. But on the other hand, it is completely trivial that there are numbers. Of course there are numbers. The number two is one, and the number three is another. And similarly for properties. On the one hand, the question of whether or not there are properties is a deep philosophical question about the nature of individuals and their features: how they go together and what the world is made from more generally. But on the other hand, it is a completely trivial question. Of course there are properties. Being a cat is one, and being a dog is another. What should we take away from this?

One lesson we might take away from all this is that something is wrong with the trivial arguments. We will investigate in some detail in chapters 2 and 3 whether this is correct. But if the arguments are good, what else could the lesson be? Besides worrying about the trivial arguments, there are two other natural reactions. First we might think that the fact that it was so trivial to answer the question shows that the question is not fully factive, in some sense to be spelled out. We can't find anything out about reality that easily, and so what we found out when we answered the question trivially was something less factive or objective than traditionally thought. Such a line will lead to a larger anti-realism about ontology, in particular in cases like the ones discussed here that seem to allow for a trivial answer to the ontological questions. How such an anti-realism could be understood more precisely, and whether it is even coherent, is a further question, but the trivial arguments can certainly be seen to motivate such a position.[2]

A second reaction is to wonder whether we articulated the ontological questions properly. Above we articulated what we wanted to ask, the ontological question about Fs, with

(10)   Are there Fs?

But maybe that isn't right. Maybe what we are trying to find out when we do ontology and metaphysics is not whether there are such things, but something else? Maybe we were misled into expressing the question we want to ask in a certain colloquial way, and that misarticulated question was trivially answered. But how else should we express the question we want to ask? Most naturally, we should look back at how ontology was motivated. Here we used expressions like "make up reality," or "exists," and thus maybe we should instead try to ask the question as follows:

(11)   Are numbers part of what makes up reality?

---

[2] See Chalmers (2009) or Kraut (2016) for such broadly anti-realist positions.

or

(12)   Do numbers exist?

or maybe simply

(13)   Are there *really* numbers?

But it is not so clear how much progress that would be. The first question seems almost a bit strange (what else would they be part of if there are any in the first place?), and the second and third are not clearly different from what we had before. How could it be that there are numbers, but they don't exist, or that there are numbers, but there aren't really numbers? It is not clear that by putting things this way we get any more substantial questions. Furthermore, couldn't we have stated Euclid's Theorem above just as well as the result that infinitely many prime numbers exist? Such attempts at a reformulation of the question might be on to something, but the above proposals so far won't be the solution. What we can conclude, however, is this: philosophers try to ask certain questions that naturally get grouped together in the discipline we call "ontology." We can call these questions *ontological questions*. Ontological questions generally are directed at certain cases, like numbers, objects, properties, or the like. We can call them the *ontological question about numbers*, the *ontological question about objects*, and so on. What we have to find out is thus this: what are the ontological questions (about numbers/properties, etc.)? That is to say, how should we properly express these questions that we want to ask? To make progress on this we should in particular see whether or not the trivial arguments are indeed valid. If they are not valid, then maybe the question we wanted to ask is properly stated as (10). But if they are valid, then it would seem that either ontology is trivial (for these cases), since (10) is trivially answered, or the question has to be stated in some other non-trivial way, or some form of anti-realism is true about ontology. To solve the first puzzle requires in part to find out what question we are trying to answer when we do ontology.

1.4.3 The Second Puzzle: How Important?

Our second puzzle about ontology arises from there being two apparently equally good, but contrary, answers to the question of what hangs on the answer to the ontological questions turning out one way or another. There are two apparently equally good, but contrary, answers to the question:

(Q2)   How important are ontological questions?

That is, how important is it, not just for philosophy or metaphysics, but in the larger scheme of things, for what the answers to these ontological questions turn out to be.

ANSWER 1: NOT VERY IMPORTANT

Although ontology is a central part of metaphysics, and metaphysics is a central part of philosophy, we should admit that in the larger scheme of things philosophy is not the most important part of inquiry. Whatever one in the end thinks of the value of a moral philosophy, or political philosophy, metaphysics won't have that importance, and ontology certainly will be even less important. Even those among us who work in metaphysics see some justification for the lack of multi-million dollar research funding for our field, even though such sums are frequently given to such projects as finding out how viruses attack cells, what subatomic particles there are, the Hubble Space Telescope, or even finding out how to factor large numbers. Philosophical questions might be questions that force themselves on to all human beings, but pursuing them is a luxury that we are fortunate enough to be able to afford. But like all luxuries, they are not very important.

ANSWER 2: VERY IMPORTANT

In the introduction to his book *Psychosemantics* Jerry Fodor discusses whether or not belief–desire psychology, i.e. the explanation of our behavior in terms of our beliefs and desires, is correct. He writes:

> The main moral is supposed to be that we have, as things now stand, no decisive reason to doubt that very many common sense belief/desire explanations are—literally—true. Which is just as well, because if common sense intentional psychology were to collapse that would be, beyond comparison, the greatest intellectual catastrophe in the history of our species; if we're that wrong about the mind, then that's the wrongest that we've ever been about anything. The collapse of the supernatural, for example, didn't compare; theism never came close to being as intimately involved in our thought and our practice—especially our practice—as belief/desire explanation is. [. . .] We'll be in deep, deep trouble if we have to give it up. Fodor (1987: xii)

Fodor has a point here. If it turns out that we don't have beliefs and desires, and thus the proper explanation of what we do does not involve them, then this would shake our conception of ourselves in a profound way. It really would be an intellectual disaster. But what if it turns out, in ontology, that there are no propositions? Then there are no such things as contents. In particular, there is no such content as the content that snow is white. But if there is no such content, how can anyone have a thought with the content that snow is white? You can't, literally, have a thought for which its content is missing. But all beliefs and desires, the building blocks of belief–desire psychology, have to have contents. If there are no propositions, then there are no contents, and thus there are no beliefs and desires. And so belief–desire psychology cannot be true. If ontology goes one way, the greatest intellectual catastrophe of our species is right around the corner.

Motivating the threat of disaster this way is, of course, not philosophically innocent. For example, there are attempts to separate the ascription of beliefs and desires with taking them to be relations to contents or propositions.[3] But the threat of a disaster is quite real since such approaches are generally considered to be a minority view,

---

[3] Some examples of such theories are the syntactic relational theory of Davidson (1968) and the multiple relation theory of Russell (1913) and Moltmann (2003b).

one which might well turn out to be incorrect. If belief relates us to propositions, as it seems to, then the negative answer to an ontological question leads to the greatest intellectual catastrophe in human history, as Fodor put it. Ontological questions can thus be extremely important.

Things are even worse when it comes to ordinary objects. If it turns out that there are no ordinary objects, then not only are there no tables and chairs, there are none of us either. If all there is is just atoms in the void then we don't exist, since we are not atoms. Even though the non-existence of propositions might lead to the greatest intellectual catastrophe in human history, the non-existence of ordinary objects would be worse, since it guarantees that there are no humans and thus there is little human history. And similarly, although less dramatically, for numbers. If it turns out in ontology that there are no numbers after all, then there are no prime numbers, and no even numbers, and no odd numbers, and so on. And if there are no prime numbers, then it is false that there are infinitely many prime numbers, something Euclid thought he established with his famous proof. And similarly almost all of mathematics is false. And if mathematics is false, then so is a lot of physics and economics and biology, since they all rely on mathematics. Thus if the ontological question about numbers goes one way, then everything else goes down the drain. The ontological question is thus of unparalleled significance. Everything depends on it.

UPSHOT OF THE SECOND PUZZLE

The second puzzle raises the issue of whether the truth of ordinary and scientific statements depends on the ontological questions having a certain answer. It seems that the truth of belief–desire psychology requires a certain position in ontology: the view that propositions exist. Without propositions belief–desire psychology can't be true, since it depends on the truth of belief and desire ascriptions, which in turn seems to depend on the existence of propositions. And so a certain ontological position seems to be required for it to be true. Similarly for mathematics, and the mathematical sciences. If there are no mathematical objects, then there are no numbers, and thus mathematics can't be true. And neither can be any of the other disciplines that rely on mathematics. So, it seems that all these sciences depend on the ontological question to turn out one way rather than another.

The second puzzle raises the question of whether this is indeed true. Does our ordinary, everyday talk as well as our more systematic theorizing require the truth of a certain position in ontology? On the one hand, it is tempting to think that ontological questions are somewhat independent from our ordinary way of describing the world. It is tempting to think that we can accept all of mathematics and still have the ontological questions about mathematical objects left open. The ontological question can plausibly be seen as not being settled within mathematics itself as it was intended. Whether this intention pans out remains to be seen.

How then should we understand the relationship between our everyday and scientific ways of describing the world and the philosophical project of ontology? To find out more about this we need to solve the second puzzle.

The second puzzle is also closely related to the question of the source of the objectivity in a particular domain of discourse. If arithmetic, for example, is about an ontology of mathematical objects, the numbers, then they could be the source of mathematical objectivity. They might exist independently of us, and adjudicate impartially all mathematical disputes. But if our arithmetical talk in everyday life as well as mathematics does not require an ontology of numbers, and if its truth is independent of their being such an ontology, then it won't be the source of arithmetical objectivity. To solve the second puzzle requires us to also make headway on this issue.

1.4.4 The Third Puzzle: How Philosophical?

Finally, there is a third puzzle, which is not unrelated to the other two, of course, but brings out another aspect of how we should understand ontology as a philosopher's project, one that in a sense is the hardest one, and that we will only be able to resolve towards the end of this book. This puzzle arises from there being two apparently equally good, but contrary, answers to the question:

(Q3)   How philosophical are ontological questions?

ANSWER 1: VERY PHILOSOPHICAL

Ontology is a central part of metaphysics, which is paradigmatically a philosophical discipline. Although the question of whether there are any electrons might well not be one for philosophy to figure out, nonetheless settling the ontological questions about numbers, properties, or propositions one way or another is to take sides in a long-standing philosophical debate. These questions are closely tied to traditional philosophical questions like the problem of universals, the source of mathematical objectivity, the limits of thought, and so on. Solving them would be real progress in philosophy. And the best attempts to solve them have been in philosophy. The existence of universals, say, has been debated in philosophy for a long time, and isn't debated anywhere else. These questions are thus philosophical through and through.

ANSWER 2: NOT AT ALL

The question of whether or not there are numbers is easily settled in mathematics, not philosophy. An answer to this question is implied by, among any others, what is widely regarded as one of the most securely established results in all of inquiry: that there are infinitely many prime numbers. This implication is equally beyond question, and thus that there are numbers, even infinitely many of them, has been established long ago by mathematical means. It is a result of mathematics, not a question for philosophy.

And similarly for the other questions. Materials science has established, by scientific means, that there are some features, or properties, of alloys that make these metals more resistant to corrosion, but more susceptible to fracture. What these features, or properties, are is known in this field. But there being such properties of alloys implies that there are properties. Thus that there are properties has been established in materials science, as well as basically any other scientific discipline. It is a result of science, not philosophy. And similarly for there being ordinary objects, like bars of metal. The question of whether or not there are numbers, objects, or properties is not a philosophical one at all. It is one that has long been settled in the sciences.

UPSHOT OF THE THIRD PUZZLE

The third puzzle makes vivid that there is a real issue about why the ontological questions we are trying to ask in philosophy aren't already answered elsewhere. If the question just is "Are there numbers?" then why isn't this question already answered in mathematics? And since mathematics establishes its results with close to certainty, why would anyone think there is still an open question about the ontology of numbers? Are philosophers so confused that they are trying to answer questions whose answers are immediately implied by the most securely established mathematical results? Although it might be tempting to some to take this line, it doesn't quite capture how the question about the ontology of numbers was intended. And it doesn't resolve whether or not this intention gets at something. What we wanted to ask when we ask about an ontology of numbers was supposed to be a question from the outside, about mathematics as a whole. The question we intended to ask was not intended to be answered within mathematics. It was intended to be a question that is still open even after we accept all of mathematics as it is. But you can't always get what you want and we have reason to think that this intention is illusionary. The question "Are there any mathematical objects?" seems to be settled within mathematics, since it establishes that there are numbers, and numbers are mathematical objects. Ontology and metaphysics were intended to be ambitious in that they have questions of fact left over for themselves, ones that aren't answered in mathematics or elsewhere. Whether ontology and metaphysics, seen as philosophical projects, can live up to this ambition is the issue raised by the third puzzle.

1.5 Towards a Solution

To make progress on all this we need to solve the three puzzles ontology gives rise to. But it won't simply be enough to take sides on the puzzles. It isn't enough to insist, for example, that ontology is trivial, not important, and not philosophical, even if this were true. What we need to understand is how these puzzles arise and how ontology naturally gives rise to this puzzling situation. After all, many philosophers feel compelled to articulate the question they want to ask when they wonder about the source of the objectivity of arithmetic with the words "Are there numbers?," even though that seems to be a completely trivial mathematical question. We need to understand not only how we should articulate the question we want to ask in ontology, but also why we naturally articulate it the way we do in light of its having an apparently obvious answer. Taking sides on the puzzles is not enough, we need to understand why and how both sides arise.

It should seem promising that if we could solve these three puzzles, and understand how they arise, then we would be in a good position to find out what role ontology should play in metaphysics, and whether metaphysics can be ambitious in the way we hoped: that it asks further questions, not immediately answered in other parts of inquiry, but still significant questions of fact. A promising strategy to achieve this is thus to look at the three puzzles, to see how they arise and to try to solve them. This is the strategy I will pursue in this book. I hope to make the case that it is the right strategy by showing that it indeed bears fruit: a solution to these puzzles illuminates not just ontology, but also the extent of the proper ambitions of metaphysics and the place of ontology in it.

The three puzzles increase in difficulty in the order given above, and a solution to the later ones will depend on a solution to the earlier ones. We will need to solve the first puzzle first, before we can attempt the second one, and we will need to solve that one before we can move on to the third one. We need to know first what the ontological questions are before we can see how important they are, and what depends on their answer. And only after we know that can we hope to see what the philosophical project can be that is connected to them.

I will thus start with the first puzzle. To understand how it arises we should look at the trivial inferences that apparently allow us to conclude from ordinary statements that there are numbers, properties, and propositions. These trivial inferences can be broken down into two parts. The first is the part where we conclude, for example, that the number of moons is four from the premise that there are four moons. The second is the part where we draw the conclusion that there is a number from that the number of moons is four. Both of these inferences deserve a rather detailed discussion. This will be our point of departure for the next two chapters: chapter 2 will discuss the first inference, chapter 3 will discuss the second one. At the end of chapter 3 we will be able to solve the first puzzle, and we will have made substantial progress towards understanding how to articulate an ontological question properly. It will then be time to move on towards solving the second puzzle and to understand how ontological questions relate to the objectivity of a domain of discourse. The middle part of this book will deal in some detail with what we do when we talk about numbers, objects, properties, and propositions. We will see what the solution to the second puzzle is for our four cases by the end of chapter 9, and it will be different for different cases. It will then be time to think about the status of ontology as a philosophical discipline, and about how metaphysics relates to other parts of inquiry. We will see the solution to the third puzzle in chapter 12. By then I hope to have made the case that metaphysics can indeed have the ambition to ask further questions of fact, that these questions sometimes are ontological ones, that they are, just as we originally thought, just questions like "Are there numbers?," what the answer to that question is, and why it all matters. But in order to get there we need to first try to solve the first puzzle.

2

Innocent Statements and Their Metaphysically Loaded Counterparts

2.1 The Trivial Arguments

Our first puzzle about ontology is a puzzle about how easy it is to answer ontological questions.[1] Such questions are questions like

(14)   Are there numbers?

which are taken to be deep, substantial, and important philosophical questions. But, on the other hand, it seems that there are trivial arguments that immediately imply an answer to that question. In particular, it seems that this question can be answered so trivially that even if one wasn't surprised by the answer, one should be surprised by just how trivial it was to get it. There are arguments that answer question (14) in the affirmative that rely only on completely uncontroversial premises, and they establish this conclusion in two apparently completely trivial steps. These arguments are the main topic of this chapter. There are a variety of lessons that could be drawn from these arguments. One is that the question (14) is indeed trivial, but it is not the one that we should ask in ontology.[2] Or one might hold that the arguments are not valid.[3] Or one might hold that ontology is trivial after all.[4] Or one might hold that the whole project of ontology is confused in some way.[5] I will draw a different conclusion below. To see what one should conclude from these arguments, we will have to look at them in more detail. In doing this we will have to start investigating what we do when we talk about numbers, objects, properties, and propositions. We will at first, in this chapter, only look at a very limited range of cases of such talk, a range that will expand significantly in later chapters. The cases we will look at now are especially significant

---

[1] This chapter is based on Hofweber (2007a). The overlapping material in it has, however, been revised, updated, and reorganized.

[2] See, for example, Fine (2001), Dorr (2008), Fine (2009), Schaffer (2009b), and many others.

[3] See Field (1989a) or Yablo (2000).

[4] See, for example, Schiffer (2003) for properties and propositions, Frege (1884) and Wright (1983) for numbers, and Thomasson (2007) for ordinary objects, and Thomasson (2015) for ontology more broadly.

[5] See, paradigmatically, Carnap (1956).

for the question of how to solve the first puzzle and how to understand the trivial arguments. We will look at the general case in the chapters to follow, which will be central for solving the second puzzle about how important ontology is.

The simplest and most striking arguments that there are numbers, properties, and propositions, all have the same form. They start out with only one premise: a completely trivial and everyday statement that seems to have nothing to do with metaphysics. I will call such statements *innocent statements*, since they are apparently completely free of any metaphysical baggage. Examples include:

(Fdog)   Fido is a dog.[6]

(J4m)   Jupiter has four moons.

and basically every other ordinary, everyday statement. The argument then moves to one or more of the what I'll call *metaphysically loaded counterparts* of the innocent statements. These are statements that are apparently equivalent to them, and obviously so, but that introduce things that don't seem so metaphysically innocent any more: numbers, properties, propositions, and truth:

(Fprop)   Fido has the property of being a dog.

(trueF)   It's true that Fido is a dog.

(Nis4)   The number of moons of Jupiter is four.

Once these are established it seems to require only a further trivial inference to bring in quantification. Once we know that Fido has the property of being a dog then we know that there is a property that he has, namely being a dog. So, after the inference from an innocent statement to one of its metaphysically loaded counterparts a second step takes us from the metaphysically loaded counterparts to a quantified statement:

(15)   There is a property Fido has, namely being a dog.

(16)   Something is true, namely that Fido is a dog.

(17)   There is a number, which is the number of moons of Jupiter, namely four.

And once we have established that, it seems completely trivial to conclude that there are numbers, properties, and propositions.

If the ontological question about numbers just is the question of whether there are numbers then it would seem that this ontological question is indeed trivially answered. Before we draw any conclusions about what the ontological questions really are, or whether ontology is trivial in these three cases, let's have a closer look at the trivial arguments. These arguments are of interest quite independently of the

---

[6] Examples that are used frequently will receive mnemonic labels, to make their identification easier below. Other examples will simply be numbered.

larger issues about the status of metaphysics, since they do give rise to a number of puzzles that are independent of metaphysics. These are puzzles about language, and understanding them, I will argue, is the key to understanding the puzzles about metaphysics. Stephen Schiffer has called the inferences from innocent statements to their metaphysically loaded counterparts *something-from-nothing* transformations in Schiffer (2003). Whether they involve a transformation of some kind or merely an inference is up for debate, but somehow we seem to be able to infer, from nothing, that there are such things as numbers, properties, or propositions. The question is whether, and how, we can get something from nothing.

We need to carefully look at the trivial arguments to make progress on this issue, and we will do so one step at a time. First we will investigate the inference from an innocent statement to one of its metaphysically loaded counterparts in this chapter. Then we will look at the second step of the trivial arguments in chapter 3: the inference from the metaphysically loaded counterpart to a quantified statement. After that we will be in a position to solve the first puzzle.

2.2 The Standard Solutions to the Metaphysical Puzzle

The metaphysical puzzle is to say how we could possibly get such easy answers to such substantial questions, which is closely connected to the question of how we could possibly get something from nothing in the trivial arguments. This question has a number of straightforward attempts at an answer. The first answer is this:

A1   The trivial arguments are not valid. They appear valid, however, for one of several reasons.

Among those reasons that make the arguments appear to be valid, even though they are not, could be a connection to fictionalism. If the metaphysically loaded statement is not literally true, but only true given that there are numbers/ properties/propositions at all, then this might well explain why the argument seems to be valid. We might slide between what is literally true, the innocent statement, and what is only true given the fiction that there are numbers. A version of this was defended by Hartry Field in Field (1989a), and by Stephen Yablo in Yablo (2000).[7]

On the other hand, one might think that the trivial arguments are valid, and draw one of several possible conclusions from that. One might conclude that since the premises are clearly true, we only use innocent statements, thus the questions that ontology tries to ask are indeed trivially answered:

A2   The trivial arguments are valid. The questions of ontology are thus easily answered.

---

[7] Yablo changed his mind about this. A more recent account of his is in Yablo (2006). His most recent view is in Yablo (2014).

This line is in effect taken by Stephen Schiffer in Schiffer (2003) for the case of properties and propositions. For the case of natural numbers it can be associated with certain ways of understanding Frege's Frege (1884) and various neo-Fregeans, like Crispin Wright in Wright (1983). The general approach to ontology congenial with this is defended by Amie Thomasson in Thomasson (2007), Thomasson (2008), and, in particular, Thomasson (2015).

One could hold that the arguments are valid, but that this, in part, shows that the questions that ontology should ask are not the ones that are simply stated as

(14)   Are there numbers?

Instead the questions should be asked with different terms in them, ones that are not tied via trivial arguments to innocent statements. This line is popular among a number of metaphysicians. They hold that the question ontology is trying to ask needs to be expressed with heavier duty metaphysical notions, like those of what is fundamental or what reality itself contains. We will discuss several proposals along these lines of how the ontological question might be stated instead in chapter 13. The third possible reaction to the trivial arguments is thus:

A3   The trivial arguments are valid. This supports that the ontological question has to be stated in different terms, relying on metaphysical notions in the question itself.

Another possibility, in fact a modus tollens to the second answer's modus ponens, is to hold that the trivial arguments are valid, but that this casts the same doubt that we had over the ontological questions onto the innocent statements. Since (Fdog) immediately implies that there are properties, the apparently innocent statement is not so innocent after all, and in fact should be seen as controversial. The controversies of ontology infect all of our apparently innocent statements. I don't know of anyone who holds this view, but it certainly is another option, our fourth and so far final one:

A4   The trivial arguments are valid. This shows that the innocent statements are not as innocent as they seemed.

To see which one, if any, of these options is the right one we will need to look at the trivial arguments in more detail. This way we can hope to find out if they are indeed valid, and what to conclude from them. In particular, we should see whether one of the standard answers A1–A4 is to be accepted.

2.3 Some Syntactic and Semantic Puzzles

The above solutions to the metaphysical puzzle treat it as such, a puzzle in metaphysics. Although each one has its own problems when seen as a metaphysical solution to a metaphysical puzzle, we will not focus on these problems here. We will rather concentrate on a number of puzzles in the philosophy of language that are also closely associated with the innocent statements and their loaded counterparts, and their apparently obvious equivalence. In this section we will first look at a common semantic assumption in all of the above solutions. Then I will raise some prima facie problems for this assumption. After that I will present the main argument that this assumption is mistaken, offer a different solution to the semantic puzzles, and finally offer a different solution to the metaphysical puzzle as well.

All of the above solutions agree that in literal uses the metaphysically loaded counterparts contain new and different semantically singular terms, compared to the innocent statements. By a *semantically singular term* I mean a singular term that has as its semantic function to pick out, refer to, or more generally denote some entity. This notion is taken loosely enough here to allow for definite descriptions to be semantically singular terms even if Russell is correct and they are quantifiers. On such a Russellian view the definite description would nonetheless require the existence of a unique object for it to occur in a range of true sentences. Thus it can be seen as being about or denoting such an object in a loose sense, even if the way in which it requires the existence of such an object is very different than that of a referring term. Semantically singular terms might thus not be a homogeneous group of expressions. In particular, on what is probably the most widely held view of such singular terms they are not a homogeneous group, since names refer, while descriptions are quantifiers, which do not refer. The details of all this don't matter for now, but we will return to this issue below. What matters now is that all of the above accounts hold that in the loaded counterparts the phrases "the number of moons," "the property of being a dog," and "that Fido is a dog" have the semantic function to denote or pick out some entity. The standard accounts disagree on whether these phrases succeed in picking out what they have the function to pick out and thus whether the corresponding entities exist, and what the metaphysical natures of these entities are if they do exist. But the semantic function of these phrases is not in dispute among the standard solutions. And there is good reason to hold that these phrases are semantically singular terms. After all, "the number of moons" looks like a definite description, just like "the composer of *Tannhäuser*." And all of them interact with quantifiers in just the way that one would expect of a referring or denoting expression. But this assumption has its problems. Some of them might be overcome. I will first mention a few of these, but I won't claim here that they are devastating and require a rejection of this view. However, I will then also present some problems which do require the rejection of this assumption.

Before we look at the prima facie arguments against this assumption we should clarify what is at issue here. The issue is not whether these phrases can be used as referring or denoting expressions, or even whether they sometimes are used that way. This would be hard to deny. After all, speakers can use an expression that semantically denotes one thing with the intention to denote another, and they can use an expression that semantically is not in the business of denoting with the intention to denote a particular thing. I certainly don't deny this. Furthermore, since "number" is a common noun it can be used to form a definite description "the number," which is perfectly meaningful, and a semantically singular term, and similarly for "the number of moons." What is at issue is rather what is going on in the common usage of the loaded counterparts. That is to say, what is the semantic function of these phrases in common usages when we make the trivial inferences from an innocent statement to their metaphysically loaded counterparts. It might well be that "the number of moons" can be and sometimes is used as a definite description, but the question is how is it used in a common usage of the loaded counterparts when making the trivial inferences. Compare this to the question of what the semantic function is of "the tiger" in a common usage of

(18)   The tiger is fierce.

Since "tiger" is a common noun it can in principle be used to form the definite description "the tiger," which aims to denote the one and only tiger (among a contextually restricted set of things, in a normal case). But this is not how it is used in a common utterance of (18). An ordinary utterance of (18) more likely involves a generic use, comparable to

(19)   Generally, tigers are fierce.

Although "the tiger" can be used as a definite description to denote a particular tiger in (18), it generally isn't used that way. Maybe it denotes the kind tiger, or maybe it doesn't denote anything at all, but rather makes a generic claim analogous to a quantificational claim about individual tigers and their fierceness. We need to consider the analogous question for the loaded counterparts: what is the semantic function of the relevant phrases in common uses of the metaphysically loaded counterparts when we make the trivial inferences?

The relevant expressions in the loaded counterparts might well be singular terms, as this notion is commonly used in philosophical discussion. But this is not a clearly defined notion understood syntactically. Being a singular term is not a category in contemporary syntactic theory and it doesn't correspond to any of the notions employed there like that of a singular noun phrase or the like.[8] It is instead a loose grouping of various expressions together that has proper names, definite descriptions, and maybe others, as their paradigmatic instances, and that has similar behavior like interacting with quantifiers and being able to take the position of an argument of

---

[8] The notion of a singular term has gotten quite a bit of attention in the philosophical literature nonetheless, since it plays a central role in neo-Fregean programs in the philosophy of mathematics. In this tradition there are attempts to give a more precise characterization of this notion, in particular Dummett (1973), Hale (1987), and essays 1 and 2 in Hale and Wright (2001), among others. Neo-Fregeans usually attempt to characterize this notion purely syntactically, and they argue that all singular terms have a uniform semantic function, namely to denote objects. Our discussion here is very relevant for this debate, but since we are not primarily concerned with neo-Fregean approaches to mathematics now we won't discuss it in this chapter. We will, however, return to this in chapter 5 and propose an alternative philosophy of arithmetic in chapter 6.

a predicate, etc. Understood this way it is innocent to take the relevant expressions in the loaded counterparts to be singular terms syntactically. What matters is what their semantic function is. Are singular terms, understood as a syntactic classification, semantically homogeneous? Do they all have the semantic function of denoting? This is not at all clear. First, as we already saw above, even the paradigm cases of singular terms are not semantically homogeneous, with names and descriptions belonging to completely different semantic categories, although both denoting in our broad sense. But more importantly, some examples of singular terms, broadly understood as a loose grouping, seem on the face of it to be rather different than denoting expressions. Consider:

(20)   What Cheney eats is classified.

(21)   I love what you have done with this room.

(22)   How Mary made the chocolate cake is detailed in her book of recipes.

These can be seen as singular terms broadly understood. They satisfy conditions that are often believed to be essential for singular terms, for example occurring in true "identity" statements, as in:

(23)   How Mary makes a chocolate cake is identical to how my grandfather used to make it.

But does this show that "how Mary makes a chocolate cake" aims to pick out an entity? Maybe it does, maybe it's a way, or a property of events, or what have you. But maybe it does something else, and although some singular terms pick out entities, others are completely different and don't aim to denote something.

Just pointing out that the loaded counterparts have more singular terms might thus give that notion too much weight. Maybe the issue about singular terms is a red herring. Instead one could simply rely on the better understood notions of a definite description and that of a proper name, which is often taken to be the paradigm of a referring expression.[9] These seem to be all that is needed for the standard accounts to understand the numbers case. The standard accounts hold that semantically

(Nis4)   The number of moons of Jupiter is four.

is just like

(CTisW)   The composer of *Tannhäuser* is Wagner.

Since "the composer of *Tannhäuser*" is pretty clearly a description, and "Wagner" is pretty clearly a referring expression, the standard accounts can thus be taken to

---

[9] On a direct reference theory of names, such names are referring expressions. On some of the alternatives they are not referring expressions. I take names to be referring expressions on many, but not all, of their uses, a topic we will discuss more in chapter 7.

hold that "the number of moons of Jupiter" is a description, and "four" is a referring expression, as they are used in (Nis4). However, there are good reasons to doubt that this is so. We should look at them, one at a time.

2.3.1 The Number of Moons of Jupiter

It is not at all clear whether or not "the number of moons of Jupiter" is a description in common uses. In fact, it is in important respects quite different from paradigmatic descriptions like "the composer of *Tannhäuser*." A description "the F" has a close relation to another quantifier: "a F." With "a F" we claim that there is at least one F; with "the F" we also claim that there is at most one. In particular, "the F is G" implies "a F is G." Thus

(CTisW)   The composer of *Tannhäuser* is Wagner.

implies

(24)   A composer of *Tannhäuser* is Wagner.

and (CTisW) furthermore claims that *Tannhäuser* had one composer. This relationship holds generally between definite descriptions "the F" and particular quantifiers "a F." However, the relation between "the number of F" and "a number of F" can be interestingly different. Most strikingly, we get a difference in plural or singular agreement with them on their most natural readings. So,

(Nis4)   The number of moons of Jupiter is four.

is perfectly fine, but

(25)   A number of moons of Jupiter is four.

is quite awkward and seems to involve an agreement violation with respect to singular or plural. However, with a plural verb as in

(26)   A number of moons of Jupiter are covered with ice.

it is perfectly fine. In addition,

(27)   The number of moons of Jupiter are covered with ice.

is quite clearly an agreement violation. So, "the number of moons" requires the singular, whereas "a number of moons" seems to require the plural on its most natural reading. We get no similar phenomenon with standard cases of descriptions. There the agreement is the same with "the F" and "a F."[10]

---

[10] This connection between "a number of Fs" and "the number of Fs" is hardly discussed in the philosophical literature. It does, however, surface in a passage by W. F. R. Hardie discussing Aristotle, who seems to have claimed that one is not a number. Hardie comments: "A man might give one as the number of his sisters, the answer to 'how many?' but also, in answer to the question of whether he had a number of sisters, might say 'no only one'." Hardie (1968: 54). Thanks to Steve Darwall for this reference.

Of course, none of this is a decisive objection to "the number of F" being a description. There are several avenues available for trying to explain why the inference from "the F" to "a F" doesn't seem to work in the numbers case. One reason might be that it is already assumed that there is only one number, and thus saying "a number" gives rise to some sort of an infelicity. This would be analogous to inferring "a richest philosopher is G" from "the richest philosopher is G." But there is more to this story even if one takes this line of explaining the data. The failure of the implication highlights that "a number of Fs" has a reading on which we do not quantify over numbers with "a number." The question now is whether "the number of Fs" similarly has a reading that is not a description of a number, and whether such a reading is the prominent one in the trivial inferences. So far we are merely warming up to that possibility.

2.3.2 Is Four

Besides the question of whether "the number of moons" in a standard use of (Nis4) is a definite description, there is also a question about the semantic function of "four" in this example. In fact, it gives rise to a puzzle all by itself, and this puzzle is not a metaphysical puzzle, but a semantic and syntactic one. Frege observed in Frege (1884) that number words in natural language occur in two quite different syntactic positions. On the one hand, they appear to be singular terms, as in (Nis4); on the other hand they appear to be adjectives, or determiners, as in

(J4m)   Jupiter has four moons.

The paradigmatic cases of words that occur in each of these syntactic positions have quite different semantic functions. But how can it be that one and the same word is syntactically both a singular term as well as a determiner? Other determiners, like "many," "some," or "the," never occur as singular terms. And other singular terms never occur as adjectives, or determiners. To mention just one example, the apparent singular term "the number of moons of Jupiter" could not be used as a determiner or adjective without resulting in immediate ungrammaticality. If we were to replace "four" in (J4m) with it we would get nonsense:

(28)   Jupiter has the number of moons of Jupiter moons.

This ungrammaticality results even though in (Nis4) "four" and "the number of moons of Jupiter" are both supposed to be singular terms standing for the same object. How "four" can occur in these different syntactic positions is in need of an explanation.[11]

2.3.3 Substitution Failure

The problems in the last two sections are specific to the numbers case. The cases of properties and propositions also raise problems for the standard solutions. The assumption shared among these solutions that the loaded counterparts in the cases

---

[11] This puzzle not only applies to the relationship between (J4m) and (Nis4), but to number words in natural language in general. The general case is discussed in detail in chapter 5.

of properties and propositions involve more semantically singular terms has several well-known problems, only one of which I want to repeat here. It is the so-called *substitution problem*. If "that Fido is a dog" and "the property of being a dog" are semantically singular terms, then they should in general be substitutable for other terms that stand for the same entity. Unless they occur in special indirect contexts, it should not matter with what term the entity is denoted, only that it is the same entity that is denoted. But this does not seem to be the case. If "that Fido is a dog" denotes a proposition, then it should denote the same proposition as "the proposition that Fido is a dog." But the following two differ in truth conditions:

(29)   Fred fears that Fido is a dog.

(30)   Fred fears the proposition that Fido is a dog.

The first is a fear about Fido, the second is a case of proposition phobia, fear of propositions themselves. Similarly for property nominalizations like "being a philosopher" and "the property of being a philosopher":

(31)   Being a philosopher is fun.

(32)   The property of being a philosopher is fun.

These are well-known examples, and I don't want to claim now that they refute a view that takes the relevant phrases to denote entities. But they do raise a prima facie problem for this view, a problem we will discuss in more detail in chapter 8.

2.3.4 The Obviousness of the Equivalence

Besides the above mentioned prima facie worries, there are also two general concerns about the relationship between the innocent statements and their metaphysically loaded counterparts. One puzzling fact about their relationship is how obviously they are equivalent. To be sure, some people think they are not equivalent, but this is usually motivated by some philosophical considerations, for example the consideration that they seem to be about different things, and thus can't be equivalent. But if we judge them as regular English sentences, and ignore metaphysical considerations, it is very hard to deny that they are equivalent.[12] Furthermore, their equivalence is obvious to anyone, in particular those not familiar with the related metaphysical debate. In fact, the more one makes oneself aware of possible metaphysical consequences of their equivalence, the less equivalent they can seem. But the obviousness of their equivalence asks for an explanation. How can they be obviously equivalent given that they seem to be about different things? One way such an explanation might go is via metaphysics: it is the nature of these things that explains why the equivalence is

---

[12] It is assumed throughout here, of course, that we are reading (J4m) as "Jupiter has exactly four moons," not "Jupiter has at least four moons." Both readings are available, but the question of the equivalence of (J4m) with (Nis4) assumes the former. For the latter, we would have to modify (Nis4) to something like "The number of moons of Jupiter is at least four."

obvious, and in fact, this is a common way to try to explain this. But this strategy is problematic, since the equivalence is obvious to everyone who understands them. It will take some further work to say how the nature of the entities in question, which isn't clearly obvious to everyone who understands the sentences, can help explain this. But there might be another, non-metaphysical way to give an explanation of the obviousness of the equivalence, one that ties it to linguistic competence, not metaphysical facts about entities. In any case, some explanation has to be given here.

2.3.5 The Puzzle of Extravagance

The key to solving the above puzzles is to solve another puzzle, which I'll call *the puzzle of extravagance*. It becomes vivid if we grant that the innocent statements are equivalent to their metaphysically loaded counterparts. It should be beyond dispute that one of the core functions of language is to communicate information. There are other things we do with language, like joking and flirting, but these uses of language seem to be derivative on the main function that it has, namely to get information from the speaker to the hearer, or to request information from the hearer. With this in mind, we can ask ourselves, why should it be that our language (and many, perhaps all, others) has systematically two ways to say the same thing? If I want to get the information across of how many moons Jupiter has, why do I have two ways to do it, by either saying that Jupiter has four moons, or that their number is four? Why would our language have these extra resources?

We can look at the same puzzle from a more Gricean and pragmatic point of view. All the metaphysically loaded counterparts of the innocent statements involve more words and more elaborate sentence structure. They are, in a word, more complicated. But according to Grice's maxims I should be as cooperative as possible, and communicate the information I have in as simple, relevant, and elegant a way as I can.[13] If I don't do this, and if I use more complicated ways to communicate than necessary, then I will often try to do something else besides communicating the information I have. In particular, the hearer's recognizing that I violate one of the maxims will make the hearer aware that I do more than just try to communicate what the sentence uttered literally means. If that is so, I might be trying to do something over and above just communicating a certain information when using the loaded statements instead of their innocent counterparts. But what?

Any language as complex as ours will have some redundancy in it. For many cases we will be able to explain why we have it. It might well be that the loaded counterparts are merely a by-product of something else. To illustrate, the conjunction of two sentences is always a sentence, even if the two sentences are the same. Therefore

(33)   Water is wet and water is wet and water is wet and water is wet.

---

[13] See Grice (1989).

is a well-formed, although artificial, sentence that is truth conditionally equivalent to

(34)   Water is wet.

A general fact about our language accounts for this case of extravagance. We have a need for having conjunction in our language and it guarantees that sentences like (33) are well-formed and meaningful sentences. Thus this case of extravagance can be explained. Such sentences are simply a by-product of something else, and that something else is what we really have a need for. Are the loaded counterparts merely a by-product of something else for which we have a need in communication? Or do they themselves play a role in communication? And if so, what is it?

In the literature on philosophy of mathematics or metaphysics the authors usually take sentences like (J4m) and (Nis4) and reflect on their truth-conditional equivalence and difference in singular terms. However, they don't discuss whether or not these have a different use in actual communication, and if so, what this difference is. This is a bit surprising because it seems to be quite obvious that their use in communication is quite different, despite their apparent truth-conditional equivalence. How do they differ? And do the loaded counterparts have a use in ordinary, everyday communication? We should look at these puzzles in the philosophy of language first, before we return to the metaphysical puzzle.

2.4 Solving the Syntactic and Semantic Puzzles

An account of the relationship between the innocent statements and their metaphysically loaded counterparts should solve the puzzles we mentioned above, both the syntactic and semantic ones, as well as the metaphysical puzzle. In the following I will propose a solution to the syntactic and semantic puzzles first. Then I will argue that this solution also gives us a solution to the metaphysical puzzle. To be more precise, an account of the relationship between the innocent statements and their metaphysically loaded counterparts has to include the following:

1. An explanation of our intuitive judgments of their truth-conditional equivalence.
2. An answer to the question of whether the meaningfulness of one of the counterparts is a by-product of something else and whether the counterparts have a function in ordinary communication. This should solve the puzzle of extravagance.
3. An account, at least in outline, of the case specific syntactic and semantic features, for example, how "four" can occur both as a determiner in the innocent statement and as a singular term in the loaded counterpart.
4. A solution to the metaphysical puzzle. We need to find out whether in these cases we get something from nothing.

As we will see, meeting the first three of these tasks will allow us to meet the fourth.

2.4.1 The Function of the Cleft Construction

There are a number of fairly well-known cases in natural language that are analogous to the relationship between the innocent statements and their loaded counterparts. They don't play much of a role in philosophy, but they give rise to some similar puzzles, in particular the puzzle of extravagance, and it is instructive to see how these puzzles are solved in their case. In fact, I will argue that they are rather close to our main concern. One good example of this is the so-called *cleft construction*: "it is X that Y." With it one can say what one says with an ordinary sentence like

(Jlike)   Johan likes soccer.

also with two different, but truth conditionally equivalent sentences, namely

(Jcleft)   It is Johan who likes soccer.

(Scleft)   It is soccer that Johan likes.

Here, too, we have (apparently) truth-conditional equivalence, but we use more words and a more complex sentence in the latter cases. So, what is the difference? When would we use one, but not the other? Is the meaningfulness of these sentences a by-product, or do they play a role in communication? What is the function of the cleft construction?

The answer is quite straightforward. Even though we communicate the same information with (Jlike), (Jcleft), and (Scleft), we do so in different ways. In an ordinary utterance of (Jlike) the information that Johan likes soccer is communicated neutrally. No particular aspect of the information is given a special status. In an ordinary utterance of (Jcleft) or (Scleft) this is not so. Some aspect of what is communicated is given a special status. It is stressed. The common term for this phenomenon is *focus*. The focus of an utterance of a sentence is the aspect of what is said that is given a special stress or importance. The clefted sentences do just that. The above ones focus either on Johan, or on soccer, respectively. They contrast what was said with other alternatives: that it was Johan and not someone else that likes soccer, and that it is soccer and not something else that Johan likes, respectively.

Clefted sentences are by no means the only way in which one can achieve a special stress. An utterance of (Jlike) can be used to communicate the information in a non-neutral way, too. A speaker could phonetically stress one aspect or another, and this way of doing it is certainly the commonest one. A speaker could utter

(SOC)   Johan likes SOCCER.[14]

---

[14] The capital letters here represent phonetic stress of the right kind. There are many different ways to phonetically stress a part of a sentence, and there are ways to annotate them in written text, but since the subtleties of all this won't play a central role in the rest of this chapter, I trust that the reader will make the proper emphasis when reading these capitals. A more complete treatment of focus in speech will not just consider intonation, but also the rhythm of a spoken sentence, the spacing between words, and so on. Although the details won't be central for us here, they can be found in prosody, a part of linguistics.

or

(JOH)   JOHAN likes soccer.

An utterance of one of (SOC) or (JOH) would not present the information neutrally. It would rather stress the fact that it is Johan who likes soccer, and not someone else, in the case of (JOH). Or that it is soccer that he likes, and not something else, as in the case of (SOC).[15] Thus (SOC) is a lot like (Scleft), and (JOH) is a lot like (Jcleft). In each case, they both can be used to communicate the same information, and in addition, they both have the same focus. It thus seems that in English we have at least two ways to achieve focus. We could raise our voice, or we could use a clefted sentence, where the focused item is put in a distinguished position. In written English, however, we only have the cleft and similar constructions, not the focus results through intonation (unless, of course, one introduces representations for phonetic stress in written English, like capital letters).

In communication we take recourse to the cleft construction for just that purpose, to achieve a certain focus effect. And we do have a need in communication to do so (I will elaborate on this in a minute). We can thus say that the function of the cleft construction is to focus on the clefted item. The clefted item is the one that was "extracted" or "displaced" and given a distinguished position, for example "Johan" in (Jcleft).[16]

The cleft construction is not the only construction we have for doing this. There are a number of constructions where a certain item in a sentence gets extracted and put in a distinguished position. The truth conditions of the sentence are unaffected by this, but a certain focus effect is achieved with it. Consider, for example, these pairs:

(35)   John ate a sandwich.

(36)   A sandwich is what John ate.

and

(37)   Mary spoke softly.

(38)   Softly is how Mary spoke.

(39)   Softly Mary spoke.

---

[15] This kind of focus is called contrastive focus. See, for example, Rooth (1985), Rochemont and Culicover (1990), Herburger (2000), Büring (1997), or Beaver and Clark (2008) for much more on focus and its relation to syntax, intonation, semantics, and pragmatics.

[16] Talk about extraction or displacement should not be seen as bringing with it an endorsement of any particular syntactic theory about how this phrase ends up in the place it ends up in. It is a natural and common way to articulate that a phrase appears away from its usual position in a subject predicate sentence, and the cleft construction is a good example where this is so. In particular, talk of extraction does not endorse anything like transformational grammar where a whole sentence is transformed into another one. Displacement might be the preferred, less loaded, while still metaphorical, notion.

Cleft constructions and similar constructions show that there are certain syntactic constructions that have as their function to present information in a certain way. That is why we use them in communication. There are other, simpler, sentences that could be used to communicate the same information. But we don't use them because we want to achieve a certain focus, and to present the information with a certain structure. Using an equivalent clefted sentence is one way to achieve this.

Thus we can say that there are at least two ways in which a focus effect can be brought about. One is *intonational focus*. There the particular intonation of a sentence uttered results in a focus effect. The other one is *structural focus*. Here the syntactic structure of the sentence brings with it a focus effect. The focus that results can be the same, but these are two different ways to achieve it.

There is a real and substantial question how these two sources of focus relate to each other. And it relates to the question of how the different aspects of the study of language—syntax, semantics, pragmatics, phonetics—relate to each other. Fortunately, we do not need to resolve this issue here. The conclusions we will reach in this chapter will all be example driven, not theory driven. That there is focus, that it can be achieved through syntactic structure or intonation, these are uncontroversial, although it is unclear what theoretical framework best accommodates all the complex data, a bit of which we will see below. Fortunately we don't have to give an account of the role of focus in language in general here, but simply exploit some rather uncontroversial connections for our present concern. For a much more detailed discussion of focus and its place in the study of language see, for example, Beaver and Clark (2008).

2.4.2 Focus and Communication

In communication we get information from one person to another, either through making an utterance with the right truth conditions, or by requesting certain information from one of the participants in the communication. But for this to work effectively not just any sentence with the right truth conditions will do. A number of aspects have to be considered that will be relevant in choosing among different sentences to utter. For example:

- What is the shared background knowledge of the participants in the communication?
- What of the information communicated is new, and what part of it is old, and shared background?
- Is there any misinformation held by one of the participants in the communication?
- What is important and what isn't for the present purpose?

To communicate effectively it is important to make clear what is new, what is important, and what is supposed to be a correction of an earlier misunderstanding. We do this by communicating information in a certain way, and this, among other things, is what focus contributes to. Focus is an important aspect of communication.[17]

Focus is not merely an added, secondary, pragmatic feature, it can also affect the truth conditions of what is said. Consider the following examples:[18]

(40)   John only INTRODUCED Jim to Jack.

(41)   John only introduced JIM to Jack.

(42)   John only introduced Jim to JACK.

(40) is false if John also took Jim and Jack fishing, but (41) and (42) can remain true under these circumstances. And (41) is false if John also introduced Jane to Jack, but (40) can remain true under these circumstances. Similarly, if John introduced Jim to Jill, then (42) will be false, but (40) and (41) can be true.

One prominent proposal to deal with the semantics of focus and the effect it can have on the truth conditions of an utterance is the so-called alternative semantics.[19] Simply put, we can think of focus as invoking alternatives to what was said. For example, when I focus on Johan in (JOH) I am invoking alternatives which can be collected in an "alternative set," like {Peter likes soccer, Sue likes soccer, etc.}. When I add "only" to this and say

(43)   Only JOHAN likes soccer.

then this is true just in case there is no true member in the alternative set, other than possibly (Jlike) itself. Thus to give a compositional semantics accommodating focus we could assign besides the usual semantic values also extra and additional semantic values: alternative sets. Focus sensitive expressions like "only" can then contribute to the truth conditions that a certain relationship between the standard semantic value and the alternative set holds. In our case involving "only" above, an utterance of (40), (41), or (42) is true just in case the proposition that John introduced Jim to Jack is true and in addition this is the only true proposition in the contextually restricted set of alternatives. What the set of alternatives is will differ depending on which item is focused, and thus the truth values of (40), (41), or (42) can differ.

---

[17] Focus also has clear relationships to presuppositions, but it is not simply to be understood in terms of presupposition. See Beaver and Clark (2008) for more on their relationship.

[18] These are standard examples for the extensive literature on the semantics of focus. See, for example, Rooth (1985) for a well-known proposal on how to deal with such examples, which will also be outlined shortly.

[19] See Rooth (1985). One alternative to it is the account that uses structured meanings instead of alternative sets. These differences do not matter for our discussion here, and the two approaches are variants of each other for simple cases like the ones we are considering in this chapter. For more on this, see von Stechow (1991) and Krifka (2004).

2.4.3 Questions, Answers, and Focus

Focus also has a close connection to questions and answers, one that we will rely upon in an argument below. This connection is tied to the relationship that focus has to differentiating new information from old, background information. Take a simple question like

(44)   Who likes soccer?

This question has a number of possible appropriate answers, basically sentences of the form

(45)   X likes soccer.

Some answers with the same truth conditions as "X likes soccer," however, are not appropriate, in particular

(46)   It is soccer that X likes.

However,

(47)   It is X who likes soccer.

is perfectly appropriate and correct. The inappropriateness of (46) is not one of ungrammaticality, since clearly the sentence is grammatical, but it is an inappropriateness in discourse. It is a failure of what is called *question–answer congruence*. The failure is basically that the focus in the answer is on the wrong thing. What is focused in (46) is what was part of the background information, namely that soccer's being liked is under discussion. For an answer to be congruent to a question it has to have the proper focus, it has to focus on the new information, the one that was requested by the question. The focus alternatives in the answer have to be the possible answers to the question. This is exactly the case with question (44) and answer (47).

   And this makes perfect sense. When one requests information with a question some part of the information requested will be shared background information, say that someone likes soccer, or that soccer is under discussion. But other parts are not shared, and are the ones that the person asking the question wants to get. When one gives an answer with a focus in it one presents the focused item as new information, and thus such an answer is only appropriate to certain questions that have certain background information. These are different with (44) and

(48)   What does Johan like?

Congruent answers to these questions will have to have a different focus. Both of these questions can have

(Jlike)   Johan likes soccer.

as their answer, but this sentence has to have a different intonation to be a congruent answer in these cases. Here the focus will be intonational, and not structural as in the clefted sentences above.

What is background also affects what elliptical answers are appropriate. For example, to (44) one can simply answer

(49)   Johan

and to answer (48) one can simply say

(50)   soccer.

This phenomenon, often called *background deletion*, reflects that what is requested is just the new aspect of the information, and that the assumed background of the question, basically "x likes soccer" and "Johan likes x," can be dropped as optional.

This relationship between questions and answers gives rise to a test of whether or not a particular utterance of a sentence comes with a focus, and background deletion allows us to determine what the focus is. We can simply see if the sentence would be an appropriate answer to a certain question, and what sub-sentential phrase would be an equally appropriate answer. The first part of this test will determine whether there is a focus, the second part will determine what the focus is. We can now apply this test to our cases of interest.

2.4.4 Another Look at the Loaded Counterparts: The Numbers Case

The above considerations allow us to see quite easily that the loaded counterparts have a focus effect. To do this we should see how they interact with questions in ordinary communication, and how an utterance of an innocent statement differs from that of the corresponding loaded counterpart.

Consider the following situation:

I visit your town for the first time, don't know my way around well, and would like to get a quick lunch. You suggest a pizza place and a bagel shop that are close. Half an hour later you see me again and ask me

(51)   What did you have for lunch?

I reply

(52)   The number of bagels I had is two.

Obviously, this is odd. To be sure, what I said might be true and it gives you the information that I went to the bagel shop. But in putting it the way I put it I did not bring out directly whether I had bagels or pizza. What I stress is how many bagels I had. Since this is not of importance here, and not what you asked me about, it makes my utterance somewhat odd. If I had said

(53)   I had two bagels.

then this wouldn't have been odd. Here how many bagels I had isn't focused on (assuming no special intonation). I told you that I had bagels, and furthermore that I had two of them. With (52) I tell you that I had two, and furthermore that it was bagels. But that isn't what matters here.

Suppose I utter (53) instead of (52) and you misunderstand me as having said that I had twelve bagels, and you ask me:

(54)   Oh my God! You had twelve bagels?

To this I could reply

(55)   No, I had TWO bagels.

or

(56)   No, the number of bagels I had is two.

This is now perfectly appropriate. What matters now is how many bagels I had, not what I had. What I had for lunch has been communicated correctly. What is still at issue is how many I had.

As simple as it is, this is the different use we have for (52) and (53) in ordinary communication. They have a different effect on a discourse and they interact differently with questions. The reason for all this is now easy to see. (52) has a focus effect, it focuses on "how many," and that is why it is not an appropriate answer to the question of what I had for lunch. (53) on the other hand does not by itself come with this focus, it has to get it through intonation. This situation is quite analogous to the clefted sentences, where the focus effect is coming from the syntactic structure of the sentence and the special positioning of the clefted item. Just as in the cleft construction, we use the loaded counterpart in communication for its focus effect. This will help us solve our syntactic and semantic puzzles.

2.5 Explaining the Focus Effect

We have seen in the above section that, in the case of numbers at least, the innocent statement and its loaded counterpart have quite a different effect on discourse. One of them brings with it a certain focus effect, independent of intonation, the other one doesn't. Now we have to ask ourselves: why is this so? What about these two statements can explain the difference in focus effect that they have? To ask this question is not to ask a metaphysical question, but simply one about natural language. As for any question that asks for an explanation of a certain phenomenon about natural language, this is not an easy question to answer in full detail. But nonetheless, often it is easy enough to see how such an answer will have to be given in outline, and this is also the case here. We will see that the common assumption of the standard accounts of involving extra semantically singular terms in the loaded counterparts is incompatible with the occurrence of the focus effect we get without special intonation. In this section we will present the argument that this is so, and propose a different account instead. We will first discuss the examples involving talk about numbers in this section, and then discuss the cases of talk about properties and propositions.

2.5.1 The Argument Against the Standard Solutions

For the standard accounts the loaded counterpart in the case of numbers is a statement where we claim that what two semantically singular terms stand for is identical. That there is a tension between the standard accounts and the focus effect can be seen simply from the fact that there ordinarily is no focus effect in identity statements. Just consider what appear to be ordinary identity statements like

(57)   Cicero is Tully.

(CTisW)   The composer of *Tannhäuser* is Wagner.

Such statements simply identify an object picked out one way with one picked out another way. Unless there is a phonetic stress in the utterance of the sentence, utterances of these sentences do not have to have a focus effect, and thus these sentences do not bring with them a structural focus effect. But according to all of the standard accounts (Nis4) is an identity statement, analogous to (CTisW). In particular, "the number of moons of Jupiter" is a singular term that attempts to stand for an object, and stands for the same object as "four," assuming (Nis4) is true. But there is a real problem with this account since there is a tension in (Nis4) giving rise to a structural focus effect, because identity statements in general do not give rise to a structural focus effect. Even in cases like (CTisW), which is truth conditionally equivalent to

(58)   Wagner composed *Tannhäuser* (by himself).[20]

we do not get a difference in focus effect, unless through intonation. Of course, we can say that he is the COMPOSER, just as we can say that he COMPOSED it. But by themselves, without special intonation, neither (CTisW) nor (58) have a structural focus effect. We will discuss in the appendix to this chapter in more detail how identity statements can affect a discourse, and what effects they can have that are similar to, but different from, a focus effect. This appendix contains a more detailed defense of the claim that identity statements do not have a structural focus effect.

The tension between (Nis4) being an identity statement and it giving rise to a structural focus effect speaks against the claim that in (Nis4) "the number of moons" and "four" are denoting or referring expressions. If the latter do denote or refer then (Nis4) would be an identity statement. Is there any option to save the standard solutions to the metaphysical puzzle in light of this? All of the standard solutions

---

[20] I am adding "(by himself)" to get the equivalence to Wagner being *the* (one and only) composer.

agree that "the number of moons" and "four," as they occur in (Nis4), are referring or denoting terms. How then can they explain that the focus effect arises?

One possible explanation compatible with the standard accounts is a pragmatic explanation. We saw above that there is a puzzle about the innocent statements and their loaded counterparts arising simply from the fact that they seem to give us systematically two ways to say the same thing. But if there were no difference at all between them, wouldn't one get a violation of a pragmatic principle (like one of Grice's maxims) if one were to utter a loaded counterpart rather than an innocent statement, since they are more complicated and involve more words? We now know that there is a relevant difference between them, namely a different focus effect. But maybe a believer in the standard accounts can turn this around and try to explain the focus effect as arising from pragmatic principles. This could, very roughly, go something like this:

When a hearer hears an utterance of a loaded counterpart they will recognize that the speaker also had the innocent statement available to communicate the same information. Thus the speaker is not communicating as efficiently as possible. From this the hearer reasons that the speaker attempts, besides communicating a particular information, also to focus in a particular aspect of the information.

Sketchy as this is, there are several problems with such an account. First, it isn't clear why the hearer should conclude from the violation of the pragmatic principles that the speaker attempts a certain focus and not something else. And, second, it isn't clear how the hearer could determine what it is that is supposed to be focused on. Why should the hearer think that the focus is on how many moons Jupiter has, in our example, as opposed to that Jupiter has that many MOONS, or that it's Jupiter that has four moons, or any of the other options? A purely pragmatic account has to bridge the gap from there being a violation of some pragmatic principle to that the hearer can conclude that the speaker attempts to achieve a focus on a particular aspect of the information communicated. And this has to be done without locating the source of the focus effect and what is in focus somewhere else, like in the syntactic form directly. I can't see how this could be done. Intuitively it is clear what settles this issue. It is the syntactic structure of the sentence that makes it clear to us that the focused item is a particular one.

2.5.2 The Explanation

Focus, as we have seen, can arise either from the intonation of the sentence uttered or from the syntactic structure of the sentence uttered. The only reasonable explanation of the relationship between syntactic structure and focus, as well as intonation and focus, is that these are basic parts of the language, and that implicit knowledge of these relations is part of linguistic competence. A competent speaker of a language will not only have knowledge of which sentences are well-formed, what the words of the language mean, and how the meanings of the simple parts determine the meanings of the complex parts. They will also know how the intonation of a sentence as well as its syntactic structure affects focus. This does not dodge trying to give an explanation, but it is to assert that such an explanation can't be gotten from any other parts of the linguistic competence of speakers, like purely pragmatic mechanisms, or purely semantic mechanisms.[21] Intonational focus and structural focus are basic areas of competence that speakers of a language have to have. How this works more precisely is, of course, a substantial question. What intonation or syntactic structure leads to which focus effect? We can't begin to get into this here, though I'd like to point to a number of related cases of structural focus that will be informative for our overall discussion.

Even though it is hard to give an account of when exactly we get structural focus, there are some paradigmatic cases that are worth looking at. Structural focus is often associated with what is called "extraction." Extraction constructions are constructions where a phrase appears in a position contrary to its canonical position, or, more metaphorically, away from where it really belongs. For example, in

(59)   Quietly is how Mary entered.

or

(60)   Quietly Mary entered.

the adverb "quietly" appears outside of and away from the verb phrase it belongs to. What the precise syntactic mechanism is that allows for these sentences isn't important for us here. In particular, it is not assumed that the sentence as a whole gets transformed from one where the adverb is part of the verb phrase into one where it appears away from it, in the style of the now discredited transformational grammar. Talk of "extraction" or "displacement" or "movement" is a theory-neutral metaphor that we don't need to spell out now. What is crucial for us instead is that constructions of this kind give rise to a syntactic focus effect, not how precisely the syntactic connection to focus is to be understood. One way to illustrate the connection of focus and extraction is to rely on the metaphor of movement that is prominent in syntactic theory. How this metaphor is to be spelled out in the end is largely an open question, but we can join the practice of using it nonetheless. And with it we can say that adverb was moved or extracted to the front, away from the verb phrase. And such extraction brings with it a focus effect. That's why (59) has a focus effect without special intonation, but

(61)   Mary entered quietly.

doesn't. And exactly this seems to be the case with (J4m) and (Nis4), repeated here:

(J4m)   Jupiter has four moons.

(Nis4)   The number of moons of Jupiter is four.

---

[21] See also Rochemont and Culicover (1990: 148f.).

In (Nis4) "four" is "extracted" or displaced from its canonical position next to the noun together with which it forms a quantified noun phrase, and appears in a position contrary to its canonical position. That's why (Nis4) has a focus effect without special intonation, but (J4m) doesn't. In fact, that the focus effect is in (Nis4) rather than (J4m) can be seen as evidence that the canonical position of "four" is as being part of a quantified noun phrase like "four moons" rather than being all by itself. Its being displaced from its canonical position explains the focus effect. If its canonical position is the one it occupies in (Nis4) or if neither is a canonical position, then it will be hard, if not impossible, to see why we get a focus effect in (Nis4) but not in (J4m).[22]

This, in rough outline, explains why we get a focus effect without special intonation with (Nis4), and why we don't have such a focus effect with (J4m). It is so far just an outline, and not a detailed account of the syntactic mechanism behind the displacement of "four" in (Nis4), nor is it an account of how syntactic structure relates to focus and prosody or intonation. Nor is it an account of how syntactic focus relates to intonational focus. All these are important further issues, but that doesn't take away from the basic motivation of the present account. We know that there is a connection between syntax and focus, and between syntactic displacement and focus in particular. Noting that (Nis4) has a syntactic focus effect motivates that "four" in it is displaced, while it isn't displaced in (J4m). And this basic connection allows us to solve the syntactic and semantic puzzles that we started with, first for the case of numbers, and then for the cases of properties and propositions.

The mere fact that there is structural focus is most instructive for our general discussion. It shows that a simple view of the relationship between syntactic form and truth-conditional semantics is mistaken. Syntactic form is not only related to the truth conditions of a sentence. It is also related to what focus an utterance of such a sentence will have. What part of a sentence has syntactically special position should not directly lead one to conclude that this part has a special semantic or truth-conditional function. In particular, what "singular terms" occur in a sentence is not directly linked to how many referring or denoting terms occur in that sentence. A particular part of a sentence might have been extracted or displaced into singular term position for a structural focus effect, but this does not have to have an effect on the truth-conditional interpretation of the sentence. It might have an effect on the truth conditions if the sentence contains focus sensitive expressions, like "only" and several others, but in general it won't. Two different syntactic structures can have the effect of presenting the same information in a different way, with a different structure, or focus.

---

[22] That number words have as their canonical position that of a "higher type" expression like a determiner or modifier, but not that of a name, is defended in detail, considering a much wider range of cases, in chapter 5.

2.6 A Solution to the Syntactic and Semantic Puzzles

To solve the semantic puzzles about the relationship between the innocent statements and their loaded counterparts one has to do the following things, as listed in section 2.4: one has to explain the obviousness of their equivalence, one has to solve the puzzle of extravagance, and one has to account for the particular syntactic and semantic features of the loaded counterparts. The last of these three tasks is different for the different cases, but the first two will have the same answer for each of the cases. We will start with numbers, which we focused on above, and then discuss the other two cases.

2.6.1 Numbers

The task of explaining our judgments of equivalence is met as follows: the loaded counterpart involves structural focus, which is closely connected, at least metaphorically, to extraction or displacement. As competent speakers of our language we are competent with the structural focus that our sentences give rise to, that is, we are competent with what syntactic structures give rise to what focus effect. As always with linguistic competence, this competence does not manifest itself in explicit knowledge of what structures do what, but rather in judgments of what utterance of what sentence is grammatical, appropriate, equivalent, and so on. We have seen such judgments about the loaded counterparts at work above in the bagel example in section 2.4.4. Thus just like our judgments of the equivalence of a simple subject predicate sentence and its clefted counterpart, our judgments of the equivalence of the innocent statements and their loaded counterparts are explained by our linguistic competence. This, in particular, explains why the two sentences (J4m) and (Nis4) are generally judged to be equivalent, except possibly by those who think they have overriding philosophical reasons for a different conclusion.

The puzzle of extravagance is also easily answered. The loaded counterparts do play a different role in communication. They give rise to a focus effect, which is an important aspect of communication. Thus the loaded counterparts are no case of extravagance. They have a function in communication, and we can see why it is good for our language to have the loaded counterparts besides the innocent statements.

This leaves us with the syntactic and semantic puzzles about number words. Here we will only discuss the occurrence of number words in our main examples:

(J4m)   Jupiter has four moons.

(Nis4)   The number of moons of Jupiter is four.

A more detailed discussion of number words in other cases, in particular as they occur in arithmetic, will follow in chapter 5. Right now we are only trying to understand what is going on in the trivial arguments that lead to the first puzzle about ontology. In particular, I don't want to claim that all occurrences of "the number of F" are to be understood along the lines that I propose the occurrence of it in (Nis4) should be understood as it is used in the trivial arguments.[23] Our focus so far is narrow, but on an especially important example. Later, in chapter 5 I will make a proposal about what we do when we talk about natural numbers in general, which is a more ambitious task tied to solving the second puzzle about ontology.

We noted above that in (J4m) "four" appears in a syntactic position that is commonly occupied by adjectives or determiners. However, in (Nis4) it appears to be in the position of a proper name. With the structural focus account of their relationship given in this chapter we can now see why that is so. In (Nis4) "four" appears away from its canonical position to achieve a structural focus effect. This focus effect is brought about, somehow, by "four" being displaced from its canonical position that it occupies in (J4m).[24] It appears outside of the phrase it properly belongs to to achieve this. "Four" thus is a displaced determiner in (Nis4), and its unusual syntactic occurrence is not in the service of reference, but in the service of focus. In particular, (Nis4) is not an identity statement where two semantically singular terms are claimed to refer to the same thing. Neither "four" nor "the number of moons of Jupiter" refer to or denote objects on a standard use of (Nis4).[25]

And this, finally, solves the metaphysical puzzle of how we can have something from nothing in these cases. (J4m) and (Nis4) have the same referring or denotational terms. There is no new referring term coming out of nowhere in (Nis4). (J4m) and (Nis4) are indeed equivalent, as we suspected, but we don't get something from nothing. No new reference is attempted or occurs. But given that the structural focus is tied to new syntactically singular terms it is understandable why someone might think that it does occur.

2.6.2 Properties

The case of properties is different from the case of numbers. In the numbers case the word which is the new singular term, "four," was already present in the innocent

---

[23] Although I hoped to make this clear in Hofweber (2007a), on which this chapter is based, as well as above, I would like to stress it again, since some authors have taken the view proposed in Hofweber (2007a) to have a much wider scope than merely to understand the uses of the loaded counterparts as they appear in the trivial inferences. For example, Friederike Moltmann says in Moltmann (2013b: 33) "Hofweber maintains that *the number of* generally has the function of a place-holder for a focus-moved numeral." This is not correct, both that I maintain it and that it is so. That phrase clearly can be used as part of a definite description and in other ways that are unrelated to focus.

[24] This does not mean that the whole sentence (J4m) gets transformed into the sentence (Nis4) via some syntactic transformation rule, as Brendan Balcerak Jackson takes it to mean in Balcerak Jackson (2013). See Hofweber (2014a) for a more detailed discussion of this issue, as well as Balcerak Jackson (2014).

[25] In chapter 9 of Thomasson (2015), Amie Thomasson maintains that one can accept my account of the focus effect of (Nis4), but maintains that "four" is referential in addition to contributing to a syntactic focus effect. I disagree with this, and would like to briefly make clear again why. First, if "four" is referential then (Nis4) would be an identity statement, and identity statements do not have a structural focus effect, as mentioned above and argued for in more detail in the appendix to this chapter. Second, if "four" is displaced in (Nis4), but not in (J4m), where it is a non-referential determiner or adjective, then, since displacement is a syntactic phenomenon, "four" is not referential in (Nis4) since syntactically displaced words don't acquire completely new semantic functions. How all this fits into a unified picture of number words in natural language is discussed in more detail in chapter 5.

statement, although at a different syntactic position. In the properties case this is not so. The expression "the property of being a dog" isn't already present in the innocent statement, but its foundation is. We can see the loaded counterpart being connected to the innocent statement since "being a dog" is the nominalization of the whole verb phrase "is a dog," with the former being the central part of "the property of being a dog." This suggests an analogy with the numbers case, in that the information that is carried by a particular aspect of the innocent statement, in this case the verb phrase, is displaced into singular term position, to achieve a structural focus effect. But to see if this is indeed so we have to see if there is a focus effect in the loaded counterparts in the case of properties, if it is a structural focus effect, and what the focus is, if there is one.

We know that there is a good test to determine whether there is a structural focus effect in a certain sentence: we have to see how utterances of the loaded counterparts interact with questions. And to do that we have to see to what questions a common utterance of the loaded counterpart would make a congruent answer. However, there is a difficulty here in that if we want to test for structural focus we have to make sure that the focus effect does not arise from intonation. This is difficult to control since certain answers to certain questions invite a certain intonation. Often an utterance of a sentence will have a focus, since a natural way to intonate that sentence at this point in the discourse will give rise to that focus. We have to make sure to distinguish structural from intonational focus in the following, and it isn't easy to do this. But by carefully considering the examples given below we can see that there is a structural focus effect in the loaded counterparts. To make this clear, let's look at some question–answer exchanges. Here we will also use background deletion as a test for what is in focus. Background deletion, discussed above, is the phenomenon that what isn't focused in an answer, what is the background, can be left out of the answer, with the resulting sub-sentential answer nonetheless being appropriate. That which can't be deleted is in focus, and thus this allows us to see what the focus is.

Consider a first example. To the question

(62)   I know gold is very valuable, but it's soft. I need something very hard. Which one of these metals is very hard?

any of the following answers are appropriate:

(63)   a. Titanium.
       b. TITANIUM is very hard.
       c. It's titanium that's very hard.

Clearly the focus is on titanium, and that is exactly as we would expect it. The contrast class is the class of these other metals that are salient in the context of the question. Now compare this with the second exchange:

(64)   I know gold is very valuable. Titanium is supposed to be special, too. But what's special about it?

(65) a. Being very hard.

    b. Being very hard is what's special about titanium.

    c. Titanium is VERY HARD.

    d. Titanium has the property of being very hard.

Here the focus is on being very hard. All of these answers focus on this, rather than titanium. That this is so can be seen by carefully considering these discourses, and by assuring oneself that a focus effect arises, and that it arises not from intonation, except, of course, in (65c). The examples above and their interaction with questions show that this is the case for the loaded counterparts in the case of properties.

Given that there is a structural focus in the loaded counterparts about properties, we can see quite easily that the solutions to the puzzle of extravagance and the problem of the obviousness of the equivalences carry over to this case as well. The loaded counterparts have a function in communication, and the equivalence is obvious from our linguistic competence. In the loaded counterparts involving properties we "relocate" the information that was carried in the verb phrase and put it in a special nominalized position. This is done to achieve the above focus effect.

2.6.3 Propositions

In the case of both numbers and properties the loaded counterparts give rise to a structural focus, and what is focused on is only a part of the information communicated. In the case of properties, the focus is the information carried by the verb phrase in the corresponding innocent statement, in the case of numbers it is the information carried by the number determiner. We could say that what is in focus is information carried by a phrase below the sentential level, that is, it is *sub-sentential focus*. But in the case of propositions this is not so. There we do not have a structural focus effect where only a part of the information communicated by the innocent statement is in focus. In the propositions case it is not the case that some of the information from the innocent statement is put into focus, while the rest is moved into the background. Instead all of the information is put into focus. All the information carried by the innocent statement is focused on, and thus what is in focus is at the sentential level: it is *sentential focus*. This makes the propositions case unique and interestingly different from the other two cases. But that we get a focus effect in this case as well can again be seen by considering some sample exchanges of questions and answers. I will list two questions, and a range of answers, all of which seem to be congruent to each of the questions.

(66) a. I have heard that Fido is a dog. Is he?

    b. Is Fido really a dog?

(67) a. Yes.

    b. Yes he is.

   c. Fido IS a dog.

   d. FIDO IS A DOG.

   e. That Fido is a dog is true.

   f. It's true that Fido is a dog.

Reflection on the question–answer exchanges above shows that our main concern here, (67e), has the same overall effect on the discourse as (67d). It has the effect of simply emphasizing the content of the innocent statement. That is, it emphasizes the whole content, not just some part of it at the expense of some other part. This is what we called sentential focus, focus at the level of a whole sentence, not merely a part of it. The emphasis does not affect the information communicated, it just emphasizes it. The speaker simply puts some extra weight on what they say, but they don't communicate different information when they do this. Emphasis on all of the information does not turn it into different information. (67d) gives us sentential focus through intonation. (67e) gives us the same focus effect without special intonation. It simply emphasizes the information also communicated with the innocent statement.

There are some subtle differences between (67c) and (67d), on the one hand, and similar ones between (67e) and (67f) on the other. (67c), depending on subtleties of how the capitals are intonated, can either have a contrastive stress on the auxiliary verb, or result in sentential focus. In the former case the contrast class might be something like {Fido seems to be a dog, Fido looks like a dog, ...}. These two are different, and it is hard to say which precise intonation gives rise to which one of them. Fortunately, we don't have to do this. It is interesting to note that a similar and somewhat neglected difference also arises between

   (67e)   That Fido is a dog is true.

   (67f)   It's true that Fido is a dog.

(67e), without special intonation, gives rise to a sentential focus of the information carried by the that-clause. (67f), on a standard occasion of utterance, on the other hand, gives rise to a contrastive focus on "true," with the contrast class, as always, depending on the situation of utterance, but likely to contain members like {It's false that Fido is a dog; it's unlikely that Fido is a dog, ...}. Although (67f) is not quite a clefted sentence, it is very similar to one, and it does have a similar contrastive focus effect. Contrastive focus on "true" and sentential focus on all the information carried in the clause are similar, but not quite the same.[26]

---

[26] That ascription of truth to a proposition has the effect of emphasis has been noted before, for example, in Strawson (1964), but I hoped to show here how this fits into a general picture, and how it can be subtly different depending on how it is done. No particular philosophical view about truth is assumed in any of this.

2.6.4 Content Carving Without Tears

Frege said that we can carve up content in different ways.[27] For Frege, the very same content can be carved up such that sometimes it is about how many things there are, and sometimes it is about numbers, which are themselves particular things. But Frege left us with a mystery: how to make sense of this position more precisely. In particular, how is it possible that the same content can be carved up one way and be about certain objects, but also carved up another and be about different objects, or no objects at all. This suggestive metaphor for the relationship between the innocent statements and their metaphysically loaded counterparts can now be spelled out, and this mystery can be resolved. We can indeed carve up content in different ways. Contents can be presented in different ways. Different aspects of the same content can be brought out, emphasized, and stressed. The syntactic form of the sentences used to communicate that content matters for what is emphasized and brought out. In this way we do carve up content differently. We can give the same information a different structure, and the sentences we use to communicate this structured information reflect this. But we do not communicate the same content by talking about different objects. We communicate the same content by focusing on different aspects of what we say. There is no mystery about it, and we can see how one can be led to consider content carving to bring out different objects, since the sentences that express the newly carved content contain new and further singular terms. But these singular terms are in the service of focus and information structure, not reference. Carving content is a great idea, and we do it all the time. However, it isn't a deep metaphysical mystery, but something rather mundane. It is the result of raising one's voice, or the syntactic equivalent of it.

2.7 On to the Second Step

The view presented in this chapter gives us an answer to the question about the relationship between innocent statements and their metaphysically loaded counterparts. According to this answer they are indeed equivalent, and obviously so. But they nonetheless have a different function in communication. This gives us a different answer to the puzzle of how we can get something from nothing than the standard answers A1–A4 listed above. Our answer is

A5   We don't get something from nothing. The loaded counterparts are equivalent to the innocent statements, but they do not contain more referring expressions. They focus on different parts of the information instead.

---

[27] See Frege (1884: §64).

This account so far does not tell us everything we need to know about the trivial arguments. It merely gives us an account of the first step in the trivial arguments, and according to it the first step is valid. What we need to see next is whether the second step is valid as well. It might seem excessive to spend so many pages already on just one step of an apparently trivial argument, in particular since in the end it turned out that it indeed is a trivial step. But I hope the effort is justified by the explanation of how this step indeed can be a trivial inference in light of the considerations that speak against this. The next step will require just as much attention, I am afraid.

If what we saw in this chapter is indeed correct, then there seems to be a real tension between the proposed account of the metaphysically loaded counterparts and the apparently trivial quantifier inferences that use these loaded counterparts as their premise. According to the present account of the loaded counterparts the relevant singular terms are not referential as they occur in standard uses of the loaded counterparts, i.e. in particular those that figure in the trivial arguments. But if the terms are not referential, and instead in the service of focus, how could it be that the quantifier inference is valid? After all, if

(Nis4)   The number of moons of Jupiter is four.

does not contain any expressions that aim to refer to numbers, then how could it be that this sentence implies, apparently trivially, that

(68)   There is a number which is the number of moons of Jupiter, namely four.

The latter sentence seems to require that there is such a thing or entity which is a number, but the former sentence seems to say nothing about that. In fact, on the account given it would seem that the former sentence explicitly does not require this. It only talks about Jupiter and its moons, and presents the information about them in a certain way so that we get a focus effect on how many moons Jupiter has. How then could it be that there is a trivial implication to a statement that explicitly makes such a requirement on what things exist? Does the quantifier inference, assuming it is indeed valid, speak against the account of the metaphysically loaded statement given in this chapter?

To see if there indeed is a tension between the account of the first step of the trivial inference offered in this chapter, and the second step, is the main starting point for the next chapter. To properly answer this question we have to investigate quantification in natural languages more closely, and this will lead to considerations that go beyond the question of a tension, and beyond the trivial inferences. Quantification has traditionally been closely tied to questions about ontology, and there is very good reason for making such a connection. To understand quantification and its tie to ontology is crucial for anyone puzzled by the philosophical project of ontology, and at the end of the next chapter we hopefully will have made progress towards this larger understanding, besides resolving the tension.

2.8 Appendix: Focus and Identity Statements

This appendix discusses the relationship between identity statements and focus in more detail, in particular whether identity statements can have an effect on a discourse similar to a focus construction. The details of all this won't matter for what is to come, so it can be skipped by anyone who has seen enough on this issue for now and wants to move on instead. But anyone who finds my claims about identity statements and focus above dubious, or would like to see more in general on this topic, will hopefully find this appendix helpful.

My argument above against the standard accounts of the loaded counterparts relied on the claim that identity statements do not come with a structural focus effect. That is to say, the syntactic structure of the sentence uttered does not determine a focus effect in that utterance. Since the loaded counterparts like (Nis4) do have a structural focus effect it follows that they are not identity statements. But there are some examples of what appear to be identity statements that seem to be in conflict with this claim. Berit Brogaard uses an example like the following in Brogaard (2007) to argue against the view defended in Hofweber (2007a), on which this chapter is based. In particular, she maintains that (Nis4) is an identity statement, and that identity statements can have a focus effect. I disagree for the reasons to be given shortly.

Take a paradigmatic identity statement, like

(57)   Cicero is Tully.

This sentence can be uttered as an answer to both of the following questions:

(69)   Who is Tully?

(70)   Who is Cicero?

However, as an answer to (69) it has to have a different intonation than as an answer to (70), which one can easily see by just going over these question–answer pairs in one's head. If we stick to one of these intonations in both cases then it becomes an incongruent answer to one of these questions. This is no surprise. Different questions require answers that present information with a different structure, even if it is the same information. For example, (69) asks about Tully, and thus makes Tully the topic of the conversation. The answer should say something about him, in particular who he is. What the topic of a sentence is is usually marked phonetically. A particular intonation of the sentence brings out what the topic of that particular utterance of this sentence is. In our case here we have the same sentence, but a different intonation either makes the first or the second term the topic of the sentence. Commonly the topic is in the subject position of a sentence, but this doesn't have to be so. It isn't, for example, when (57) is given as an answer to (69). A classic example in the linguistic literature that makes the same point is the following.[28] Consider the sentence

---

[28] See Jackendoff (1972: 258 ff.).

(71)   Fred ate the beans.

It can be a congruent answer to each of the following two series of questions:

(72)   What about Fred? What did he eat?

(73)   What about the beans? Who ate them?

But to be a congruent answer the intonation has to be different in each case. The intonation has to mark the topic of the sentence properly, which is either Fred, or the beans, and it has to focus on the other part, again, either Fred or the beans. The sentence topic does not have to be what is in subject position, as it isn't when (71), properly pronounced, is given as an answer to (73). In this case the beans are the topic, as is required by the question, and that Fred ate them is what is said about them, and so Fred is put into focus. This focus and topicalization is simply the result of the intonation of the sentence. We won't have to discuss here how intonation does that, and what intonation does what. This is the topic of an extensive debate in the linguistic literature, but all we need here is that this happens, not how precisely it happens.[29]

Examples like (57) suggest that the topic and the focus of an identity statement are simply a result of intonation. It seems that the term in pre-copula position as well as the one in post-copula position can be either one of these, with proper intonation, and some proper intonation is needed besides what is determined by the syntax to have it one way or another. Thus the syntactic structure of the sentence does not bring with it a particular focus, and thus there is no structural focus effect. And this is just the position relied upon in our main argument above that (Nis4) is not an identity statement.

However, this symmetry does not always seem to hold. Some identity statements seem to be asymmetrical in that one of the terms has to be in subject position. These are examples that might suggest that there is a structural focus effect in an identity statement after all. Thus these are the cases we will have to look at more closely. I will discuss two such cases in the following. For both of them, I will argue, the explanation for the asymmetry has nothing to do with a structural focus effect, but there is a different explanation of the asymmetry in these cases.

Our first case of an asymmetry concerns answers to questions like "Who is X?" When we ask "Who is X?" then identity statements formed with proper names can be used symmetrically, given proper intonation, to form a congruent answer. The examples (57), (69), and (70) illustrate this. The same seems to be true when we ask a question involving a description, and give answers that involve a description and a name. Consider:

(74)   Who is the composer of *Tannhäuser*?

(CTisW)   The composer of *Tannhäuser* is Wagner.

---

[29] For more on how topic is marked phonetically, see, for example, Büring (1997).

(WisCT)   Wagner is the composer of *Tannhäuser*.

Again, proper intonation is required for these answers to be congruent with the question, but the order of terms does not seem to be essential. But now consider the following. Suppose you ask

(75)   Who is Wagner?

Then the order of terms seems to matter. It is perfectly fine to answer with (WisCT), but it is awkward to answer with (CTisW), even when one tries to straighten things out with intonation. Why is that?

I won't be able to give a full explanation of all the relevant data here, but what is of crucial importance to us is to see whether the explanation could be that an identity statement involving a description in subject position comes with a structural focus effect, one that somehow makes (CTisW) an incongruent answer to the question (75). But this explanation we can rule out right away. Notice that sometimes identity statements involving descriptions in subject position are perfectly fine answers to questions like (75). Consider, for example

(76)   The man to your left is Wagner.

or

(77)   The conductor is Wagner.

These are perfectly congruent answers, given proper intonation. If identity statements involving such descriptions had a structural focus effect it should apply to all such cases, but it clearly doesn't. The explanation of why (CTisW) is an incongruent answer to (75) thus has to be something else. Why are some descriptions OK, while others are not? In outline, the explanation goes along the following lines.

For a sentence to be a congruent answer to (75) it has to have Wagner as its topic, since Wagner is the topic of the question. Usually, in English at least, the topic is in subject position, but it does not have to be in that position. The topic can be in post-copula position, as in the examples above, but for that to be possible it has to be phonetically marked as the topic. But when the pre-copula material is long and complicated, as in (CTisW), it is hard to get the reading where Wagner is the topic, since it is so late in the sentence, and thus we naturally take the pre-copula term to be the topic. When trying to understanding the sentence we naturally try to give it a topic, and if the pre-copula material is complex it is hard to make the post-copula name the topic. This explains why (77) is OK, and similar sentences like it, but (CTisW) is not. However, the exception to this seems to be if the object denoted by the pre-copula material is salient to the speaker, as in (76). This somehow makes it easier to parse the sentence as having a post-copula topic. The details of how this works aren't clear, of course, and we won't be able to provide them. What matters to us here is that the explanation for these asymmetries in identity statements does not arise from identity statements having a structural focus effect, as these examples make this clear.

There is also a second case of asymmetries in identity statements, which has a different explanation than the above cases. It can be seen when we consider answers to questions like

(78)   What has Wagner ever done?

(CTisW) is not a congruent answer to it, but (WisCT) is. In this case the contrast is even stronger than in the above cases. How should we explain this asymmetry?

As a first step we should notice that here it is more puzzling that (WisCT) is a congruent answer than it is that (CTisW) isn't one. Why would an identity statement be a congruent answer to a question about what Wagner has done? After all, we are not asking who Wagner is, but what he did. An identity statement seems to tell us who he is, but not what he did. In fact, this is where the difference in the two answers lies. (CTisW) is not a congruent answer since it is an identity statement, and thus doesn't address the question. But (WisCT) is a proper answer in part because it is in fact not an identity statement when used as an answer to the question. The description in (WisCT) is in predicate position, and it functions as a predicate that specifies a property the subject is claimed to have.[30] Thus on a standard reading (WisCT) does not say that Wagner is identical to the unique thing which composed *Tannhäuser*. Rather it says that Wagner has the property of being the composer of *Tannhäuser*. And this property is closely tied to the property of having composed *Tannhäuser*. Composing is an activity, and thus something Wagner has done. This is why (WisCT) addresses the question and is a congruent answer to it, but (CTisW) is not.

The fact that the description occurs as a predicate in (WisCT) then raises the question of why it can be a congruent answer to (75). If the post-copula term does not denote an object with which Wagner is identical, then why does (WisCT) answer the question? Questions like (75) have two kinds of congruent answers.[31] The first is one where the topic is identified with another object which is familiar to the person addressed. The second is where the speaker specifies a discerning property of the topic. Thus I can answer either with an identity statement, or with a subject predicate sentence which specifies some special properties of Wagner. The latter is the case when I use (WisCT). The former is the case when I answer with (CTisW).

All this is in no conflict with our claim that identity statements do not have a structural focus effect. That there is an asymmetry in such identity statements, i.e.

---

[30] See Partee (1987) and Fara (2001) for more on predicative uses of descriptions.
[31] See Boër and Lycan (1986).

that sometimes they are a congruent answer to a question when one term is in subject position, and sometimes when the other one is, does not show that there is a structural focus effect in identity statements. We have seen above how, in outline, these asymmetries are to be explained, and that a structural focus effect couldn't explain the data.[32]

[32] In Hofweber (2007a) I also discuss at some length whether uncontroversial focus constructions sometimes are identity statements. This was proposed by Phillipe Schlenker in Schlenker (2003) with a proposal that specificational sentences are concealed questions, which in turn are identity statements. However, I argue there that they are only identity statements in a different sense of the phrase, not the one relevant for us. For discussion of proposals of treating sentences like (Nis4) as specificational sentences and as concealed questions, see Romero (2005), Brogaard (2007), Moltmann (2013b), Felka (2014), and Schwartzkopff (2015). See also Barlew (2015) for more on this general topic.

3

Quantification

3.1 The Significance of Quantification

Quantification has played a large role in debates in ontology, and apparently for good reasons. That quantifiers play a central role in ontology is implied by the following two claims:

1. The ontological question[1] about Fs is "Are there Fs?"
2. The possible answers to the ontological question about Fs, i.e. "There are Fs" and "There are no Fs," are quantificational statements.

To answer the ontological question about Fs is thus to determine the truth value of a quantified statement.

However, both of these claims are controversial. The first claim is explicitly rejected by metaphysicians who hold that ontology instead is concerned with what there fundamentally, or ultimately, or really, is. We will discuss such approaches in detail in chapter 13. It is also rejected by Jody Azzouni, who holds that neither a quantificational sentence, nor any other sentence, in natural language is able to state the ontological questions, at least not without some further commitment of the speaker which itself can't be put into words. On such a view an articulation of an ontological question requires the help of a secret pragmatic ingredient.[2]

The second claim is usually considered to be trivial. Of course "There are numbers" is a quantificational statement, since it contains the quantifier "there are," the plural version of the quantifier "there is." However, this is mistaken. "there is" is not a quantifier. Although philosophers often call "there is" the particular or existential quantifier in natural language, this is not correct, and we will discuss this issue shortly.

Despite these concerns, I hold that both of the above claims are true, and thus that quantificational statements do have a central role to play in ontology. But this role is more complicated than is generally assumed, in part because of a widely held, but mistaken, view about quantification in natural language. This chapter is devoted to spelling out how quantification is to be understood and how it relates to ontology.

---

[1] Recall that an ontological question is just whatever question is properly to be addressed in ontology.
[2] See Azzouni (2004) and Azzouni (2007). John Burgess has criticized this aspect of Azzouni's view in Burgess (2004), I have done the same in Hofweber (2007b). See Azzouni (2010a: 82f.) for a brief reply to both.

This will be essential for understanding what is going on in the second step of the trivial inferences discussed above, for solving the first puzzle about ontology, and for seeing how to properly state ontological questions.

3.2 The (Fairly) Uncontroversial Facts About Quantifiers

3.2.1 The Structure of Quantified Noun Phrases

Quantifiers are a class of expressions in natural language that include "something," "everything," and many more. These expressions famously figure in valid inference patterns, and are a central part of modern logic. Logic, however, is mostly concerned with quantifiers that do have clear inference patterns, but the class of quantifiers in natural language is a much larger one. In first-order logic, for example, we restrict ourselves to the quantifiers "something" and "everything," and understand some other quantifiers, like "nothing," in terms of them. But the treatment of quantifiers in first-order logic has clear limitations. First, not all quantifiers in natural language can be understood in terms of just "something" and "everything." For technical reasons, quantifiers like "most things" or "finitely many things" can't be expressed with just the quantifiers and other resources of first-order logic. Second, and more importantly, the treatment of quantifiers in first-order logic clearly misses out on some crucial features of quantifiers in natural language. Quantifiers are complex expressions in natural language. They are built up from expressions like "some," "all," "most," etc. and common nouns like "thing," "dog," and so on. In fact, there are various ways to modify a noun to make even more complex quantified expressions, such as "some tall, hairy dog." In the study of natural language this complexity among quantifiers was investigated with great success, and one compelling theory about how quantifiers work in natural language is the so-called Generalized Quantifier Theory (GQT). According to GQT quantified noun phrases are built up from determiners (Det) and nouns (N), which together with a verb phrase (VP) form a sentence. Determiners include expressions like "some," "all," "most," "many," "a," "four," and so on. The semantics of quantifiers can then be given in a very general way. Considering just the simple, extensional case, we can understand the truth conditions of a quantified sentence of the form

(79)   Det N VP

with instances like

(80)   Some dog barks.

as involving a claim about the relationship between the extensions of the VP and the N. In our case, with the determiner "some," the claim would be that the sentence is true just in case the intersection of the extension of the N with the extension of the VP is not empty. If the determiner is "all" it is the claim that the extension of the N is a subset of the extension of the VP. If the determiner were "four" it would be that the intersection of the two extensions has four members in it, and so on. A nice survey of GQT is given in Gamut (1991); one of the classic early papers about it is Barwise and Cooper (1981). We will discuss GQT in some more detail below.

3.2.2 "There Is" Is Not a Quantifier

One common, but mistaken, statement about quantifiers is that the phrase "there is" is a quantifier, for example as it occurs in

(81)   There is a number which is the sum of 2 and 4.

This sentence does contain a quantifier, but it is not "there is." It is instead "a number." The latter is built up from the determiner "a" and the common noun "number." The "there is" is part of what is called the "existential construction," which is discussed, for example, in Barwise and Cooper (1981) and McNally (1997). The existential construction, in English, is of great interest to linguists since it has a number of puzzling features. For example, not every determiner can occur following the existential "there is." Consider:

(82)   a. There is a dog in the garden.
  b. *There is every dog in the garden.
  c. *There is the dog in the garden.

The last two are ungrammatical, even though the following apparently equivalent statements are grammatical throughout:[3]

(83)   a. A dog is in the garden.
  b. Every dog is in the garden.
  c. The dog is in the garden.

Even David Lewis called "there is" a quantifier in his Lewis (2004). I am sure he knew that this is strictly speaking not correct. Although many philosophers call "there is" a quantifier, I still count the fact that this is not so among the uncontroversial facts about quantifiers. I don't think any of them would aim to defend it once the relevant facts are pointed out. And, in fact, "there is" in the existential construction is generally, if not always, followed by a quantifier. This point is important, since it relates to the question

---

[3] This feature of the existential construction is called the *definiteness effect*, and is discussed in detail in, for example, McNally (1997). Although there is a clear difference between the starred and unstarred examples in (82) it is not completely uncontroversial whether examples like (82c) are always ungrammatical. For example, consider an exchange like this: A: "Who would want to live next to that noisy bar?" B: "Well, there is the guy who runs the bar." or maybe even B: "Well, there is everyone who works at the bar." We don't have to settle this issue here. McNally (1997) also discusses various other puzzling features of the existential construction, which are also not relevant for our discussion here.

whether or not statements like "There are numbers." are quantificational. If "there is" or "there are" would be a quantifier, then they clearly would be. But since they are not quantifiers there is a real issue here whether this is a quantificational statement. But as we will see shortly, "There are numbers." is a quantified statement, but not because "there are" is a quantifier.

3.2.3 Singular and Plural Quantifiers

Quantifiers can be plural or singular. Plural quantification is a well-known aspect where standard, singular, first-order logic fails again in the representation of ordinary natural language quantification. Plural quantifiers have been studied in both logic and linguistics, and they will be a part of our discussion in the following, even though we will mostly focus on singular quantification. Some determiners combine with nouns to form a singular quantified NP, some form plural quantified NPs. For example, "all" leads the plural "all dogs," whereas "every" leads to the singular "every dog." Furthermore, plural quantifiers can occur without having a determiner explicitly present next to the noun. Bare plurals like "dogs" have quantificational readings:

(84)   Dogs are eating my lunch.

has a generic reading (in general dogs are eating my lunch) and a quantificational one (some dogs are eating my lunch right now).[4] The quantificational reading of bare plurals can be used in the existential construction:

(85)   There are dogs in the garden.

This, as above, is simply truth conditionally equivalent to

(86)   Some dogs are in the garden.

The existential construction can also occur with an "empty predicate," where the predicate is simply vacuous or an unpronounced trivial predicate. One example is:

(87)   There are dogs.

which might be seen as being of the form "There are dogs Φ," with Φ being the empty predicate. This is again equivalent to "Some dogs are Φ," which on the understanding of the vacuous predicate is just to say that some dogs are some way or other. In particular, (87) is a quantificational statement, but the quantifier in it is not "there are" but rather "dogs." And the same holds for sentences often used in debates about ontology. "There are numbers" is a quantificational statement, but the quantifier in it is not "there are" but "numbers." This supports claim 2 in section 3.1, but it so far leaves claim 1 open, i.e. the claim that the ontological question about Fs is the question "are there Fs?" To make progress on that we need to look more closely at what we do with quantifiers in communication.

---

[4] It is widely believed that the generic reading is also quantificational, involving generic quantification. I will sideline this here.

3.2.4 The Domain Conditions Reading

Quantifiers quite clearly have at least one role in ordinary communication, a role that is closely tied to how our language represents the world in its most general features. Whether this way to represent the world is in the end accurate, and captures how the world really is, is, of course, another question, one we will discuss in detail in later chapters. But even if it is inaccurate, there are certain quite uncontroversial aspects of how natural languages represent the world. Quantifiers have a crucial connection to this general way of representing the world.

Our language represents the world as containing objects which have certain properties. Some of these objects are picked out by terms which keep track of particular objects, paradigmatically names. In the simplest subject–predicate sentences we have a subject term which picks out an object in the world, a predicate which expresses a property, and the sentence as a whole asserts that the object has that property. But the simplest case does not always obtain, and sometimes we make claims not about a particular object, but about the domain of objects more generally. We want to say something about these objects in general, whatever they may be, rather than about a particular one of them. To do this is one of the functions of quantifiers. When I say

(88)   Something fell on my head.

in an ordinary use of that sentence, I want to make a claim about the domain of all objects and say that among them is one that has a certain feature, falling on my head. I am not using the quantifier to pick out the domain of all objects, in the way that "Fido" picks out the dog. I use the quantifier to make a claim that is true just in case the domain of all objects is a certain way. Or in other words, I am making a claim about the objects, whatever they may be, and so that one of them is a certain way. Quantifiers play at least this role in a language like ours that represents the world as containing objects. It not only represents it as containing some particular object or other, but as containing a domain of objects. Sometimes we make claims that concern that domain as a whole, and quantifiers are used to do just that.

To talk about a domain of *objects* should be understood in the most neutral way possible, without bringing in already some substantial notion of what an object is. We could equally well have used other, related notions like entity, individual, or thing. Similarly when we describe the other aspect of how we represent the world as containing objects that have *properties*, this, too, should be seen with as little philosophical baggage as possible. We could just as well have said that the objects are represented as having features, or characteristics. Furthermore, no assumption is made here that we represent the world as containing only objects, or that the domain of objects exhausts what reality is like. All these claims are very substantial and controversial. What should be uncontroversial is that we represent the world, correctly or not, as at least containing objects and sometimes we make a claim about the domain of these objects, whatever they may be, when we use quantifiers.

This aspect of what we do with quantifiers is part of their use in ordinary, everyday communication. We make claims about the domain of objects, whatever they might be. What is controversial, however, is the question of whether this is the only use of quantifiers. Maybe quantifiers have besides the uncontroversial use also another one. This is exactly what I take to be the case, and I hope to make it clear in the following. In ordinary, everyday communication we use quantifiers in another way as well, besides making a claim about the domain of objects. This further use is just as closely tied to our needs in ordinary communication, but it serves a different need and is also otherwise importantly different. We will see what this need is, and how it is different, in the following.

As is customary, let's call a *reading* of a sentence a way the sentence can be correctly understood. A sentence has more than one reading if there are two ways to understand the sentence that don't have the same truth conditions. In this sense, the sentence

(89)   Everyone likes a famous movie star.

has two readings. One where everyone likes some famous movie star or other, the other where everyone likes the very same movie star, say Sean Connery. In this example the standard explanation is that the two readings arise from different relative scope that the quantifiers "everyone" and "a famous movie star" can have. On the first reading "everyone" has wide scope, on the second one it has narrow scope. The question for us then becomes whether quantified statements in general have more than one reading, not necessarily arising from issues of relative scope, but from the quantifiers themselves. Given the uncontroversial facts about quantifiers we can say that quantifiers have at least a *domain conditions reading*, or more briefly a *domain reading*, or as we will also call it, for reasons to be made explicit below, an *external reading*. On this reading the quantified sentence makes a claim about the domain of objects, whatever they may be. This is how quantifiers are commonly used, and it is clear that we have a use for them in this way in ordinary communication. What is at issue is whether there is another reading as well.

3.3 Semantic Underspecification

Semantic underspecification is a general phenomenon in natural language. It occurs when the semantic content of a sentence does not determine a unique content of a literal utterance of that sentence. We will in this section first look at the phenomenon in general, while trying to sideline controversial issues that won't matter for our main points to follow. After that we will discuss to what extent it applies to quantifiers in particular.

To understand the thesis of semantic underspecification more clearly one has to make several notions more precise, and depending on how they are made precise the thesis becomes more or less controversial. For example, depending on what one understands by a sentence it can become either trivial or controversial. If a sentence is simply a sequence of words, then clearly many sequences of words can be used to make utterances with different truth conditions. Standard examples of syntactic ambiguity, like

(90)   He gave her cat food.

can be read as him giving food to her cat, or as him giving her food intended for cats. Similarly

(91)   Sue hit a man with a wooden leg.

which might specify what the man was hit with, or what the man hit was like with respect to his leg. A more restrictive use of a sentence would take a sentence not merely to be a sequence of words, but such a sequence together with basic syntactic structure, in particular argument structure. In this sense of a sentence the obvious two ways to read (91) are not readings of the same sentence. An even more restrictive way of understanding a sentence is as its full syntactic tree, with all relevant information that syntax, the linguistic discipline, might attach to it. This way of understanding a sentence makes it a highly theoretical notion and brings in all kinds of further issues that are not relevant for our discussion here, related to the relationship between syntax and semantics. We will thus investigate semantic underspecification in the sense in which a sentence is understood in the second way: a string of words with basic argument structure. Given this understanding of a sentence, it gives us a particular notion of semantic underspecification. In this section we will see that such underspecification is widespread, and that there is one particularly interesting case of it.

This notion of underspecification relies on a particular notion of a sentence, which is understood as a sequence of words with basic syntactic structure. But when are two strings of letters the same word? This is another complicated question, which can be made vivid with our example (90). Is the "her" in both readings the same word? One is the possessive pronoun, the other one is just a regular pronoun. If you tried to do this sentence with the male pronoun it would not work, with one reading requiring "his" and the other "him." This might be taken as evidence that in the female pronoun case the words are different, they are just spelled with the same letters.

The details of how to understand the phenomenon of semantic underspecification are controversial within the philosophy of language, but fortunately they won't matter for our discussion.[5] We only need to consider the general phenomenon, not its precise theoretical description. The phenomenon can be illustrated with this simple picture: when we make an assertion we utter a sentence, and that utterance has certain truth conditions. These are truth conditions of an act of uttering, or of the product of that act, a particular token of letters or sounds. What we uttered, however, was a sentence, part

---

[5] For further discussion of the general issue of semantic underspecification and its place in the philosophy of language, see for example Stanley (2000), Recanati (2001), or Bach (2005).

of a language, and this sentence has a certain semantic content in that language. The semantic content of the sentence comes from what the words mean in the language and how they were put together. The question for us now is about the relationship between the semantic content of the sentence uttered and the content of the utterance of that sentence. To keep things simple, all we need to look at is content understood as truth conditions, not more fine-grained notions of content. Does the content of the sentence determine the content of the utterance? Are they simply the same?

Although it might be tempting to think that the content of the sentence uttered and the content of the utterance of that sentence are most of the time the same, this is quite clearly false. First, there are the obvious cases where the contents are not the same, but they seem to be closely related. Most prominently, the content of the English sentence

(92)   I am hungry.

isn't the same as the content of my utterance of it right now. However, it plausibly determines the content of the utterance together with general features of the event of the utterance. The content of the sentence might be something like that whoever utters this sentence is hungry when they utter it, but the content of the utterance is that TH is hungry at t. The former content settles what the content of the utterance will be on any given occasion of utterance. The details of this are, of course, up for grabs. We are instead interested in whether there are also other cases of a gap between semantic content and utterance content. Are there cases where semantic content does not determine utterance content, but only puts some constraints on it? Are these cases widespread or only marginal? And do they occur with quantifiers? To get a sense of the situation, let's first look at some well-known cases of what appears to be semantic underspecification.

Genitive A genitive construction like

(93)   John's car is fast.

can make a variety of different contributions to the truth conditions of an utterance, and the semantics of the sentence containing the genitive doesn't seem to determine which one it is. (93) can be uttered with the truth conditions that the car John owns is fast, or the car John is driving, or the car John is sitting on, and so on. Almost any relation between John and a car can be made salient enough so that an utterance of (93) has the truth conditions that the car John has that relation to is fast. The sentence by itself is silent on what relation might be the one in a particular utterance.

Plural Another well-known case of semantic underspecification arises with plurals. Plural expressions systematically have at least two readings, which in turn give rise to different readings of sentences in which they occur. Consider:

(94)   Four philosophers carried two pianos.

The commonest way to read this sentence is that four philosophers together carried two pianos one after the other. But this sentence also has a reading where four philosophers each carried two pianos together. Or four philosophers each carried two pianos one after the other. Or four philosophers together carried two pianos together. These readings are generated by two readings that the plural phrases "four philosophers" and "two pianos" have. They can either be understood *collectively*, as being about a group of people or pianos. Or *distributively* as being about each one in that group. Plural phrases have these two readings, but the semantics of the plural doesn't say which one it has on a particular occasion of an utterance. The semantics of plurals simply tells us which readings such a sentence can have; it sets some limits on how it can be read, but not how it is to be read on a particular occasion of utterance.

Reciprocals Another interesting class of expressions that leads to semantic underspecification are reciprocal expressions, like "each other." These expressions are interesting because more so than the cases above, it is not at all clear to ordinary speakers that they do give rise to underspecification and what different readings they give rise to. In fact, they can give rise to a very large number of readings, something that has only recently been uncovered in linguistic research. Take some simple examples:

(95)   The exits on the Santa Monica freeway are less than a mile from each other.

(96)   The professors live less than a mile from each other.

On the preferred reading of the first example, it is not required that all exits are less than a mile from any other one, all that needs to be the case is that there is another exit within a mile. On the preferred reading of the second example, however, something else is required. It is not enough if every professor lives within a mile of some other professor. Rather it is required that all the professors live within a one mile circle. However, both examples admit of both kinds of readings. There are also many other readings that reciprocals can give rise to, which are discussed in Dalrymple et al. (1998).

Polysemy Possibly the commonest source of semantic underspecification of sentences is polysemy, the phenomena that particular expressions have more than one of a group of closely related readings. This is almost universally true of verbs in English, in particular, and to an extreme, of verbs like "get" or "do." Consider, for example

(97)   Before I get home I have to get some beer to get drunk.

"get" here has three slightly different, but related, readings. And this phenomenon is very general. Almost any verb can be used with more than one reading, but the readings that it can have generally are closely related. This aspect distinguishes polysemy from ambiguity, which here would be lexical ambiguity. Ambiguity in this sense is the phenomenon that different words are written and spoken in the same way. "bank" is the standard example. Ambiguity does not count as a case of semantic underspecification for us here because the sentences both written as

(98)   I'll meet you at the bank.

are different sentences. They are built up from different words that happen to be written and pronounced the same. But in cases of polysemy it is the same word that has the different readings. That "bank" and "bank" are written and pronounced the same way is just an accident, but that "get" is pronounced and written the same way in its different occurrences in (97) is not an accident. Here we have one word, but a word that allows for different readings. Similarly

(99)   My parrot expired on Wednesday, my drivers license expired on Thursday.

It is one word "expire" with two different, but related readings. Polysemy is not only common with verbs but many other parts of natural languages: adjectives like "hot" and "cold," nouns like "rat" (animal or snitch), and so on and so forth. The phenomenon of polysemy is everywhere in natural language. Overall, then, we have good reason to think that semantic underspecification is widespread and semantic content underdetermines utterance content in a variety of ways.

3.4 Quantifiers as a Source of Underspecification

In this section I will argue that quantifiers are polysemous. They have more than one reading and they thus give rise to different readings of sentences in which they occur. These different readings have different truth conditions, as is required for the readings to count as different. We will see how the readings differ in truth conditions, and what evidence we have for there being different readings in the first place. We will tackle these tasks one at a time.

3.4.1 How to Settle the Issue

When we wonder why we should think that there is more than one reading of quantifiers, we should assess this question from the point of view of the ordinary investigation of natural language. That is, we should find out whether or not there are different readings to quantifiers not on metaphysical grounds, but on empirical grounds, as a result of the investigation into our own language and what use we make of it in ordinary situations of communication. When we look at our language and what we do with it, can we see that quantifiers have more than one reading? Whatever might follow from that for philosophy, however desirable or undesirable this might be for philosophical reasons, that should take second place. First we should look at our language, then we should find out what follows from it for philosophy.

But how do we decide whether or not quantifiers, or any expressions, have different readings? One simple way is to consider a sentence, and see if it can be read in two different ways. This is how we can easily see that sentences like (93) or (94) have different readings. Once the different readings are pointed out it will generally be agreed upon that they can be read in these different ways, and thus we have evidence that there are these different readings. Unfortunately when it comes to quantifiers the issue is so controversial that philosophers in particular firmly hold their ground by denying that they see any but one particular way to read the sentence. And often they disagree what the one and only way to read the sentence is. In the debate about quantification in the past, to be discussed in more detail below, this method has proven to be ineffective. A method of cases, by looking at individual examples, won't be good enough. I will thus propose to proceed differently. In the following I will look at the use we have for quantifiers in ordinary, everyday communication more generally, and motivate that we have a need for quantifiers to play two related, but different, roles. These two roles correspond to two readings quantifiers have which go hand in hand with two different contributions they can make to the truth conditions. Thus instead of focusing on individual examples of quantified sentences we will focus on the need we have for quantifiers in general in ordinary communication, and on the basis of this we will be able to see that quantifiers should play two different roles in ordinary communication, answering to two different needs we have for them. Only once we have seen what these communicative needs are, how quantifiers allow us to meet those needs, and why they require different contributions to the truth conditions, will we illustrate the difference with individual examples that bring out these two readings.

3.4.2 The Communicative Function of Quantifiers

Quantifiers, uncontroversially, have at least one function in communication. They are used to make a claim about the domain of objects. This reading of the quantifiers we labeled above the domain conditions reading (or alternatively the domain reading or external reading), since what the quantifier contributes to the truth conditions is to impose a certain condition on the domain of objects, together with the rest of the sentence, that has to be fulfilled for the sentence to be true. In the simplest case

(88)   Something fell on my head.

in its domain conditions reading imposes the condition on the domain of objects, whatever they might be, that one of them fell on my head. This use of the quantifier is tied to the way we represent the world as being like in natural language, correctly or incorrectly. We represent it as containing objects, and sometimes we make claims about these objects, whatever they may be. In the above case we simply say that among these objects is one that has a certain feature. The question is whether quantifiers are always used in their domain conditions reading.

I will argue that quantifiers have besides their external, domain conditions reading, also another one. This other reading will be labeled the *internal*, or *inferential role*, or simply *inferential*, reading, for reasons to be made explicit below. To do this I will argue for the following:

1. We have a need in ordinary, everyday communication to have expressions that have a certain inferential role as well as the syntactic features of quantifiers.

2. The expressions that we use to meet that need in ordinary communication are quantifiers.
3. Quantifiers in their domain conditions reading don't have the inferential role, and so don't meet our expressive need in these cases.

From these I argue we can reasonably conclude that quantifiers in English are polysemous and have at least two different readings that make different contributions to the truth conditions of utterances of sentences in which they occur. Once we have seen the argument from a general communicative need for quantifiers with a different reading than their domain conditions reading I will present several examples that illustrate the different readings further. At this stage individual examples will hopefully be less controversial.

We use language to represent information that we gather about the world. Sentences, for example, can express the contents of our beliefs. And when we try to communicate our beliefs we utter sentences that get the content of our beliefs across to the hearer. Different parts of the sentence uttered carry different parts of the information. In a simple subject–predicate sentence the subject terms tell us what the information is about, the predicate tells us what that thing is supposed to be like. In straightforward cases of communicating information, where I have the information I want to communicate, and I have a sentence available that communicates that information, things are easy. But sometimes things aren't so easy. I would like to look at one particular case where an ordinary obstacle can make communication more complex, and where our second reading of quantifiers makes an appearance. It is *communicating under partial ignorance*. Sometimes I don't have all the information any more that I was hoping to communicate. I might still have some of it, but lost some other part. I am partially ignorant, in that I have lost some, but not all, of the information I wanted to communicate. Still, I have kept some information, and what I can still hope to do is to communicate the information that I have left. I just need a sentence that allows me to do that. Let's look at an example.

Suppose you are writing a biography about Fred, an elusive and mysterious figure. One day you learn the very useful and revealing information about Fred that he admires Ronald Reagan, the American president who is also admired by many Republicans, members of Reagan's political party. This, you hope, will allow you to conclude all kinds of things about Fred. But when you hope to tell me about it, you forgot who precisely it was that Fred admires. All you remember is that whoever it was, that person is also admired by many Republicans. This is not quite as good as what you hoped to tell me, but it is still better than nothing. And you can communicate the information you have left to me quite easily by uttering the following, which also tells me that you only have incomplete information:

(100)   There is someone Fred admires who is also admired by many Republicans. But who was that again?

Of course, if you had not forgotten who it was then you could have done better. You could have told me this:

(101)   Fred admires Ronald Reagan. Reagan is also admired by many Republicans.

But you didn't have all the information left, so (100) will have to do. The quantifier in (100) takes the place of the part of the words that would communicate the information that you had lost. Had you known who it was you could have used (101) instead. But since the "Reagan" part of your information was lost, all you had left was that whoever it was that Fred admires is tied to someone many Republicans admire as well. The quantifier does just that for you. It replaces the "Reagan" part of expression of the information, and it ties who is admired by Fred to who is admired by many Republicans, i.e. it is the same one.

This situation is completely general. It doesn't matter for the example that it is Reagan who is admired. It could have been anyone. All that is required is that there is some useful information about Fred, and that you lose part of it, but you still are able to communicate the rest, what you have left after losing part of the information. Suppose instead then that the information you learn about Fred is that he admires Sherlock Holmes, the super-rational crime solver in Conan Doyle's stories, who is also admired by many detectives. But, again, you forget who it was that Fred admires before you get to tell me. All you have left is that whoever it is is also admired by many detectives. And that information you can easily communicate to me with

(102)   There is someone Fred admires who is also admired by many detectives. But who was that again?

This is still useful, and it lets you draw conclusions about Fred in various ways. And when you remember, you might say "Of course, Sherlock Holmes!" Then you can tell me what you were hoping to tell me all along:

(103)   Fred admires Sherlock Holmes. Sherlock is also admired by many detectives.

Quantifiers in ordinary communication sometimes play the role of placeholders for parts of information that are missing, either because they were forgotten, or because they were never acquired. That is not the only thing quantifiers do, but it is one of the things they do, something that we have a need for in ordinary, everyday communication. Quantifiers play that role, among others, when we try to communicate information. And for them to do that they have to have a certain inferential role: they have to figure in certain patterns of valid inferences. To see this, we have to note two things. First, when you forget part of the information you had then the expression of the incomplete information you have left is implied by the expression of the complete information you used to have. That is to say, (101) implies (100). When you lose part of the information then this is a loss in at least this sense: what you had before implies what you have left, but not the other way round. The second thing to note is that there are many different information states you could have started out with; it might have been Reagan, or Superman, or Leonardo da Vinci, etc., who is admired by both Fred and many Republicans. But when you forget who it was then in each case you end up with the same information, namely (100). It doesn't matter who it was that is admired, which is the part of the information you forgot, you are left with the same information after you forgot that. So, no matter what you replace "Reagan" with in (101), it also implies (100). And this tells us that for "someone" to play the role we need it for in (100), it has to have a certain inferential role, namely the inferential role that

(104)   "Fred admires t." implies "There is someone Fred admires."

where "t" is any term for a person, like "Reagan," "Sherlock," or any other one. More generally, the inferential role of the particular quantifier "something" is this:

(105)   "F(t)" implies "Something is F."

for any term "t" and any predicate "F." This is what we need in ordinary communication.[6] I will argue next that "something" in its domain conditions reading does not have that inferential role, and thus quantifiers need to make two different contributions to the truth conditions to play their two roles in communication, i.e. quantifiers are polysemous.

3.4.3 Inferential Role and Empty Terms

In its domain conditions reading "something" makes a claim about the domain of objects that are part of reality. The inference from "F(t)" to "Something is F" in the domain conditions reading of the quantifier is thus valid just in case "t" picks out an object in the domain that the quantifier ranges over. Thus the particular quantifier "something" in its domain conditions reading has the inferential role discussed above just in case every term "t" in the language picks out an object in the domain that the quantifier ranges over. But this seems to be false for English and other natural languages. There appear to be many examples of terms that fail to pick out an object, and there appear to be many examples of *empty names*, names that do not stand for anything. For example, there are names from fiction like "Sherlock Holmes," there are names from failed scientific theories, like "Vulcan," the alleged planet responsible for the particular orbit of Mercury, and many more. It might thus seem clear that inferential role and domain conditions come apart in truth conditions in a language like ours.

---

[6] It is common in philosophical debates to object to, or at least be suspicious of, using examples involving "admire" or similar intensional transitive verbs, since they create "opaque" contexts, have an unclear semantics, and make everything more complicated. But this is all irrelevant here. All that matters is that there is a part of the sentence that carries a certain aspect of the information, in this case the information who is admired. That that part occurs in a sentence that behaves differently when it comes to substituting co-referential terms is irrelevant for this. We will also see other examples below.

But this reasoning is problematic in its reliance on examples. Each alleged example of a term that fails to pick out an object is controversial within the philosophical debate. Some philosophers have argued that "Sherlock Holmes" does succeed in referring after all, although not to ordinary material, but to an abstract object.[7] Such theories are not unproblematic, but it would be a mistake to try to refute or argue against them here. We don't have time now to get bogged down trying to argue against every philosopher who denied that there are any truly empty terms in our language. Single examples are thus not the way to go.

Just as in the case of quantifiers, it is hopeless to try to convincingly motivate that some terms are empty by giving examples. Since there are high philosophical stakes here, every example is contested. Instead we can look again at general features of communication, and in particular general features of having names in our language, that will give us enough reason to think that we do have empty names in our language. This way we can avoid controversies about particular examples, and focus on what matters for us: that we have good reason to think that there are examples of empty terms. In particular, let's in the following consider first empty names, and then look at other terms which can be empty as well.

A name is a word that is used to pick out an object. With the use of such names we can represent information about that object. We have many names that, somehow, succeed in doing this. How names manage to stand for the object they name is not at all clear, but it should be clear that something has to happen for them to do this. They, somehow, must be attached properly to the object, and this attachment must, somehow, be preserved in a community of speakers. Many names do succeed in working this way. But it should be clear that this is an achievement when it works. Maybe it is a common achievement, maybe it is a fairly easy thing to do and it works most of the time, but it is an achievement nonetheless. But when things can go right, they can go wrong. Sometimes an attempt to attach a name to an object doesn't succeed. And sometimes such an attachment isn't successfully preserved over time. It is hard to tell when this happens, and with which names, since we don't know yet what has to happen for it to succeed. But no matter how we manage to get names attached to objects, it is an achievement when it happens. And as an achievement, it is something that can fail to be achieved. Even if it is easy to achieve, it is hard to imagine that it is so easy to achieve that we have reason to think we always achieve it. In our language we have an incredibly large number of names that stand for an incredibly diverse group of objects. Although we don't know how names attach to the objects they name, the diverse body of names and objects named we have in our language gives us plenty of reason to think that something will have gone wrong with some of them. We have good reason to think that sometimes we fail to achieve what they are trying to do. And thus we have good reason to think that some names in our language

---

[7] For example, Peter van Inwagen has argued for such a position in his van Inwagen (1977) and Amie Thomasson has defended a detailed version of it in Thomasson (1999).

are not attached to any objects. They are empty names, names that aim to pick out an object, but, for some reason or other, fail to do so. Whether particular fictional names are among them is controversial, whether "Vulcan" is among them might be less controversial, but whether there are any names of this kind at all should not be controversial.[8]

We have a need in ordinary communication for quantifiers to have a certain inferential role. And we have good reason to think that some terms or other in our language are empty. These two facts together tell us that quantifiers in their domain conditions reading can't do all we need them to do for us, in ordinary communication. In their domain conditions reading quantifiers don't have the inferential role they are supposed to have. "F(t)" does not imply "something is F," on that reading. "t" might be one of these empty terms, and if so we would move from truth to falsity. No object in the domain of quantification is denoted by "t," since "t" is empty. The domain conditions reading does not give one the inferential role we need in communication.

But we do use quantifiers in just that inferential way when we want to communicate in ignorance, and in other situations as well. In actual communication quantifiers do play just that role. And this is just fine since quantifiers are polysemous, just like many other expressions. They have a domain conditions reading and an inferential role reading. These two readings are different with respect to truth conditions, but they are also related, in that they coincide in simple cases, to be discussed shortly.

I thus hold that quantifiers are polysemous, they have at least two different readings: one is the domain conditions reading, where they make a claim about the domain of objects in the world; the other is the inferential role reading, where they are inferentially related to their instances. Above we have focused on the particular quantifier "something," which in its inferential role reading has the inferential role

---

[8] Another group of empty terms are terms other than names. All we need for our main argument is that there are terms that can carry some aspect of information that don't succeed in picking out an object. But "picking out" doesn't have to be reference, it can be what is often called designation or denotation, the relationship that holds between a definite description and the object, if any, that uniquely satisfies the description. There clearly are cases of descriptions that don't designate an object, as for example "the president of Texas" (Texas has a governor, but not a president). Such terms can still figure in true sentences, as in

(106)   Fred admires the president of Texas.

But descriptions by themselves are attached to a large body of literature, and many subtle issues, which we should also sideline here. Maybe the above description "the president of Texas" is used referentially to pick out the governor of Texas? Maybe Fred admires whoever would be the president of Texas, even though no one in fact is? We won't have to get into these issues, since all we need is that some terms are empty, and this should be clear from the discussion of names.

Another fairly uncontroversial case of empty terms are demonstratives that fail to refer. For example, I might say

(107)   Did you see that deer?

while failing to refer to a deer, since there isn't one there. I was merely tricked into thinking so by some flickering light. Any such example would do, but I'll stick to names in the following.

(108) "F(t)" implies "something is F."

A parallel case is the universal quantifier, which in its inferential role reading has the inferential role

(109) "Everything is F" implies "F(t)."

We will see below more about what an instance of a quantified sentence is, i.e. what expressions "t" can be used in the above inferences. And we will discuss, in the appendix to this chapter, how this account of quantification can be extended to generalized quantifiers.

3.4.4 Some Examples

Above we motivated the need for an inferential reading of quantifiers with a general communicative need. This motivation for such a reading did not depend on any particular examples of quantified statements. In this section we should look at some examples of quantified statements, whether or not they have different readings, and what their readings are. Consider:

(110) Everything exists.

This example is subject to persistent debate within philosophy. Some philosophers take it to be obviously true. After all, they reason, all the things that there are, out there in the world, have one thing in common: they all exist. "Everything exists" is thus quite trivially true. On the other hand, some philosophers have argued that (110) is quite trivially false, that we can easily refute it with counterexamples. It can't be that everything exists, since Santa doesn't exist. If *everything* exists then it should follow that Santa exists as well. But Santa doesn't exist, so not everything exists. Both of these arguments are very good, but these two ways of reasoning for and against the truth of (110) are simply based on the two readings that the quantifier in (110) has. On the inferential reading it does trivially follow from (110) that Santa exists. The latter is false, and so on that reading (110) is false. On the domain conditions reading, on the other hand, no such inference is justified. The domain conditions reading does not give one the inferential role that from "Everything is F" any instance "F(t)" can be inferred. On the domain conditions reading of (110), however, the motivation for its truth is perfectly compelling. On the domain conditions reading (110) is true, and on the inferential role reading it is false.[9]

(110) is, of course, a philosophically loaded example. We can similarly see the two readings at work in perfectly ordinary examples. Consider, in particular, quantification into the position of a complement of a verb that has a "notional" reading:

(111) I need an assistant.

_____

[9] More on quantification and existence will follow in section 3.5.2.

This sentence has a reading where I need some assistant or other, any one will do, and a reading where I need someone in particular, who happens to be an assistant. Similarly

(112) There is something we both need, namely an assistant.

has a reading where we both need someone to be our assistant and a reading where we both need someone who happens to be an assistant. It is easy to see how these two readings arise from the two readings of the quantifier, but hard how they can be acceptable if quantifiers only have one reading. The inferential reading of the quantifier gives us the first reading of (112), while the domain conditions one gives us the second.[10]

3.4.5 Inferential Role and Truth Conditions

To say that quantifiers have an inferential role reading is not to endorse a substantial approach to semantics. It does not favor an inferential role semantics over a truth-conditional semantics, and it is neutral with respect to the larger debates about what the goals and methods of a semantics theory should be. An inferential role semantics is generally seen as a large-scale alternative to a semantics based on reference and other language–world relationships. But to say that quantifiers have an inferential role reading is not to endorse any side with respect to such larger debates. Inferential role semantics could certainly accommodate inferential role readings of quantifiers. And a truth-conditional semantics can deal with it just as well. Not only is an inferential role reading of the quantifiers compatible with a truth-conditional semantics, we can furthermore see what truth conditions give it the inferential role it has. This section will investigate the contribution to the truth conditions that quantifiers make in their inferential role reading.

The inferential role reading of the quantifier is just another reading it has, one, as I will spell out shortly, that makes a different contribution to the truth conditions of the utterance of a sentence in which it occurs, but it makes a contribution to the truth conditions, just like the domain conditions reading of the quantifier. The question now is simply this: what contribution to the truth conditions gives the quantifier the inferential role it is supposed to have? When we want to know what the contribution to the truth conditions of the inferential role reading of the quantifier is, we must try to find out what contributions to the truth conditions would give it that inferential role. And here, it turns out, there is a compelling candidate for what the truth conditions should be.

A quantified sentence like

(SomeF) Something is F.

---

[10] Friederike Moltmann uses examples like (112) in Moltmann (2003a) to motivate that "something" is a special quantifier, in a technical sense, and proposed a semantics for it that is distinct from the usual semantics for other quantifiers. I hold, to the contrary, that quantifiers in general are polysemous and that the "special" uses of "something" can be seen as resulting from one of the readings that quantifiers in general have.

in its inferential role reading has to have truth conditions such that each instance "F(t)" implies it. That is, it has to have the following inferential role:

(113)   F(t)
        _____

        Something is F.

And no matter what instance you start out with, it has to imply the very same sentence (SomeF), in its inferential role reading. The simplest way this can be is the following: (SomeF) in its inferential role reading is truth conditionally equivalent to the disjunction of all the instances "F(t)" which imply it:

(114) $F(t_1) \vee F(t_2) \vee F(t_3) \vee \ldots$

If the quantified statement has those truth conditions, then indeed every instance will imply it, since a disjunction is implied by each one of its disjuncts. If the internal reading of the particular quantifier gives the utterances of sentences in which it occurs truth conditions equivalent to the disjunction of all the instances that imply it, then this would lead to exactly the inferential role for which we need this reading of the quantifier. The question thus becomes: what is an instance of the quantified statement, and how many of them are there?

When we talked about an instance of (SomeF), we meant a sentence of the form "F(t)" where "t" was any term. But what is a term? In simple formal languages it is easy to say what a term is: a (closed) term is either a simple name, or the result of applying a function symbol to things which are already terms. Terms are thus inductively defined, as the smallest set containing all the names and being closed under "term forming operations," that is, getting more complex terms from simpler ones, generally with the use of function symbols. But this picture of terms is based on the idea of terms being denoting expressions, ones that aim to pick out objects in the domain, in either a simple way or a complex way. In natural language, on the other hand, we have many categories of expressions that are in certain ways similar to paradigmatic cases of terms, but completely different in other ways. Take expressions like "the tiger" in their generic use, as in

(115)   The tiger is fierce.

This expression appears syntactically in exactly the position of paradigm terms, but it is completely unclear whether it has the semantic function of picking out an object. Does the particular quantifier in its inferential role reading interact with this term? It seems that our reasoning about communicating in partial ignorance would carry over to this case just as well. You might learn (115) but then forget who or what it is that is fierce, and communicate simply what you have left by

(116)   Something is fierce. (But what, again?)

And when you remember you could simply say: "Of course, the tiger!" It seems irrelevant for this use of the quantifier not only whether or not this term succeeds in referring, but even whether it is in the business of referring. All that matters is that some part of the information is carried by some part of the sentence that can be properly replaced with the particular quantifier. And this is true here no matter what precisely is going on with the semantics of generic noun phrases like "the tiger." Thus what should be considered a term has to be understood more broadly than in artificial formal languages, which reflects the extra complexity that natural languages have. To say what precisely the instances of quantified statements are that are relevant for their inferential role reading, or equivalently, what terms are in this particular sense, is a substantial task. It is not sufficient simply to take names or similar expressions, as seen above. Nor is it sufficient to understand as terms all expressions that syntactically can occupy the position of the quantifier. The latter won't do since the quantifiers themselves can occupy that position, leading to unintended results. For example, it doesn't seem to capture a reading of the quantifier when one reasons:

(117) Something is fierce. But what was that again? Oh yes, of course, it's something.

Even more extreme, suppose someone were to assert that

(118) Nothing is fierce.

intending it in the sense where neither Fred, nor the tiger, nor anything else is fierce. It would be strange to contradict that person by saying that they can't be right because if nothing is fierce then there is something which is fierce, namely nothing. (118) doesn't seem contradictory, as it would be on an inferential role reading that allows any grammatical instance to be inferentially relevant. To make the inferential role precise one needs to make precise what terms are and thus what grammatical instances are inferentially relevant. Although it would be desirable to do this precisely, this precision will not be required for anything that follows, nor is it required to appreciate the motivation and function of the inferential role reading of the quantifiers. We will thus rely on our ordinary judgments about what is and is not a good inference involving this particular reading of the quantifier. For most cases, these will be clear, and when complications arise below they will be made explicit.[11]

Suppose then that we have an understanding of what an instance of a quantified statement is. How many such instances are there? Even though we don't have a precise account of instances, we can be sure that the number of instances will be countably infinite. The most paradigm cases of a term alone guarantee that it will be infinite. We have, for example, in our language a term for every natural number (just regular

---

[11] To specify the class of terms for a fragment of English precisely one would first have to make that fragment precise, which for any fragment that contains at least generic NPs and non-trivial examples of terms is itself a substantial task. Once we are given a precise characterization of a fragment then specifying the terms in that fragment should present no substantial further difficulty, I conjecture, but the characterization of such a fragment is a substantial task.

arabic numerals would count), and we have terms like Mary, the mother of Mary, the mother of the mother of Mary, etc. But since our instances are tied to our language and our language has a finite basic vocabulary, only countably many instances can be constructed. That means that the truth conditions of a quantified statement in its inferential role reading are equivalent to a disjunction (in case of the particular quantifier) or conjunction (in case of the universal one) of infinitely many instances. But this should not be disturbing. The claim is not that these sentences somehow are infinitary conjunctions or disjunctions, only that they are truth conditionally equivalent to them. My proposal is, of course, not that English is an infinitary language.[12] That is pretty clearly not so. Every sentence of English is finite. What the proposal is instead is that the truth conditions of some readings of some sentences in English are the same as the truth conditions of some sentences of an infinitary version of a fragment of English. Thus the truth conditions of

(119)   There is someone I admire.

on its internal reading are equivalent to the disjunction

(120)   I admire Fred Dretske, or I admire Fred Flintstone, or I admire Fred Astaire, or . . . .

where we have one disjunct for every instance of (119). To be a little more precise, the inferential role reading of (119) is truth conditionally equivalent to the disjunction of all instances "I admire t," where "t" is one of our terms. This is commonly written as

(121) $\bigvee_{t \in Term}$ I admire t

where "Term" is the class of all terms in our language, the class we didn't specify precisely above. But it should be clear that what the terms are is central to what the

---

[12] Infinitary languages are technically unproblematic. Disjunctions and conjunctions of infinite length are just as well behaved as those of finite length. In formal logic languages with infinite disjunctions and conjunctions have been studied and used for a long time. Such languages can have not just countable disjunctions or conjunctions, but uncountable ones as well. It is common to use the following way to describe what kinds of disjunctions can be formed in a formal language. If we start with an artificial language $\mathcal{L}$, we call the language $\mathcal{L}_{\infty,\omega}$ the language that allows for arbitrary conjunctions or disjunctions, with no restriction over how many sentences can be conjoined at once into a single conjunction or disjunction. We call $\mathcal{L}_{\omega_1,\omega}$ the language that allows for countable conjunctions and disjunctions only, and more generally, we call $\mathcal{L}_{\kappa,\omega}$ the language that allows for conjunctions and disjunctions of κ many sentences at once, whereby κ is some infinite cardinal number. When a language allows for any infinite disjunctions or conjunctions, then we call it an *infinitary language*. The second "ω" in the above subscripts is always the same, and is used to describe the complexity of quantification, i.e. how many variables we can bind with a single quantifier. This level of complexity in the description of infinitary languages is irrelevant for us now, but it will be important later on in chapter 9, and we will return to it then.

We will mostly consider only single quantifiers here, not iterated ones. However, sentences with iterated quantifiers pose no further problem. They are truth conditionally equivalent to, for example, infinitary disjunctions where the disjuncts are infinitary conjunctions, and so on for deeper iterations. For all the details on the more technical side, see Keisler (1971).

truth conditions of the quantified sentence are, and that any indeterminacy about what the terms are leads to indeterminacy about what the truth conditions are. This is not to say that there is any indeterminacy in what the terms are, i.e. which expressions figure correctly in certain inferences. But it is to say that if there is any indeterminacy here, then it will affect the truth conditions of the quantified statements in their inferential role.[13]

Thus, for the case of the inferential reading of the particular quantifier, having the truth conditions of being truth conditionally equivalent to disjunction over all the instances in our own language gives (SomeF) the inferential role that we want it for. This is the solution to the problem of what truth conditions give (SomeF) this inferential role, and it is the optimal, although not the only, solution. Other truth conditions would also give the quantifier the inferential role where every instance implies (SomeF). Nonetheless, the truth conditions outlined above are the optimal solution in that they exactly give (SomeF) this and no other undesired inferential behavior. The truth conditions of (SomeF) need to be such that it is implied by any one of its instances. Thus it needs to be logically weaker than each of them, and the strongest truth conditions with this feature are simply being truth conditionally equivalent to the disjunction of all the instances. However, even weaker truth conditions would also give (SomeF) this inferential role. For example, if we add, besides the disjunction of all its instances, a further arbitrary disjunct D then those proposed truth conditions of (SomeF) would still have it implied by all its instances, besides by D as well. But (SomeF) could then also be true even though all of its instances are false, simply because D might be the only true disjunct. These truth conditions would give (SomeF) the inferential behavior we focused on, but it would also give it further inferential behavior that was unintended: being implied by D. The optimal solution to our problem gives (SomeF) the inferential behavior we want it for and just that inferential behavior. This is, to have the strongest truth conditions that give it the inferential role for which we need it: being truth conditionally equivalent to all its instances in our own language. Similarly, for the universal quantifier, we should take it to have the weakest truth conditions that give it the inferential behavior we want it for: being equivalent to the conjunction of all instances in our own language, but no further conjunctions, and not being logically stronger by implying all the conjunctions of the instances besides some additional things as well. The inferential behavior of the internal reading of the quantifiers is thus perfectly captured in the truth conditions that generalize over the instances in our own language. These truth conditions are the optimal solution to the problem of what truth conditions give the quantifiers their inferential role, on their internal reading.

---

[13] Similarly, some indeterminacy will arise from what the language is from which we source the terms. Languages are not clearly defined things in the study of language, i.e. of speech and communication. Nonetheless, the notion of a language is clear enough for our purposes here.

For just this reason the truth conditions of (SomeF) have to be seen as being equivalent to the disjunction of the instances in *our own* language. We could get weaker truth conditions if we took the instances not just from our language, but a possible extension as well. These would, in effect, add many Ds among the disjuncts, but those would give (SomeF) an inferential behavior beyond, and contrary to, what we want. In particular, (SomeF) could be true while each instance in our language is false. And furthermore, if the instances come from terms in some extension of our language, which one would it be? Some particular extension or just any possible extension? Should anything that could possibly be a term be accommodated in a disjunct? This would be completely contrary to the inferential use we have for quantifiers as placeholders for forgotten aspects of information that we once possessed. What we want instead is an inferential reading that is truth conditionally equivalent to the strongest truth conditions that give it the inferential role we want. Thus the truth conditions of the inferential reading should be seen as reflecting the optimal solution to our problem: being truth conditionally equivalent to conjunctions or disjunctions over the instances in our language. This captures the properly language-internal aspect of the inferential reading, as motivated here. Further philosophical considerations might push in a different direction, in particular the ones we will see in chapter 9 below, but the motivation from the role of inferential readings of quantifiers in ordinary communication supports the truth conditions that correspond to the optimal solution: generalizing over the instances in our own language. Further complexities and various refinements of these and other issues are discussed in chapter 9, and we will leave them until then.

3.4.6 How the Readings Are Related, but Different

It is now time to have a closer look at why and how the two readings are different. To see that the two readings differ in truth conditions, it is first of all important to make clear that and why neither reading implies the other one. The inferential role reading of

(SomeF)   Something is F.

does not imply the domain conditions reading, since it might be that "F(t)" is true, for a term "t" which is an empty term, and at the same time nothing in the domain of discourse is such that it is F. (Example: Fred admires Sherlock, but nothing else.) Then the inferential role reading of (SomeF) is true, while the domain conditions reading is false. The inferential role reading of (SomeF) is true since it is implied by the true instance "F(t)." The domain conditions reading of (SomeF) is false since the domain of quantification contains no F. Thus the inferential role reading does not imply the domain conditions reading.

For the other way round, it might be that (SomeF) is true in the domain conditions reading, but no F is denoted by any terms in our language, nor is there any other, non-denoting, term in our language that gives us a true instance of "F(t)." Here it is harder to give a clear example, since it is not clear which objects are denoted by terms we have, nor whether there might not be some non-denoting term that gives us a true instance. But it should be clear enough that simply because there is an object in the domain of quantification that satisfies a certain predicate "F," there is no reason to think that there will be a term in our language that stands for such an object, nor that there will be some non-denoting term that gives us a true instance of "F(t)." Having such an object in the domain of quantification, and having certain terms in our language, are two different, and independent, things. In such a circumstance the inferential role reading is false, while the domain conditions reading is true. Thus the domain conditions reading does not imply the inferential role reading.

In particular, the external reading is not just a restriction of the internal reading, and the internal reading is not just a restriction of the external one.[14] Quantifiers as they are commonly used are restricted, either explicitly or contextually. In the particular quantifier case, an ordinary utterance of (SomeF) will often have a further implicit restriction to Gs. An utterance of (SomeF) will assert that some G is F. But in this case the restricted utterance implies the unrestricted one. So, no matter what the restriction might be, (SomeF) in the internal reading is not equivalent to a restriction of (SomeF) in the external reading. The unrestricted internal reading can be true while the unrestricted external reading is false, as we saw above. Since the restricted external reading implies the unrestricted one, it follows that the unrestricted internal reading is not equivalent to a restriction of the external one. If it were, then the unrestricted internal reading would imply the unrestricted external one, which it does not. And similarly for the other way round. Neither reading is equivalent to a restriction of the other. They operate along different dimensions.

Even though in a language like ours describing the world as it is, the two readings of the quantifiers differ in truth conditions, this doesn't have to be so. When the language and the world match up in an especially simple way, then the two readings coincide in truth conditions. If every object in the world is denoted by a term in the language, and every term in the language denotes an object, then the two readings will have exactly the same truth conditions. Whenever there is an object o in the domain that satisfies predicate "F," then there will be a term t in the language that denotes o, and thus there will be a true instance "F(t)." And whenever there is a true instance, then there is an object in the domain denoted by the term in the true instance. In the simplest case the truth conditions of the two readings coincide, but in the real case they do not. The two readings come apart in truth conditions when we have non-denoting terms or undenoted objects. In our language and world we have both. But the fact that the two readings coincide in the limit is nonetheless of interest. It is evidence that one and the

---

[14] Kit Fine, in Fine (2009), argued that if quantifiers are in fact used in more than one way then one is a restriction of the other. This is mistaken, as the present account of the two readings of quantifiers hopefully illustrates. We will, however, revisit this issue below in section 9.6.

same expression, a quantifier, plays both of the roles that we need some expression to play: domain conditions and inferential role. These two are different communicative tasks, but they are related. These two roles can be played by the very same expression with the same contribution to the truth conditions in the limit, in the simplest case where language and world are a close match. The actual case is not that simple case, however, and so we get the next best thing: the two communicative roles are played by two different readings of the same expression, with two different contributions to the truth conditions.

When we have polysemous expressions, be they quantifiers, verbs, or any other expression, there is a real question about what determines which reading is the proper one on a particular occasion of utterance. Is an explicit intention required on the part of the speaker? Is some implicit intention enough, or is it maybe even underspecified and indeterminate which reading is being used on a particular occasion? It is clear that speakers don't need to have the different possible readings in front of their minds, and then consciously pick one of them. Often it is a notable discovery that there are a variety of readings of an expression, and what they are. And only once it is discovered and pointed out are ordinary speakers explicitly aware that there are these readings. To use an example discussed above, it is fairly involved to find out how many readings sentences with reciprocal expressions like "each other" have, but nonetheless, competent speakers use them just fine without such an investigation or explicit awareness. What determines what the correct reading of a particular expression in an utterance is is thus not a trivial question to answer. Speakers' intentions will certainly play a central role, but we do not need to give a theory of how precisely it works here. The phenomenon of polysemy is widespread, and not in any special way tied to the proposal about quantifiers made in this section. However the correct reading of polysemous verbs is determined, it probably will be just like that in the case of quantifiers.

When I explained the external reading in terms of "ranges over the domain of objects" and similar phrases, I was not trying to define it in terms of notions like "object" or "domain." I was using these phrases to trigger the external reading in you, who as a competent speaker of English has full access to it. The readings are not defined in this chapter at all, they are merely characterized and distinguished. This is an important methodological point. The present proposal is one about what our language already contains. To disagree with it requires to disagree either on whether we have a need for an inferential reading of quantifiers, or on whether we indeed have such a reading in our language, even if we could use one. The issue is not one about defining or specifying to those who don't understand one or the other reading how it is supposed to be understood. If I am correct and our sentences have these readings, then examples might trigger them, and contrast them with the other reading. Specifying under what communicative circumstances we might draw on them might help. If someone insists that they just don't get one or the other of the readings, or that for them there is always only one reading, then there is little I can do. It would be no different than if someone insisted that plurals always have a collective reading, and they just don't get the distributive one. I can try to trigger it by adding "each" to "four dogs," but if the other person still insists, there is little that can be done. All I can hope to do is to make clear that we have a need in ordinary communication for an inferential and a domain reading of quantifiers, to try to trigger the readings by using examples or extras like "in the domain of objects, whatever they may be," and to get clearer on what the truth conditions of these readings are. I can only hope that any denial of there being one or the other of these readings is not based on prior philosophical commitments.

The two readings of quantifiers were also labeled the "internal" and the "external" reading. I use this terminology for two reasons. First, the internal, or inferential role, reading is properly metaphorically described as "internal," since it relates instances of a sentence to the quantified sentence inferentially, and thus is concerned with the relationship between sentences within the language. It is language internal, in this metaphorical sense. The external, or domain conditions, reading has its truth conditions essentially tied to something language external: a domain of objects that the quantifiers range over. Here, too, the metaphor of externality is apt. Second, there are some connections with some differences between internal and external readings of quantifiers and internal and external questions about what there is, in a sense made famous by Carnap. These connections and differences, and Carnap's view more generally, are discussed in more detail below in section 3.5.3.

Time to sum up. Quantifiers are polysemous. There is one quantifier spelled "something" in English, but it has more than one reading. On one reading, the external or the domain conditions reading, it ranges over the domain of objects that the world contains. On the other reading, the internal or inferential role reading, it inferentially relates to other sentences in our own language in a certain way. In the case of the particular quantifier, it is implied by any instance. In the case of the universal quantifier, it implies any instance. (We will look at other quantifiers soon.) We can see that there are these two readings on the one hand by reflecting on particular examples of quantified sentences, and on the other by thinking about general needs we have in communication that we fill with the use of quantifiers. With our language we are representing a world of objects, some of which we have terms for, but many we don't have any terms for. Sometimes we use quantifiers to make a claim about the domain of objects. This puts the truth conditions of the quantified statement directly in the hands of something external to the language, the domain of objects, and it thus is properly called the external or domain conditions reading. Sometimes we merely relate representations within our language to each other. This is what we do when we use quantifiers in their internal or inferential role reading. The quantifier here has a role internal to the language, by relating some sentences within the language to others within it. We have a need to do both of these things, and polysemous quantifiers allow us to meet that need. All this is simply a reflection on language, and not motivated by metaphysics or ontology.

3.5 Compare and Contrast

The proposed view of the polysemy of quantification is reminiscent of a number of positions that have been defended in the past by various authors. In this section we will look at several such cases, and investigate how the present view is similar and different from these positions. This will hopefully clarify what the present view is and isn't. In the next section we will then be able to solve the first puzzle about ontology.

3.5.1 Substitutional Quantification

One traditional debate about quantification is the debate between those who advocate for an objectual interpretation of the quantifiers and those who favor a substitutional interpretation. This debate is traditionally one about which semantic interpretation of the quantifier is the correct one. Those who favor an objectual interpretation hold that, with the case of the particular quantifier again as our example, the following gives the truth conditions of a quantified sentence:

(122) "Something is F" is true iff there is an object o in the domain of quantification such that o satisfies the predicate "F."

On the other hand, those who push for a substitutional interpretation hold that

(123) "Something is F" is true iff there is a term t in English such that "F(t)" is true.

Here I treat these as two proposals about the semantics of the quantifiers in ordinary English. Commonly these proposals are given as alternatives for the semantics of quantifiers in a formal language $\mathcal{L}$. The objectual semantics then is:[15]

(124) "$\exists xF(x)$" is true in a model $\mathcal{M}$ iff there is an object o in the domain of $\mathcal{M}$ which satisfies F.

The substitutional formulation is:

(125) "$\exists xF(x)$" is true in a model $\mathcal{M}$ iff there is a term t in $\mathcal{L}$ such that "$F[t/x]$," the result of substituting "t" for "x" in "F(x)," is true in $\mathcal{M}$.

These two proposals for the semantics of quantifiers are normally considered competing proposals about what semantics quantifiers should be considered as having. There are good examples supporting each one of them. For one, the objectual one is well supported by examples like

(126) Some things are not denoted by any term in our language.

which intuitively seems right. On the other hand, the substitutional is supported by examples like

---

[15] I am neglecting the more general way to formulate this in terms of satisfaction by sequences of objects to deal with multiple quantifiers, since this isn't central for our main issue here.

(127)   There is someone I admire, but who doesn't exist.

which seems to be true when uttered by someone who only admires Sherlock, but doesn't seem to come out true with the objectual reading of the quantifier. But which one is the right interpretation? A good part of the debate about substitutional quantification in the 1960s and 1970s concerned this question. It is a debate about which of two competing views about the semantics of quantifiers is the correct one for formal or natural language quantifiers. Each one of the competing proposals has good examples in their favor, and each one has some apparent problems.

The objectual interpretation of the quantifier corresponds to our domain conditions reading of the quantifier, and the substitutional interpretation corresponds to our inferential role reading. However, the present proposal about quantifiers is one about different readings of quantifiers, not different interpretations. This might seem like a verbal point, but it is important. The debate about interpretations is a debate about which one of two competing hypotheses about the semantics of quantifiers is correct. The answer is neither. Instead quantifiers are semantically underspecified and have two different readings. It is not the case that quantifiers as a whole either have a substitutional or an objectual semantics. Instead they are semantically underspecified and polysemous, but the readings they have are tied to these competing interpretations. The objectual interpretation of a quantifier is truth conditionally equivalent to the domain conditions reading of that quantifier. The substitutional interpretation is truth conditionally equivalent to the inferential role reading. Quantifiers have both of these readings, but neither is the right interpretation, i.e. the correct specification of the unique truth conditions.

I could have formulated my proposal in this chapter by tying it to the debate about objectual vs. substitutional quantification, but I prefer not to. This debate, as it was carried out in the 1960s and 1970s in particular, was wrongheaded in a number of ways.[16] It was wrongheaded in its approach to the semantics of quantifiers, and in its connection to questions about ontology. I would thus like to distance myself from the traditional debate about substitutional quantification. There are several worries about the traditional use of substitutional quantification.

First, substitutional quantification has often been associated with those who hold that proposing a certain semantics for quantifiers, somehow, is a route for quantifying over certain things without having to hold that there are such things. But this strikes me as a mistake. If you want to know the answer to

(128)   Is there a god?

then this question would be answered by

(129)   There is a god.

---

[16] Among the original papers on substitutional quantification and its role in philosophy are Marcus (1962), Parsons (1971), Wallace (1971), Linsky (1972), Kripke (1976), and Marcus (1993).

It is irrelevant for this what the semantics of quantifiers is. The semantics of the quantifier has nothing to do with the fact that (129) is an answer to (128), unless some expressions in these sentences are somehow used in different ways. But whether the semantics is uniformly a substitutional one, or an objectual one, is irrelevant for the fact that (129), if true, answers (128).[17] Simply making a proposal about what the semantics of the quantifiers uniformly is has nothing to do with ontology.

Second, the use of substitutional quantification in philosophy is full of uses where substitutional quantifiers are employed to avoid "ontological commitment."[18] A common move of the friends of substitutional quantification is that since we can interpret certain quantifiers substitutionally we can avoid ontological commitment to certain entities.[19] The fact that a substitutional interpretation is available for certain uses of quantifiers is taken to have a consequence for what we have to accept exists when we use quantifiers this way. But this is all wrongheaded. It doesn't matter how we can interpret various uses of quantifiers. What matters is what we in fact do with these uses of quantifiers. Even if I could interpret every single use of quantifiers substitutionally, it wouldn't solve any questions in ontology. Substitutional quantification has been rightly charged with a nominalist's attempt at a free lunch. I want to have no part of that, as hopefully will become clear when we return to discussing ontology shortly.

Third, the substitutional interpretation of the quantifiers traditionally ties the truth conditions of quantified statements to quantified statements that quantify over expressions in the language, and thus gives substitutional quantification the appearance of disguised metalinguistic quantification. But this is the wrong sense in which such (uses of) quantifiers are "internal" to the language. It is not because they quantify over linguistic items that makes them language internal, but rather their inferential relationship to other sentences, which is the defining feature of the internal reading of the quantifier. It is preferable to focus on the inferential reading instead. A substitutional interpretation turns out to give a quantified statement that inferential role, and the latter, not the former, is what matters. What matter are the different functions quantifiers have in ordinary communication, and what contributions to the truth conditions let them fulfill that function.

Fourth, substitutional semantics has frequently encountered the charge that even though it specifies the truth conditions of the quantified statements, it does not specify the meaning, and thus is defective. See Lycan (1979) and van Inwagen (1981) for such challenges. Whether this charge is legitimate against the traditional substitutional proposal is doubtful, but certainly it has no bite against the present proposal. The internal reading of the quantifier has a clearly specified function in ordinary communication

---

[17] See also Kripke (1976: 376 ff.). Kripke says on p. 380 that he is skeptical that the substitutional quantifier has any role in the interpretation of natural language. I don't think he got that one right.

[18] Or at least to hold that no issue of ontological commitment arises. See, for example, Willard Quine "Ontology is thus meaningless for a theory whose only quantification is substitutionally construed [. . .]" Quine (1969: 64).

[19] This occurs even in Schiffer's excellent Schiffer (1987b). See also Gottlieb (1980).

and it makes a precise contribution to the truth conditions. We can see why we have it, what contribution to the truth conditions it makes, and why this contribution gets us what we want it for. We can't hope to do much better than that in specifying the meaning of an expression.

I thus prefer to distance myself from talk about substitutional quantification. Too much in its history is tainted with wrongheaded debates. And its name focuses on the wrong thing and not what matters. The inferential role reading is truth conditionally equivalent to a substitutional interpretation, properly formulated, but I prefer to talk about different readings of the quantifiers. This is not to distance myself from the philosophers who pushed for a substitutional interpretation. They often were motivated by exactly the right considerations, but it often was packaged all wrong.

3.5.2 Meinong and Non-Existent Objects

Some people have considered apparently true sentences like

(130) There is someone I admire who doesn't exist.

and concluded that quantifiers range over a larger domain of objects than merely those things that exist. The domain of quantification, on this view, contains both things that exist and things that do not exist, just as it contains things that are red and things that are not red.[20] Examples like (130) involve an unrestricted quantifier. Other examples involve the quantifier implicitly restricted to things that exist. This general idea is usually attributed to Meinong, and his followers in this regard are generally called Meinongians. The basic idea of Meinongians is that sentences like

(110) Everything exists.

can be understood as involving unrestricted quantification or restricted quantification. Understood unrestrictedly it is false, since the domain of quantification contains objects which do not exist. Understood as restricted to what exists it is true. It then just, trivially, says

(131) Everything which exists exists.

In short, Meinongians can try to accommodate the motivation for an inferential role reading by understanding it as a case of quantifier domain restriction. However, this is a mistake. It is not a mistake because there is anything wrong with quantifier domain restriction. It occurs all the time. The question simply is whether quantifier domain restriction is the proper way to understand these examples and the communicative need for an inferential role reading. And this is not the case. First off, any quantifier in a domain conditions reading, no matter how inclusive the domain, will only have

---

[20] Prominent examples of proposals along these lines are Parsons (1980b) and Zalta (1999). A sophisticated but unorthodox recent form can also be found in Priest (2007).

the relevant inferential role if every term in our language stands for an object in the domain. But, as we have seen above, we have good reason to think that some terms in our language fail to denote or refer. If there are such terms, then the domain conditions reading will not be able to satisfy our communicative needs.

Furthermore, quantifier domain restriction is implausible as the source of the two readings of (110) and why one of the readings seems quite clearly true. If all that is going on is either a domain restriction or no domain restriction, then the question remains why (110) has a reading that strikes us as clearly true. Of course, the explicitly restricted (131) is true, but why does (110) strike us as true? The answer for the Meinongian must be that even though the restriction is not explicit, there is a salient class of things to which it is implicitly restricted. And that class is the especially important and salient class of the existing things. But note that nothing like this is true for similar statements which are also not explicitly restricted:

(132)   Everything is human.

(133)   Everything is alive.

These or similar sentences do not have a natural reading that strikes us as true, even though the humans or living things can arguably be seen as at least as salient as the existing ones. If Meinongianism is true, then the existing things are just some special subclass of all the things, just as the living ones or the humans. As philosophy of language, the Meinongian proposal fails to explain why utterances of (110) strike us as true, while those of (132) strike us as false. This suggests that quantification "over" non-existent objects as in (130) should not be seen as quantification of a larger domain that includes both existing and non-existing objects, but rather as arising from the inferential role reading of the quantifier. This is the proposal I favor, and argued for above.

Meinongians sometimes present the following examples as arguments for the domain restriction account. They hold that there is a difference between

(134)   There is someone we both admire.

and

(135)   There exists someone we both admire.

The former, we are told, naturally corresponds to the unrestricted reading of the quantifier, whereas the latter has the quantifier explicitly restricted to existing things. "exists" in (135) does not merely involve contextual restriction, but explicit restriction to existing things, and that is the difference when using "exists" instead of "is." But this is mistaken as a view about whether the domain of the quantifier "someone" is restricted. It is correct that we have such a restriction in

(136)   There is someone who exists whom we both admire.

But in (135) "exists" does not restrict "someone," and in fact belongs to a quite different part of the sentence. To see this, just consider the negative version

(137)   There doesn't exist someone we both admire.

This sentence doesn't have a reading where we both admire someone who doesn't exist, i.e.:

(138)   There is someone who doesn't exist whom we both admire.

But (137) should have such a reading if "exists" restricts "someone" in (135). If "exists" can restrict "someone" in (135), then "does exist" and also "doesn't exist" should be able to do this as well. But (137) clearly has no reading where such a restriction occurs, so whatever compels us to understand (135) along Meinongian lines, it isn't quantifier domain restriction. However, there is a better explanation of the difference.

For there is indeed a difference between (135) and (134). It is more natural to take a speaker who asserts (135) to say that someone who exists is admired than if they uttered (134) instead. But the reason for this is not that one involves a restricted quantifier while the other one does not. The reason for this instead is that with (135) one of two readings is being made more prominent. To illustrate this general phenomenon, consider the case of the distributive and collective readings of plurals:

(139)   Four philosophers carried a piano.

It can be read as describing what every one of four philosophers carried, or what they did together. However, by adding "together" or "each" we are forcing one reading or the other:

(140)   Four philosophers each carried a piano.

(141)   Four philosophers together carried a piano.

The addition of "together" or "each" forces one of two readings of a sentence that otherwise would have both readings. This is analogous to what the word "exists" does in our examples (134) and (135). "exists" does play a real role, but not to restrict the quantifier, rather to force or at least suggest one reading over another. These examples should thus not be taken to support the Meinongian position, which essentially relies on quantifier domain restriction as the difference between the two.

The word "exists," however, is more complicated than what the above suggests. It is an English verb, and like (almost) all other verbs it is polysemous. On certain uses it certainly has the function of suggesting an external reading of a quantifier. To take one example, consider:

(142)   Numbers exist.

"Numbers" here is the quantifier, and "exists" is the predicate. It can be read as either

(143)   Some numbers exist.

or

(144)   All of the numbers exist.

These two readings are the result of the different readings that a bare plural quantifier like "numbers" can have. The former reading is equivalent to

(145)   Something is a number.

where "something" is used in its external reading. That this equivalence holds has nothing to do with "exists" being a quantifier, or with (143) being analyzed as (145). But the equivalence of these particular readings of (143) and (145) holds nonetheless.

However, "exists" can also be used in other ways where it is not connected to external readings of quantifiers. On an ordinary utterance of either of these sentences:

(146)   Racial discrimination still exists in North Carolina.

(147)   After the sexist remark quite some tension existed in the room.

"exists" is not used to force or suggest an external reading of a quantifier. As with verbs and natural language in general, things are complicated. I won't attempt here to give a more detailed analysis of various uses of "exists." We can acknowledge the polysemy of the verb and its relation to one of the readings of the quantifiers. I furthermore conjecture that "exists" and quantifiers are in harmony in the following sense: for every reading of the quantifiers, there is a reading of "exists" such that

(110)   Everything exists.

is true on those readings. But nothing to come depends on that.

A final issue that should be briefly addressed concerns the acceptability of sentences used in our main examples above, in particular, whether

(148)   Fred admires Sherlock.

implies

(149)   There is something/someone Fred admires.

on any reading of a quantifier. I have often heard in conversation that it is acceptable to infer

(150)   Fred admires something/someone.

but not (149). But it is hard to justify the difference. As we saw above, the difference between (149) and (150) is that (149) is an existential construction, whereas (150) is the corresponding subject–predicate sentence. In general these corresponding pairs seem to have the same truth conditions. There can, however, be a difference between such pairs. Sometimes one of them has more available readings than the other, in particular when they involve intensional transitive verbs. Consider the difference between

(151)   Fred is looking for a wife.

(152)   There is a wife Fred is looking for.

(151) has two readings which (152) does not have. First the reading where Fred is looking to marry some woman or other (he had enough of being a bachelor), second the reading where Fred is trying to find some woman or other who is married (to interview her about marriage, say). However, both have the reading where Fred is looking for some particular woman who is married. The existential construction thus allows fewer readings, but remains equivalent to the subject predicate sentence on the readings that are still allowed. In particular, the existential construction rules out so-called notional readings of these sentences, ones where Fred is looking for some wife or other. This difference, however, does not apply to our examples like (149) and (150) containing polysemous quantifiers. Here both readings are available in both sentences, and they are equivalent on the corresponding readings. In particular, it is hard to read (149) as having a notional reading: he admires someone, anyone would do. Such a reading might well be ruled out in the existential construction, but it isn't there in the first place.[21]

All these arguments against a Meinongian alternative to our two readings of polysemous quantifiers have nothing to do with worries about an ontology of non-existent objects. They are simply arguments that the Meinongian alternative in terms of quantifier domain restriction is as a matter of fact incorrect about our uses of quantifiers in the relevant examples that appear to motivate it. The polysemy of quantifiers view defended here is to be favored instead.

To conclude, we can now see what the connection is between quantification, existence, and "there is" statements. Quantified statements are tied to what exists. When a quantified statement on its external reading is true then this is connected to what exists (on the reading of "exists" that is probably the most prominent one). However, not all quantified statements are directly tied to what exists, since quantifiers can be read internally as well, and on that reading no immediate connection obtains to what exists (using the same prominent reading of "exists"). This connection does not obtain because "exists" is a quantifier. It is clearly not a quantifier, it is a verb or a predicate instead. The connection obtains since what we want to make a claim about when we use quantifiers in their external, domain conditions reading, is related to what we want to say when we say that something exists. Furthermore, quantified statements are connected to "there is" statements, and generally equivalent to them. Not because "there is" is itself a quantifier, but because it is generally followed by one, and helps via the "existential construction" to make quantified claims in a particular way.

---

[21] Note that such a reading is available for "Fred admires a wife," i.e. he admires any woman who is married, but not for the corresponding existential sentences.

3.5.3 Carnap on Internal and External Questions

I have argued that quantificational claims systematically have two readings: an inferential one, also called the internal reading, and a domain conditions one, the external reading. To distinguish internal and external readings of quantified claims is reminiscent of Rudolf Carnap's view that we need to distinguish internal from external questions about what there is. In this section we will briefly look at how the two views are related and how they are different.

In his essay "Empiricism, semantics and ontology," Carnap (1956), Carnap hoped to show that talking about metaphysically sounding things like propositions or truth is perfectly fine for anti-metaphysical positivists like himself. In particular, truth-conditional semantics, which relies on such things as propositions and properties as semantic values, is positivistically acceptable. Carnap's idea for the defense of this is that there is a crucial distinction between two things we need to do when we propose a theory. We first need to pick a language in which to formulate the theory, and second we need to state the theory in that language. To accept a language to formulate a theory in is a choice one has to make, and one has quite a bit of liberty in choosing one. All that needs to be the case for such a language to be acceptable is that it has a clear syntax, clear inferential relations among its sentences, and that expressions are properly tied to experiences in their semantics in order to be meaningful. Once a certain language is accepted it then becomes trivial to draw certain conclusions in that language. Take the "numbers language" as one example, which is the language that allows for talk about numbers, and meets the outlined minimal conditions for being an acceptable language. Once I select the numbers language in the first place it becomes trivial for me to conclude that there are numbers. This triviality comes simply from the rules internal to the language, which I decided to adopt to describe the world. In order for me to even be able to ask the question "Are there numbers?" I must accept the numbers language. And then the answer is trivial: "yes." So understood the question "Are there numbers?" is an internal question, one asked within the numbers language, with all of its rules for inference, syntax, and the rest.

Carnap thought, however, that these trivial questions about what there is can't be what metaphysics is after. They are trivial questions after all. The metaphysician must be seen as trying to ask a different question. This question the metaphysicians express with just the same words: "Are there numbers?," but it is intended to ask a question about whether reality contains such things, or whether the numbers language gives a correct description of reality, and thus whether it was the objectively right choice of a language to describe it with. This other question is the external question about numbers. It is a question about the correspondence of the numbers language to reality, and not one that is settled internally, simply by the rules of the language itself.

However, Carnap thought that these external questions as intended are meaningless and devoid of "cognitive content" Carnap (1956: 209). There is no other question to ask here besides the internal, and trivial, one. The other question the metaphysician hopes to ask, the external question, was intended to be one about the numbers language as a whole, one that one might approximate with words like "Does the numbers language correspond to reality?" This external question can't, for Carnap, be seen as asking a question of fact, since the only question properly asked with the words "Are there numbers?" is the internal one, the only one that the language licenses. Since there is no question of fact to be asked here besides the internal question, the external question is devoid of "cognitive content" in the technical sense that it can't be seen as a question of fact. Instead, it can at best be seen as a practical question, a question about what to do, in particular whether it is advisable to use the number language to describe the world. But it isn't a further question of fact, and anyone who hopes to answer such a question of fact is engaged in a confused project.[22]

The distinctions between internal and external questions for Carnap, and between internal and external readings of quantified statements as defended here, are thus only marginally similar. On the present account internal and external readings have exactly the same status. They are both equally meaningful readings, making equally factive statements, both with truth conditions and both with uses in ordinary as well as metaphysical conversations. They simply fulfill different needs in communication, but otherwise they are just alike. In particular, external readings of quantified statements are just as meaningful and factual as internal ones. Furthermore, for Carnap a quantified sentence like "there are numbers" and its question equivalent "are there numbers?" have only one reading. Correspondingly, only one question can be asked with an utterance of that question sentence where the content of the question asked is the content of the sentence uttered. On the present view such sentences have two equally good readings, and thus two different questions can be asked with utterances of these question sentences. The external question, i.e. the question "are there Fs" on its external reading, is therefore a perfectly meaningful question, one we would like to know the answer to.

On the other hand, the present account agrees with Carnap that internal readings of quantified statements, in particular very general quantified statements like "there are numbers," are often trivially true. The reasons for why this is so are different, of course, but the spirit of this aspect of the two positions is the same.[23]

Besides these important differences between Carnap's view and the present account, the metaphor of internal–external is apt for both. We will see towards the end of this

---

[22] Many people have criticized Carnap's particular version of his internal–external distinction. Willard Quine has a well-known criticism in Quine (1966). Some philosophers have recently tried to defend views like Carnap's, in particular with an emphasis on various aspects of the view that can naturally be attributed to Carnap. For a recent defense of Carnap's rejection of ontology, see Yablo (1998), for his anti-realism about ontology, see Chalmers (2009), for his taking ontological questions to be practical ones, see Kraut (2016), for his deflationism about ontology, see Thomasson (2015). How Carnap's view relates to some other contemporary positions, including, in particular, that of Eli Hirsch in his Hirsch (2011), is discussed in Eklund (2009) and Eklund (2016).

[23] For a more detailed discussion of Carnap's view and how it relates to the position defended in this book, see Hofweber (2016b).

book that on the present account the internal–external distinction for questions about what there is is crucial for defending ambitious metaphysics, and not a central part of its rejection, as Carnap would have it. Still, I think Carnap's insight that there are two questions we can ask about what there is is a very deep one, although his version of articulating what the difference is was misguided. In particular, it was misguided to reject the external questions, and ontology with it. I think the opposite is true, but it is still a while until we get to this in chapter 12.

3.5.4 Lightweight and Heavyweight Quantifiers

Natural language might be seen as containing a metaphysically lightweight quantifier and a metaphysically heavyweight quantifier.[24] In the lightweight sense there are infinitely many primes, but that leaves it open whether there are numbers in the heavyweight sense. With such a distinction at hand it might seem quite easy to solve all of our above puzzles. But this would be quite unsatisfactory. There can't just be a basic distinction between lightweight and heavyweight quantification in natural language, a distinction tied to ontology and metaphysics in particular. Why would our language have both of them? It would be amazing if it contained primitive distinctions or resources mainly to carry out metaphysics, but with no role in ordinary communication. And it would be puzzling how such quantifiers should be understood and how they relate to each other.[25] If there is such a distinction to be made, it must be shown to arise from our language, and this can't simply proceed by example, pointing to our puzzles, or by wishful thinking in order to defend that there is work for ontology as part of metaphysics: to settle what there is in the heavyweight sense. If there is a metaphysically significant distinction between different uses of quantifiers, then this must be shown to be so by making two things clear: (a) we have reason to think that there are different uses of quantifiers in natural language arising not simply in metaphysical debates, but in ordinary communication, and (b) these different uses somehow give rise to a metaphysically significant distinction among quantifiers. What we have seen so far is that (a) is indeed the case. Whether and, if so, how this connects to a metaphysically significant distinction is still to be worked out in more detail. As we will see, there is such a connection, but it isn't simply that one reading of the quantifier is metaphysically lightweight while the other is metaphysically heavyweight. Instead the significance of the distinction of the two readings for metaphysics is a consequence of the function of those two readings in ordinary communication. But it isn't the distinction between heavyweight and lightweight. To see more on the significance of the two readings for metaphysics, it is now time to solve our first puzzle about ontology.

---

[24] See, in particular, Chalmers (2009) and Fine (2009) for a discussion. Both attribute this view to Hofweber (2005b), but that is not quite right, as I hope to make clear here.

[25] We will discuss in more detail whether we should think that such metaphysical resources are part of our shared conceptual repertoire in chapter 13.

3.6 The Solution to the First Puzzle

Suppose that what I said about quantification in this chapter is correct. What would it show for our puzzle about the trivial arguments? These arguments started with an innocent statement, inferred from it one of its metaphysically loaded counterparts, and finally inferred a quantified statement from that, leading immediately to the conclusion that there are numbers/properties/propositions. With what we have seen above in mind we can now analyze and understand each step in this argument.

The inference from an innocent statement to one of its metaphysically loaded counterparts is indeed valid, as we have seen in chapter 2. Thus the inference from

(153)   I have two hands.

to

(154)   The number of my hands is two.

is valid, and indeed trivial. It is just as valid and trivial as the inference from (153) to

(155)   What I have two of is hands.

These inferences preserve the truth conditions of the premise to the conclusion, but they give the information presented a particular structure or focus. The syntactic structure of the sentence guarantees this focus effect, and our competence with our language allows us to see what the difference is between the premise and the conclusion. This inference is thus valid, and its validity can be seen by anyone competent with the language in which they are given.

The next step of the trivial arguments is also valid, and also trivially so. From (154) I can conclude immediately that

(156)   There is something which is the number of hands I have, namely two.

This inference is trivial in the inferential role reading of the quantifier "something." And for this inference to be valid, and trivially so, it is irrelevant what the semantic function is of the number word "two" or any other expression in the sentence. All that matters is that the relevant expression occupies the right syntactic position for a proper interaction with the quantifier. This second step would not be trivially valid (and I argued in chapter 2, would not be valid at all) if the quantifier was used in its external reading. However, in its internal reading this is indeed a completely trivial inference, and in the way in which it seems to us that we can trivially infer this we are relying on the inferential role reading of the quantifier. In the domain conditions reading this inference is not trivial (and not even valid).

So, the trivial arguments are indeed trivial and trivially valid. The inference goes through, and it does so no matter what the semantic function is of number words or terms for properties or propositions. And this solves our first puzzle. The puzzle was that the question of whether there are numbers on the one hand seems to have a trivial answer, and on the other hand seems to be a substantial question which shouldn't have such a trivial answer. But if the trivial argument is indeed valid, then it would seem that the question of whether there are numbers has a trivial affirmative answer. This seems to favor the side who thinks that the question has a trivial answer. But things are not as simple. If the account of the trivial inferences given above is correct, then this does not only affect the trivial arguments and what is going on in them, it also affects the question that we hoped to answer. The question we were wondering about was the question

(157) Are there numbers?

This question, just like the answer

(158) There are numbers.

contains a quantifier. It is not the "are there" or the "there are," as we saw above, but rather it is the plural quantifier "numbers." To make this clearer, again, let's look at the singular case

(159) Is there a number?

(160) There is a number.

Here we have the quantified noun phrase "a number" built up from the determiner "a" and the noun "number." The plural case is no different, except that it is plural, of course, and does not have an explicit determiner. The content of the question would be no different if we explicitly added a determiner as in

(161) Are there some numbers?

(162) There are some numbers.

Here the quantifier is made more explicit: the plural quantified noun phrase "some numbers."

If what we said about quantifiers above is correct, then it implies that both (157) and (158) have one reading arising from the domain conditions reading and another one arising from the inferential role reading of the quantifier. This applies to all the versions of the question as well as the answer, be it in the singular or plural, with the determiner explicit or not. Let us focus on (159) first, which is in a sense the simplest case, since we can ignore the issue of plural there. This question can thus be read in two ways, corresponding to the two readings of the quantifier "a number." Similarly, the answer (160) has two readings as well, and the two readings of the question correspond to the two readings of the answer. The domain conditions reading of the question is answered with the domain conditions reading of the answer. And the inferential role reading of the question is answered with the inferential role reading of the answer. When we establish (160) in its inferential role reading then we have shown that there is a true instance of "t is a number." When we establish it in its domain conditions reading, then we have shown that there is an object in the domain that we quantify over which satisfies the predicate "is a number." And that is to say that the world contains such a thing as a number. Similarly, when we ask (159) in its inferential role reading, then we ask whether or not there is a true instance of "t is a number." When we ask it in its domain conditions reading, then we ask if the world contains such a thing as a number.

Now we can see the answer to our puzzle: the trivial arguments are indeed valid, and they establish trivially that there are numbers in the inferential role reading of this sentence. But the question of whether there are numbers is a substantial and difficult question in the domain conditions reading of the question. When we asked whether there are numbers among the things that the world is made from, then we were asking the question in its domain conditions reading. When we trivially concluded that there are numbers, then we established this in the inferential role reading. But the domain conditions question is not answered with the inferential role answer. The trivial argument does not establish that there are numbers in the domain conditions reading. Thus the trivial argument is indeed valid, but it does not answer the question we posed. And that is so even though the question just was the question of whether there are any numbers, and the trivial argument established that there are numbers.

What we have seen above shows that the question itself has two readings. Only on one reading is it trivially answered. And it is the other reading on which it is a substantial question. Once we keep in mind the different ways both the question and the answer can be understood we can see that the puzzle disappears. There is no tension between there being a substantial question of whether there are numbers and it also being trivial to conclude that there are numbers. The reading according to which it is a substantial question is a different one than the one according to which it gives a trivial answer. The question "are there numbers?" is a trivial question, and it is a hard and substantial question. It is trivial on one reading, and substantial on another.

What does all this mean for what we should think the ontological question is? In ontology we aim to ask a certain question, and it was very natural for us to express the question as "Do numbers exist?" or "Are there numbers?" We were originally worried that this can't quite be the proper statement of the ontological question, since these questions seem to be trivially answered in the affirmative, whereas the ontological question, as it was intended, was supposed to be a substantial question about the world. We have now seen that this worry was mistaken. Thus we can at least preliminarily draw a conclusion about what the ontological question about numbers is, at least according to how we conceived of ontology at the outset. Just as we originally thought, the ontological question about numbers just is the question

(163)   Are there numbers?

but only on its external, domain conditions reading. As we have seen, on the external reading this question is equivalent to

(164)   Do numbers exist?

on one of the readings that this sentence has. The ontological question about numbers is thus just the question that we originally thought it was. It doesn't follow so far, of course, that the ontological question is one for philosophy or metaphysics to address. It might well be that this question is answered in mathematics, and that metaphysics has no work to do here. Ontology might not be especially metaphysical, but at least it has a clear question it should try to answer. Whether ontology also should answer other questions is so far, of course, left open. We will get back to all this later.

And although our solution to the first puzzle shows that our initial statement of the ontological question was correct, it also shows that very common motivations for a metaphysical project are mistaken. A classic example is the following way to set up the problem of universals. It is based on an argument that initially has great plausibility, and it is often used. But we can now see that this argument involves an equivocation. It moves from one reading of a quantifier to another. Here is the common way to motivate the problem of universals:

(165)   a. Fido is a dog, and Fifi is a dog.
        b. Thus: there is something that Fido and Fifi have in common, namely being a dog.
        c. Thus: there is some thing or entity which they have in common.
        d. Question: what is this entity and how does it relate to Fido and Fifi?

This argument naturally leads to the project of finding out what this thing, the universal which Fido and Fifi both have, is, and how it relates to the objects that have it. Is it concrete or abstract? Is it everywhere Fido and Fifi are? How do they both partake in it? And so on. These might be good questions in the end, but the above argument doesn't support such a project. The argument makes a crucial error. The inference from (165a) to (165b) is trivially valid.[26] But the inference from (165b) to (165c) is based on a mistake. (165b) is only established with the quantifier in its inferential role reading. But in (165c) the quantifier is used in its domain conditions reading. This inference simply does not hold. The inferential role reading does not imply the domain conditions reading. However, the motivation for the metaphysical project requires the domain conditions reading. The argument given thus does not establish that there are things that Fido and Fifi have in common, and that we should try to find out what they are.

Although the motivation for the problem of universals given above is mistaken, it is important to be clear how little has been shown so far for larger-scale philosophical questions. What we have seen so far does not show that the problem of universals is based on a mistake, just that a certain motivation for it is based on a mistake. Nothing

[26] We will see more on inferential role uses of quantifiers in examples like these in chapter 8.

so far says anything about the problem of universals as such. The problem of universals will be discussed in detail in chapter 11.

Even though the question of whether there are numbers in its inferential role reading is trivial, this does not show much about the status of the question in its domain conditions reading. The trivial arguments do not directly answer the domain conditions question, but this does not mean that the domain conditions question can't be answered in other ways, nor even that it can't be trivially answered in some other way. What we should think about the domain conditions reading of the question is so far left completely open.

But what we have seen so far should help us to figure out whether the domain conditions reading of the question is answered in other ways. Does mathematics answer the question on the external reading? Are there other arguments that give us an answer to this question? To answer these questions we in effect need to solve the second puzzle. And this is what we will try to do next.

3.7 Appendix: Generalized Quantifiers

This appendix deals with extending the two readings of quantifiers to a larger class of quantifiers called generalized quantifiers. It can be skipped, since the details won't be relied upon in the following. However, it is important to show that the two readings apply to quantification in natural language in general, not just to the simple cases we saw above.

The internal reading of the quantifiers is most apparent when it is directly tied to a simple inferential role, as it is in the case of the particular and universal quantifiers, i.e. "something" and "everything." But the phenomenon of the two readings is not restricted to these cases of quantifiers. Instead, it applies throughout the quantificational apparatus as a whole. The best way to think about quantifiers in natural language as a whole is generalized quantifier theory (GQT). This theory has been developed in detail over the last couple of decades for the external, domain conditions, reading of quantifiers. And it can be easily modified to accommodate the internal, inferential role, reading of quantifiers as well. In this appendix I will outline how this can be done. I will focus on the main idea and not go into the details of GQT. We will simply focus on the truth conditions of quantified statements, in both their external as well as their internal readings.

Quantifiers in natural language are built up from determiners and nouns. Determiners include expressions like "some," "all," "most," "many," "two," and so on. These are then used to form quantified noun phrases like "some man," "all women," "most dogs," and so on. The quantifiers studied in formal, first-order logic, "something" and "everything," can be seen as just special cases of quantifiers, ones built up from the determiner "some" or "every" and the noun "thing." The semantics for sentences with quantified noun phrases in them, i.e. sentences of the form [[DET N] VP], can then be given uniformly, for example by assigning extensions to the noun N and the verb phrase VP, and associating the determiner with a condition that has to hold between the extension of N and VP.[27] To give a simple example, the sentence

(166)   Every dog barks.

is true just in case the extension of "dog" is contained in the extension of "barks." And this in turn is the case if and only if

(167) $\{x|x \text{ is a dog}\} \subset \{x|x \text{ barks}\}$

The truth conditions of

(168)   Two dogs bark.

differ from (166) in that the determiner "two" makes a different contribution than the determiner "every." Whereas the former demanded that the extension of the N is a subset of the extension of the VP, in the latter case the determiner demands that the intersection of the extension of the N and the VP contains two members:[28]

(169) $|\{x|x \text{ is a dog}\} \cap \{x|x \text{ barks}\}| = 2$

And similarly for other determiners: "some" (the intersection is non-empty), "few" (the intersection is relatively small[29]), and so on. The study of generalized quantifiers then focuses on universal features that such quantifiers have in natural language, on how to treat complex determiners, etc., all things we don't have to get into. What is crucial for us here is that the semantics of quantifiers given this way is the semantics of the external, domain conditions reading. The truth conditions are given in terms of extension of the predicates, which in turn are subsets of the domain of quantification, the domain of things that the world contains. Does the internal reading of the quantifier apply to generalized quantifiers as well, or is it restricted to the simple quantifiers "something" and "everything"? And if it applies, how can we specify its semantics?

The internal reading of the quantifiers clearly applies to quantifiers more generally, and not just to the quantifiers studied in first-order logic. We can see that the motivation for the internal reading carries over from "something" to "two things," "few

---

[27] Alternatively, but equivalently, we could assign determiners functions as their semantic value which map noun extensions and verb extensions onto truth values. More commonly one assigns functions as the semantic values also to the N and the VP, so that a determiner gets assigned a function that maps functions assigned to the N to functions that assign VP values truth values. In the end it all comes down to the same idea. In a proper treatment one would use not simply extensions, but intensions, but this, too, is something we can put aside to convey the main idea.

[28] This is for the reading "exactly two." For "at least two" it should, obviously, be modified to end in "≥ 2."

[29] "few" is a little bit more complicated, since it has a variety of readings. On one of them "few F are G" simply means that the overall number of Fs that are G is small, whereby context settles how small it has to be. This is the reading we are using here. Other readings include one where the ratio of the Fs that are G to the Fs that are not G is small.

things," "two dogs," and so on for other generalized quantifiers. For example, instead of forgetting who it was when I remembered that there was someone that we both admire, I might instead remember that there were two things we both admire, but forget who they are. Then I can communicate what I still know with

(170)   There are two things we both admire.

and here, again, there is a reading where what I said is true when you and I both admire Santa and Sherlock, but nothing else. Similarly for

(171)   There are few things we both admire.

and other generalized quantifiers. The internal reading applies to the quantificational apparatus as a whole, not just to particular examples of quantifiers. The question thus is how to give the semantics for the internal reading of quantifiers in general, i.e. generalized quantifiers.

We can see how to give the semantics for the internal reading of quantifiers more generally by thinking about (170) briefly. When we discussed above the inferential role of "something," we used the situation of partial ignorance to argue that there is a reading of

(172)   There is something we both admire.

It has to have truth conditions such that any true instance

(173)   You admire X and I admire X.

implies it. This might suggest that the truth conditions of (170) should be such any two true instances of (173) should imply it. After all, if you and I both admire X, and you and I both admire Y, then no matter what X and Y are, it should guarantee the truth of (170). But this is not quite right. There might be two different true instances of (173), but nonetheless (170) is false, on any reading. This is simply so when the two instances are instances with two terms "X" and "Y," but X and Y are identical. Both you and I might admire only Nixon, but that gives us more than one true instance of (173) since we both admire Richard Nixon, and we both admire President Nixon, and we both admire Tricky Dick.[30] So, the number of true instances is no direct guide to the number of things we both admire.

We don't admire two things here even though there are many instances of (173), since the terms that give us the instances stand for the same thing. For any two instances with terms "X" and "Y" we have to consider whether X is identical to Y. This does not just apply to terms that stand for an object in the world, but also for empty terms, ones like "Sherlock." It might be that the only thing we admire is Sherlock, but there is more than one true instance of (173), since we then also both admire Detective Holmes. Of course, Sherlock and Detective Holmes are identical. The

---

[30] "Tricky Dick" was a nickname for Nixon, with negative connotations.

identity statement "Sherlock is identical to Detective Holmes" is true, even though the relevant terms don't pick out any objects in the world. That non-referring terms can figure in true identity statements should be clear, although it is somewhat puzzling. Sherlock is indeed identical to Detective Holmes. And that is so even though there are no such things as Sherlock or Detective Holmes. Similarly for many other phrases that function grammatically like singular terms, but are not, arguably, in the business of reference or picking out any objects. For example:

(174)   How I make a chocolate cake is identical to how my grandpa used to make it.

How to understand identity statements with non-denoting terms is an interesting topic, but for now we don't have to have an account of it; we should just acknowledge the phenomenon that there are true identity statements with non-denoting terms in them, and that it is relevant for our understanding of internal quantifiers.

We can now see how to understand the internal reading in examples like (170). We admire two things when there are two instances of (173) *up to true identity statements of the terms involved*. That is to say, if we have true instances with terms "X" and "Y" and the identity statement "X = Y" is true, then they count as one thing admired, but if the identity statement is false, then they count as two things admired. And this holds whether or not the terms are referring. We can generalize this idea to give a semantics for the internal reading of quantifiers, and see how the internal reading relates to the inferential role in these more general cases.

Identity is an equivalence relation. In our setting here, however, we can distinguish two senses of equivalence relations. On each one, an equivalence relation is reflexive, symmetric, and transitive, but each of those has an internal and an external sense. For example, transitivity holds just in case for all a, b, c if a = b and b = c then a = c. And the "for all" here has an internal and an external reading. Correspondingly, we can say that identity is transitive in the internal sense just in case the above "for all" statement is true on the internal reading of the quantifiers. It is transitive in the external sense just in case the statement is true on the external reading. Similarly, we can distinguish internal and external senses of reflexivity and symmetry, and finally of being an equivalence relation itself. An internal equivalence relation is one that is reflexive, symmetric, and transitive, each in the internal sense. An external equivalence relation is one that is reflexive, symmetric, and transitive, each in the external sense. Identity is an equivalence relation in both senses. This is quite uncontroversial for the external sense, but holds for the internal one as well. For example, if how I make the cake is identical to how you make it, and how you make it is identical to how grandpa made it, then how I make it is identical to how grandpa made it.[31]

---

[31] In the literature on free logic there is some disagreement on how best to understand identity statements with non-referring terms. Some approaches, positive free logic, accept the truth of "Sherlock = Sherlock" while others, negative free logic, reject it. I side with those who accept it, not as a matter of a choice of logic, but based on a judgment of the truth of the corresponding English sentence. See Nolt (2014) for a detailed discussion.

Since identity is an internal equivalence relation it follows that terms that figure in true identity statements get divided up into separate, non-overlapping groups: they form equivalence classes.[32] And what we have to consider when we consider how many things we both admire, in the internal reading, is how many true instances of (173) there are, up to true identity statements. In other words, we are not counting how many individual instances there are, that is how many different terms can be put in the right places to make a true sentence, but rather how many different equivalence classes of terms that figure in true instances there are. What matters is how many terms, up to occurring in true identity statements, lead to a true instance.

Let's introduce some terminology. Suppose T is the set of all the terms in our language. When t is one such term we can form the equivalence class t_{IS} as the class of all terms s such that the identity statement "$t = s$" is true: $t_{IS} = \{s|\text{"}t = s\text{"}$ is true$\}$. Note that the latter condition is not that "t" and "s" are the same term, but that the identity statement formed with those terms is true. t_{IS} is thus the class of all terms such that "t" occurs in a true identity statement (IS) with this term. Now we can say that an equivalence class t_{IS} is *realized* in a formula $F(x)$ just in case there is a true instance $F(t)$, with "t" a member of t_{IS}. With this we can now specify the truth conditions of both the internal and the external readings of a generalized quantifier in parallel.

Suppose D is the domain of objects over which we (externally) quantify, and T is the class of terms within our own language L. Let F' be the extension of a predicate $F(x)$ on D, i.e. the set of objects in D that satisfies $F(x)$. Let $\bar{T}$ be the set of equivalence classes of terms formed with the equivalence relation: x and y are such that the statement "$x = y$" is a true identity statement.[33] Let $\bar{F}$ be the set of equivalence classes in $\bar{T}$ that are realized in $F(x)$. Given these definitions we can state the truth conditions for generalized quantifiers in both their internal and external readings. I put them next to each other to make vivid how parallel they are.

Table 3.1 Truth conditions of various generalized quantifiers

Sentence	External reading	Internal reading
Some F is G	$F' \cap G' \neq \emptyset$	$\bar{F} \cap \bar{G} \neq \emptyset$
All F are G	$F' \subset G'$	$\bar{F} \subset \bar{G}$
Most F are G	$|F' \cap G'| > |F' \cap \neg G'|$	$|\bar{F} \cap \bar{G}| > |\bar{F} \cap \neg\bar{G}|$
n F are G	$|F' \cap G'| = n$	$|\bar{F} \cap \bar{G}| = n$
Few F are G	$|F' \cap G'|$ is relatively small	$|\bar{F} \cap \bar{G}|$ is relatively small
Infinitely many F are G	$|F' \cap G'|$ is infinite	$|\bar{F} \cap \bar{G}|$ is infinite

---

[32] An equivalence relation on a domain D, to repeat, is a relation R that is reflexive, symmetric, and transitive. An equivalence class d_R is the set of all members of D that stand in relationship R to d, i.e. $d_R = \{x \in D|R(x, d)\}$. It follows that D gets divided into disjoint subsets $d_{1R}, d_{2R} \ldots$ which together make up all of D. In particular $d_R = e_R$ if and only if $R(d, e)$.

[33] Strictly speaking we should use corner quotes here, which quote the result of replacing the value of x and the value of y for x and y, respectively, in "x = y."

And so on and so forth for other generalized quantifiers.

Given the truth conditions of the internal reading of generalized quantifiers, we can see that the associated inferential role is determined by the truth conditions. For example, take the internal reading of

(170)   There are two things we both admire.

This sentence can't be deduced from

(175)   You admire Sherlock, and I admire Sherlock, and you admire Nixon, and I admire Nixon.

We need the further premise that Nixon and Sherlock are different. But once we add that then this guarantees that two equivalence classes are realized in the predicate "we both admire x." Since Nixon and Sherlock are' not identical it follows that the equivalence class built around "Nixon" and that built around "Sherlock" are different equivalence classes. And since the first is realized with "Nixon" and the second is realized with "Sherlock" two different equivalence classes are realized in the predicate. Then (170) follows, given the truth conditions specified above. The truth conditions of the internal reading guarantee the inferential role it gives rise to.[34]

---

[34] The above truth conditions for the internal reading are intended for extensional contexts only. It has to be assumed that "$t = s$" and "$F(t)$" imply "$F(s)$." With that restriction in place we are guaranteed that if a certain equivalence class is realized in a predicate, then every term in that class gives a true substitution instance with that predicate. If we do not make that restriction, and deal with quantification into intensional contexts, then not only will the semantics given for the internal reading not be adequate, an extensional version of generalized quantifier theory will not be adequate for the external reading as well. Intensional versions of GQT for the external reading are well-known, but we will not attempt to generalize the above semantics for the internal reading to the intensional case, in part because dealing with intensional transitive verbs brings in many other complicated issues that will just sidetrack us now. One uncomplicated issue worth mentioning is this. Suppose you admire Superman and I admire Clark Kent, and that Clark Kent and Superman are the same person. Then there seems to be a reading of

(176)   There is someone we both admire.

which is true, and a reading which is false. For the latter, it might well be that I despise Superman while I only admire Clark Kent, whereas you do the opposite. On our proposed semantics for the internal reading it would come out true, since the same equivalence class of terms is realized by "you admire t and I admire t." The false reading of (176) can be captured by not relativizing the semantics to equivalence classes of terms, but sticking with terms directly. (176) is true on that reading if there is a true instance. Relativizing to equivalence classes of terms is required in the extensional contexts. In intensional contexts there seems to be a reading corresponding to relativizing to equivalence classes, and one where we take terms directly.

4

Internalism, Externalism, and the Ontological Question

The view of quantification defended in chapter 3 suggests that the ontological question about Fs just is the question "Are there Fs?" on its external reading. The account of the trivial arguments put forward in chapters 2 and 3 shows that this question is not trivially answered in the way it seems to be answered. The question is simply left open by the trivial and sound arguments. What we need to know now is how to answer this question. And ideally we should try to find out how to answer this question not just for our examples that give rise to trivial arguments about what there is, but also those that do not. In the following chapters we will try to make progress on this question for our four main cases: natural numbers, ordinary objects, properties, and propositions. Later on we will see how much of this generalizes to other cases. In the present chapter we will look at what we need to do to make progress in finding the answer to these ontological questions. This chapter is thus devoted to the strategy for making progress. We will see what the crucial considerations are, and why they are the ones that matter. In doing this I will outline some of the central arguments that will help us not just to answer the ontological questions, but also to solve the second and third puzzles about ontology, which are so far left open.

I will argue that there are two large-scale views about how a domain of discourse in general can be understood: internalism and externalism, to be defined shortly. Finding the answer to which one of them applies to a domain of discourse, talk about Fs, is the central, but not only, consideration for answering the ontological question about Fs. To do this we will first investigate the difference between internalism and externalism about a domain, and then see why this distinction is central for finding the answer to the corresponding ontological question. Depending on whether internalism or externalism is true about talk about Fs will be closely tied to the issue whether a certain answer to the ontological question is required for the truth of what we say about Fs. And this will bring us back to the second puzzle about ontology, the puzzle about how important ontological questions are. I will once more use talk about numbers as our main example for illustration.

4.1 Internalism and Externalism

The trivial arguments in the case of numbers did not answer the external question about numbers since the number word "four" is not a referring or denoting expression as it occurs in the relevant use of "The number of moons is four." Because of this, the quantified statement "There are numbers" only follows on the internal reading, but not on the external one. However, this is a very special case of talk about numbers. The account of the loaded counterparts offered in chapter 2 was an account of a unique occurrence of number words. Most talk about numbers will have to be understood quite differently. Thus, even though the trivial arguments don't answer the external ontological question about numbers, the issue remains whether other talk about numbers implies an answer to the external question. And here there are two main cases to consider. First, it might be that ordinary talk about numbers implies an answer to the external question, even if it doesn't do so via the trivial arguments. Second, it might be that the results of mathematics imply such an answer. Mathematics has established beyond doubt that

(177)   There are infinitely many prime numbers.

Does this imply that there are numbers on the external reading? It will depend on what reading (177) itself has been established. This sentence, just as any other quantified sentence, has an internal and an external reading. Did mathematics establish it in one or the other of these readings, or even in both? If mathematics established it in the external reading, then, of course, an affirmative answer to the ontological question about numbers would immediately follow. But on which reading mathematics established it won't be a trivial thing to find out. Nonetheless, this is what we need to look at if we want to find the answer to the ontological questions. Although in the trivial inferences number words were used non-referentially, and quantifiers were used in their internal reading, this leaves open what we do in general when we talk about numbers. Do we use number words referentially, and quantifiers externally? Or are number words in general non-referring expressions, and is quantification over numbers in general used internally? And similarly for any other domains of discourse: talk about objects, properties, and so on.

Whether or not what we say about numbers, either in mathematics or outside of it, implies an answer to the ontological question about numbers is very closely tied to the question of whether or not the truth of what we say depends on the answer to the ontological question being one way rather than another. If the truth of arithmetic implies an answer to the ontological question, then arithmetic is only true if this is the correct answer. And this is what gave rise to the second puzzle about ontology: on the one hand, ontological questions seem to be of merely academic interest; on the other hand, it seems that everything depends on the answer coming out a certain way.

Does our talk about numbers, objects, etc., imply a certain answer to the respective ontological questions? Our solution to the first puzzle and the account of the trivial inferences makes clear that simply its implying that there are numbers, objects, etc. is not enough to settle this question. Basically anything implies that there are numbers on the internal reading. We need to know whether arithmetic, as well as what we accept about natural numbers more generally, implies this on the external reading as well.

It is uncontroversial that we talk about numbers both in mathematics and outside of it. But this alone does not settle the issue we are concerned with here. When we say that we talk "about" numbers we should distinguish two different senses of aboutness. One we can call *referential aboutness*, which is the sense in which a name is about the thing it names. In this sense you can't talk about Santa Claus, since there is no Santa Claus.[1] A second sense of aboutness is *topical aboutness*. In this sense you can talk about Santa Claus just by making Santa Claus the topic of your conversation. In the topical sense you can talk about aliens all night long, whether or not there are any, but in the referential sense you can only talk about aliens if there are any. Topical aboutness is thus a more neutral notion from an ontological point of view.[2] The distinction between referential and topical aboutness is not quite the same as the distinction between minimal (or non-discriminating) and substantial (or discriminating) conceptions of reference. On a minimal conception of reference, "t" refers to t in all instances. Thus "Santa" (minimally) refers to Santa. But "Santa" does not (substantially) refer to Santa. In a similar spirit one could distinguish minimal and substantial senses of aboutness.[3] The distinction I hope to draw here is different from the distinction between minimal and substantial reference. It is instead the distinction between reference and topic. The topic of what one is talking about can be something one isn't referring to in any sense. For example, when one talks about aliens all night long it is unlikely that anyone even attempted to refer to any alien. Aliens were simply the topic of the conversation. Often we do refer to things that are also the topic of our conversation, for example, when we talk about Obama. But often these two come apart, for example when we talk about aliens, or Santa.

---

[1] In what sense a name is about the object it names is, of course, controversial, as is whether or not names are referential in a stricter and more technical sense. On the latter sense a name would not be referential if it was a disguised description, but it would still be about a particular thing even in that case. It would "denote" that thing, but not refer to it. This issue won't matter for our discussion now, but we will take it up again in section 7.2. Referential aboutness only concerns reference broadly construed, in the sense spelled out shortly.

[2] Topical aboutness is also more intensional than referential aboutness. The topic of our conversation might have been the sage of Königsberg, wondering who that might be, but on a natural way of understanding the topic of our conversation was not Kant, even though he is the sage of Königsberg. Our conversation wasn't topically about Kant, even though it was referentially. This aspect of the difference between topical and referential aboutness won't play any role in the following.

[3] For examples of such distinctions, see Putnam (1981: 1) or Azzouni (2010b: 42f.). This distinction is again to be distinguished from the question of whether or not a deflationary or non-deflationary conception of reference is correct. See Båve (2009) for more on this issue.

We talk about numbers, objects, properties, and propositions, all the time, in the topical sense. We say that the number of moons is four, that there are infinitely many prime numbers, that there are some features wine has in common with beer, or that it's hard to believe that Fred ate the cookies. Many sentences that topically talk about numbers, objects, properties, and propositions we take to be true, and for good reason. What is not so clear is what we do when we talk about them. There are two questions that are crucial for getting clearer about this, and these questions apply equally, but separately, to the cases of talk about numbers, talk about objects, talk about properties, and talk about propositions. We will again discuss them using the case of numbers to illustrate.

First, there is a question about singular terms. When we use number words as singular terms, are they referring expressions on these uses? We saw above that on some uses number words are nothing like referring expressions, they are displaced determiners which for a certain reason appear in a syntactically unusual position. But what in general is going on with number words in singular term position, and other positions as well? Are they in general referential expressions on these uses? Are they never referring expressions? If they are not referring expressions, are they relevantly similar to referring expressions? In our discussion of some occurrences of number words above we saw that sometimes these words are nothing like referring expressions. But maybe on some other occasions they are very much like referring expressions while not being strictly referential? For example, if number words were, somehow, disguised descriptions, then they might denote numbers, while not referring to them (assuming again that descriptions are not referential). And thus we might talk about numbers with number words in more than just a topical sense. We should thus distinguish *strictly referential expressions*, ones that are about things in the same way that names are generally believed to be about the things they name, from *broadly referential expressions*, which are about things in a more permissive sense to include definite descriptions. Definite descriptions, on the standard Russellian view, are quantifiers that describe a unique object. Although they don't refer to that object, they are closely enough associated with it to warrant the label of denoting this object, and denotation would be close enough to reference for our present purposes. Furthermore, it might be that demonstratives, say, are for reasons yet to be uncovered sufficiently different from names and descriptions that they should be seen as neither referring nor describing the objects that they are about, but they would still be sufficiently similar to names and descriptions, and sufficiently different from determiners or adverbs, to warrant being grouped with the broadly referential expressions, ones that in the relevant sense are about something. We will in the following focus on referential expressions in the broader sense, since for us it will be the relevant one in the discussion to follow. What are we doing when we use number terms? Are we using them in a broadly referential way, aiming to pick out some things or talk about some things, or are we doing something quite different with them? Our discussion above suggests that on some occasions we do something different, but that was an isolated case. What do we do with them more generally? If we can get clearer about these questions, we will know more about what we do when we talk about numbers.

Second, there is a question about quantifiers. When we quantify over numbers, in what reading are we using these quantifiers? Quantifiers generally have an internal and an external reading, and for any particular use of a quantified sentence we can ask with which reading it was employed. We know that quantifiers sometimes are used in the internal sense when we quantify over numbers, as we do in the trivial arguments. We also know that sometimes we use external quantifiers when we talk about numbers, as when we ask whether or not there are any numbers, on the external reading of the question. But how is our more general practice of quantifying over numbers to be understood? Is there a pattern in how we quantify over numbers in our talk about numbers, and is one of these two cases an exceptional case? In general, is such quantification internal or external, i.e. are such quantifiers used in the internal reading, or in the external reading?

Before we can address these questions directly we need to refine the issue a bit more. Any sentence that contains a quantifier has two readings. In particular, it is not the case that sentences with quantifiers over numbers in them always contain either external or internal quantifiers. All quantifiers have an external and an internal reading. The question is thus not one about quantified sentences themselves, but about the uses, or utterances, of such sentences. Any quantified sentence can be used with an internal or an external reading of the quantifier. The question is how, in general, they are used, and whether there is a general pattern here at all. A speaker has the freedom to use a quantified sentence either in the internal or the external reading. We sometimes use quantifiers over numbers in the internal reading, for example in the trivial arguments, and sometimes in the external reading, for example when we discuss ontology, and similarly for number words. We sometimes use them as non-referring expressions, even when they are syntactically singular terms, as in the trivial arguments again. But someone can clearly use number words with the intention to refer to objects. A dedicated Platonist might do this with an explicit intention to pick out a Platonic object with a particular use of a number word. Our intentions are in this way up to us. What is at issue is how, in general, we use such quantifiers and number words. Is there a pattern in our use of such quantifiers that points in one direction or the other? Is the dedicated Platonist's use of number words an exception to how they are commonly used, or is it the norm?

There might be no general pattern at all. Maybe number words are used referentially and non-referentially throughout, with no primacy or pattern in favor of one or the other. But if there is a pattern among singular terms, and a pattern among quantifiers, then there are certain ways in which we should expect these patterns to combine. We can note two stable positions about how our use of number words and quantifiers over numbers can be understood, which arise from the connection between quantification and reference. If number words in general are referring expressions, that is, if they are used by a speaker with the intention to pick out an object,[4] then quantifiers in general can be expected to be used in their external reading. After all, if uses of number words like "five" are intended to pick out an object, then quantifiers like "every number" should be used to range over the domain of objects that the uses of number words aim to pick out. But on the other hand, if number words in general are not referring expressions, if they are not used with the aim to pick out an object, then quantifiers in general should not be expected to be used in their external reading. If number words like "five" do not aim to pick out an object, then it would be strange to aim to quantify over a domain of objects with quantifiers like "every number." Not that it would be impossible. But if there is a pattern at all in our use of number words and our use of quantifiers over numbers, then there are two *stable options* and two *unstable options*. The unstable options are first that quantifiers are generally used externally, but number words are not broadly referential, and, second, that quantifiers are used internally, but number words are broadly referential. An unstable option might be true, since our talk about numbers might be confused or incoherent in this way. It might also be true that some number words are used referentially and some are not. Maybe those for even numbers are and those for odd ones are not. It would not make much sense of our talk about numbers, but it might be true for all we have seen so far. But even though we can not, at the outset, rule out that our talk about numbers has some level of incoherence attached to it, we can certainly hope for and expect more. Number talk is not some defective part of discourse. We should expect, although this can't be guaranteed, that one of the stable options is true for it. Thus we will now consider the two stable options. First there is *externalism about talk about numbers*: the view that, in general, number words are broadly referring expressions and quantifiers over numbers are used in their external, domain conditions, reading. Second, there is *internalism about talk about numbers*: the view that, in general, number words are not broadly referring expressions and quantifiers over numbers are used in their internal reading. These are the two large-scale options we have on how to understand our talk about numbers in a coherent way. The crucial question for us now is which one, if any of them, is true when it comes to talk about numbers: internalism or externalism?

The case of talk about numbers was used to illustrate the difference between internalism and externalism, but this distinction applies more broadly. For any domain of discourse, talk about numbers, material objects, events, and so on, can we ask whether internalism, or externalism, or neither, is true? As with the case of numbers, it might well be that neither one is true. It might be that the domain is heterogeneous, and that internalism is true about talk about a certain kind of event, and externalism about a different kind of event. This is possible, but it certainly would be surprising. *Internalism about a domain of discourse* is the view that in general the singular terms in

---

[4] "referring expression" can be understood as a success term, where reference has to succeed, or more neutrally as a term that simply aims to refer, whether or not it succeeds. I use it in the neutral sense here, since the issue for now is what we try to do with talk about numbers, not whether we succeed in doing it.

that domain are not referential, and that the quantifiers are in general used internally. *Externalism about a domain of discourse* is the view that, in general, the quantifiers are used externally and the singular terms are used referentially. For each domain of discourse we can ask whether internalism or externalism is true about that domain.

The crucial question for deciding between internalism and externalism, if any, is to see whether *in general* quantifiers and singular terms are used one way or another. If there is no pattern of use, then neither option might hold. But if there is a pattern, then one or the other might turn out to be true. It is important to note that the issue can't be settled by looking at single cases, either as single examples of the occurrence of number words in a particular sentence, or as the use of number words by a single speaker. A particular person might be exceptional and use number words differently than is the common use in a linguistic community. And a particular occurrence of number words might be quite different from the common occurrence in natural language. In fact, our account of the function of number words in the loaded counterparts might well be a special and exceptional case. We should not conclude from the account of the loaded counterparts and trivial arguments given in chapters 2 and 3 that internalism is true. The question of the general use of number words is largely left open by this.

4.2 Towards a Solution to the Second and Third Puzzles

If one of the stable options, internalism or externalism, is true about a domain of discourse, then this is the key to a solution of our second puzzle for that domain. If externalism is true, then the truth of what we say in that domain requires an affirmative answer to the respective ontological question. Thus the truth of what we say depends on that answer to the ontological question being correct. And, of course, the evidence we have for the truth of what we say is evidence, subtleties aside, for the truth of that answer to the ontological question. On the other hand, if internalism is true, then no answer to the ontological question is implied by what we say in that domain. To make this clearer, consider again the case of talk about natural numbers. If internalism is true, then number words in general are non-referring expressions. Quantifiers in general are used internally, simply generalizing over the instances. Since number words don't refer, no referents are required to exist for the truth of what we say. Since quantifiers generalize over the instances, no more is required for their truth than the truth of the instances. Thus if internalism is true about talk about numbers, then no answer to the ontological question about numbers is implied by what we say about them. And thus no such answer is presupposed for the truth of what we say. We will see more details about all this later, when we will have made progress on the question of internalism or externalism for our four main ontological debates. But for now it should seem promising that deciding between internalism and externalism is crucial for answering the second puzzle. To pursue this question thus seems to be a good strategy towards solving the second puzzle.

And the question of whether internalism or externalism is true is also highly relevant for the questions of whether there is any philosophical work to do at all in ontology, and thus it is central for solving our third puzzle as well. If externalism is true about talk about numbers, then arithmetic implies an answer to the ontological question. Whatever philosophical work there is to do will have to be closely tied to whether there is truth in the results of arithmetic. After all, if the results of arithmetic are true, as they surely seem to be, then the question is settled and no further work needs to be done. But if, on the other hand, internalism is true, then the ontological question is not settled by the results of arithmetic, and further work needs to be done. And if this situation were to obtain, it just might justify ambitious metaphysics. If arithmetic leaves the ontological question about numbers open, then there is a real factual question about numbers left open and available for the taking by metaphysics. Metaphysics might claim that remaining question for itself, and have the ambition of answering it. Much more will have to be said about how this could go and whether arithmetic indeed does not answer the ontological question about numbers. We will look at this way of defending the ambitions of metaphysics in detail in chapter 12, but that the issue of internalism or externalism is very relevant for it is hopefully already plausible, although it will have to be postponed for a proper discussion until later. What we can see already is that the issue is central for answering the ontological question.

4.3 Internalism and the Answer to the Ontological Question

Settling the question of whether internalism or externalism is true about talk about Fs is often, but not always, the central question for finding the answer to the ontological question about Fs. The issue of internalism and externalism not only is central for finding out whether our ordinary talk about Fs depends for its truth on a certain answer to the ontological question, it also is crucial for finding out what the answer to that ontological question is. This is clear in one sense, but surprising in another. It is clear that if it turns out that externalism is true about talk about numbers, then the results of mathematics imply immediately an answer to the external question. If mathematical statements aim to refer to numbers and quantify over them in the external sense, then an affirmative answer to the ontological question about numbers is implied by these results. This leaves it open so far whether we should believe that the results of mathematics are true, but it is certainly hard to doubt that it's true that there are infinitely many primes. So, if externalism about Fs is true, then an affirmative answer to the external question about Fs is guaranteed by the truth of what we say in that domain. Whether what we say is indeed true can still be a substantial question, as we will see in the case of objects below, but it is a good part of the way towards the answer. That so far is the unsurprising part.

But what if internalism is true? As outlined in the last section, if internalism about number talk is true, then what we say about numbers does not imply an answer to the external question about numbers. But then, what could imply such an answer? After all, if what we say about them isn't enough, what else can we do? Does the external question turn into an unsolvable problem? It turns out that the opposite is true. If internalism is true about talk about numbers, then the answer to the external, ontological question about numbers is guaranteed to be "no." Internalism guarantees that the answer to the external question is negative. In the following I hope to spell out the argument for this claim in general. It will be slightly refined below, and we will come back to it in later chapters once we have looked more closely at what we do when we talk about numbers or properties. This is the more surprising part.

Suppose for the moment that internalism is true about talk about natural numbers. We haven't seen, of course, how this could be so in general, but we have seen enough to have a sense of what it would be like if it were true. It would mean that number words are non-referring expressions, both in arithmetic as well as outside of it, and number quantifiers in general are used in their internal reading. In particular, even in arithmetic statements like "$2 + 2 = 4$" number words don't refer to numbers if internalism is true. Arithmetical equations, on the internalist picture, are still true, just as the loaded counterparts were. Even though number words are not referring expressions, they nonetheless occur in true sentences. They just do not refer in these sentences, but do something else semantically. As a consequence, what we say when we talk about numbers does not imply an answer to the external, ontological question about numbers. No reference to numbers is attempted, no externally quantified statements over numbers are made, and thus these statements are true or false whether or not numbers exist. What we say about numbers leaves the external question about them open.

But the fact that internalism is true about number talk guarantees an answer to the external question about numbers, and that answer is "no." Thus if internalism is true about talk about numbers, then the external question about numbers has a negative answer. Not because anything we say about numbers implies this, but the fact that internalism is true about that domain of discourse implies it. The argument for this claim is simple and straightforward, but it is important for what is to come later, and thus we will discuss it in some more detail here, and return to it in later chapters. What is important for appreciating this argument is that its assumption, the truth of internalism, is a quite substantial assumption which we have not established in the least, nor made in any way plausible, nor even outlined how it could possibly be the case for any of our domains of discourse. But suppose internalism were true for talk about natural numbers. Then some account or other, like the one given for what is going on with the singular terms in the loaded counterparts in the trivial arguments, would carry over to number talk in general. Number words, in general, would then be non-referring expressions that appear in singular term position for various reasons without being broadly referential. A negative answer to the ontological question about numbers follows from this substantial assumption.

The simple argument for a negative answer to the external question from the assumption of internalism is the following. If internalism is true about talk about natural numbers, then number words are non-referring expressions. Thus expressions like "2" or "the number 2" do not refer, in the broad sense, to anything. Thus none of the objects in the domain of our external quantifiers are referred to by these expressions. And thus none of them is the number 2. Since "the number 2" does not pick out or denote any object, whatever objects there may be, none of them is the number 2. So among all the objects, none of them is the number 2. That's the simple argument on which I should elaborate.

The argument is analogous to the argument that if "Sherlock" doesn't refer to anything, then nothing is Sherlock. "Nothing is Sherlock," as I just used it, must be true if "Sherlock" doesn't refer since if "Sherlock" doesn't refer, then "Sherlock" doesn't stand for any of the objects over which our quantifiers range. Thus none of the things that my quantifier "nothing" ranges over is Sherlock, and so nothing is Sherlock.

If internalism about talk about natural numbers is true, then number words are non-referring expressions. They are non-referring expressions of a different kind than "Sherlock," but the simple argument just given doesn't care about what kind of a non-referring expression they are. We can distinguish two kinds of expressions that do not refer in the broad sense. First there are those that have the function to refer, but fail to carry out that function. A paradigmatic case of an expression that does not refer in this sense is an empty name. It has as its semantic function to pick out an object, but it can fail to carry out that function, for example when there is no appropriate object to be picked out. "Sherlock" is a good, although, of course, controversial, example of this kind of non-referring expression. A second kind of expression that does not refer is an expression that is not even in the business of referring. It does something different than to pick out or denote an object and it has a completely different semantic function. Take a quantifier like "nothing," or an intensifier like "very." These are non-referring expressions in that they do not even have the function of referring, they semantically do something altogether different. If internalism is correct about talk about numbers, then number words are non-referring expressions of this latter kind. They do something other than referring broadly understood.

All that we need for the simple argument above is that number words and phrases like "the number 2" are non-referring. It is irrelevant for this argument whether number words are non-referring because they try to refer, but fail, or they are non-referring because they are not even in the business of referring. Either way, there is no such thing as the number 2. Since the phrase "the number 2" is non-referring, by the assumption of internalism, and thus does not pick out, or refer to, or denote, any object, it follows that whatever objects there might be, none of them is the number 2. Since "the number 2" isn't a referring expression, and since I am using it in just that non-referring way here, it does not pick out any of the objects that are the domain of my external quantifiers. So, none of the objects in the domain of quantification is the number 2. So there is no number 2, using the quantifier externally. And similarly for all the other natural numbers. So there are no natural numbers: neither the number 1, nor the number 2, nor the number 3, etc. Internalism about talk about natural numbers guarantees that the external question about natural numbers has a negative answer.

There is a general worry about how the above argument could possibly be any good. After all, internalism is a thesis about language and its use. It is a position about what people do when they say certain things and what the semantic function of various expressions is. But how could a semantic and linguistic thesis like internalism imply a metaphysical thesis like the thesis that there are no natural numbers? It would seem that one could never bridge the gap between language and the world and draw conclusions about the latter simply from results about the former. Internalism is simply a view about the semantic function of various expressions and their use. It doesn't even include which sentences are true or false. But the semantic function of number words, say, should not be enough to conclude that there are no numbers. We will need more to draw that conclusion, or so the worry goes.

Although the general spirit of this worry is right-headed, it does not undermine the above argument. To see this, it helps to get clear on what the argument does and does not show. It is true that non-referring expressions of the first kind, those that have the function to refer, but fail to carry it out, can't be seen to be non-referring by semantic considerations alone. All that considerations about language tell us here is whether they have the function to refer. We need to know about the world to see whether they carry out that function successfully. But it is nonetheless true that once we know that they are non-referring expressions then we can know that the object they have the function to refer to does not exist. If we know that "Betty Crocker" is a name for a person, and that name does not refer, then we can conclude that Betty Crocker does not exist. What we won't be able to find out from considerations about language alone is whether that name fails to refer. But once we know that it fails to refer then this is all we need to know to figure out whether Betty Crocker exists.

The issue with non-referring expressions in our second sense is in certain ways similar, and in other ways different. These expressions are expressions that don't even have the function to refer—they do something completely different semantically. Here, too, it is sufficient to know that an expression is non-referring to find out about what exists, but here, in contrast, we can find out that an expression is non-referring by considerations about language alone. We can find out by the study of language alone that an expression is not even in the business of referring, but does something completely different instead. Take the example of a quantifier. Quantifiers are not referring expressions, they do something else, for example range over a domain of objects. "nothing" is such a quantifier. But suppose someone wanted to find out what *The Nothing* is, which, by definition, just is whatever object our phrase "nothing" refers to. Where is The Nothing? What are its features? We can conclude from semantic considerations alone that The Nothing does not exist. Since it, by definition, is whatever "nothing" refers to, and since we know that "nothing" is not a referring expression at all, but a quantifier, we can conclude that there is no such thing as The Nothing. We don't have to look at all the objects and figure out whether any one of them is The Nothing. We can tell in advance that for any object o, o can't be The Nothing since "nothing" doesn't refer to o. So no object is The Nothing. The argument for a negative answer of the external question about Fs from the assumption of internalism about Fs is analogous. If internalism is true about talk about Fs, then all the relevant singular terms that are about Fs, in the topical sense of "about," are like "nothing." They all do something other than referring. And so internalism guarantees that there are no Fs. Semantic considerations are sufficient to allow us to conclude this. This then is the simple argument for a strategy to answer ontological questions with semantic considerations. We will elaborate various aspects of this issue in more detail in chapters 6 and 8.

Although internalism about number talk implies that there are no numbers, it leaves completely open that there are other things, even other things that some people thought are what numbers are: certain sets, or positions in an ω-sequence, or classes of possible inscriptions of numerals, or what have you. Internalism about number talk does not rule the existence of any of these out, but it guarantees that our number talk is not about any of them, and that none of them are numbers. If internalism about number talk were true it would most likely have substantial consequences for the philosophy of arithmetic. I will argue in the following that this is indeed the case. We will have detailed discussion of what we do when we talk about natural numbers in chapter 5 and how this affects the philosophy of arithmetic in chapter 6, where we will also revisit the arguments presented here. After that we will consider other domains of discourse.

4.4 How to Settle the Issue

Although the above considerations were fairly general, without giving us any clear sense of how internalism could possibly be true for a whole domain of discourse, they hopefully made the case that the issue of internalism or externalism is a central issue for the project of finding the answer to the ontological questions. In the next couple of chapters we will aim to answer the question of whether internalism or externalism is true for the four domains that are our main focus here: talk about natural numbers, objects, properties, and propositions. We will start with the case of talk about numbers. To make progress on this question in this case we need to see whether, in general, talk about numbers involves attempts of reference to objects and the use of quantifiers in their external reading. This is, first and foremost, a question about the semantics of number words. Are number words, words like "four," names for objects? Are they like "Fred," or are they completely different? This question might seem easy, but as we will see shortly in chapter 5, this question is quite a bit more complicated than it seems. But it should be clear that this question is central for the question of whether internalism or externalism is true about number talk. If number words are names for objects, then it could be expected that they, in general, are used with the intention to refer to such objects. On the other hand, if number words are nothing like names for objects, then we should expect that, in general, they are used quite differently than names are used. The semantics of the language we use in communication should match up, in general, with what speakers do when they utter sentences in that language. Looking at the semantics of number words will thus be a crucial step for finding out what we do when we talk about numbers. But the semantics of number words by itself might not be enough. It might be that the semantics of number words by itself leaves the issue open, just as the semantics of quantifiers does. Quantifiers are semantically underspecified and have internal as well as external readings. Number words, too, might be semantically underspecified, and have readings that are more in line with an internal use, and others that are more in line with an external use. To really understand what we do with talk about numbers we might have to look beyond the semantics. But looking at the semantics of number words is a natural start.

In the next chapter we will investigate what we do when we talk about natural numbers. Our focus will be on the natural numbers only, in part because this is difficult enough, and in part because the case of talk about natural numbers is special, in ways to be spelled out, that don't carry over to talk about other numbers like the real numbers or complex numbers. We will see more about other numbers in chapter 6. After we have looked at talk about numbers we will be able to draw a number of conclusions for the philosophy of arithmetic. We will then repeat this process for objects, properties, and propositions. Once we see what we do when we talk about them we can draw a number of conclusions about philosophical problems tied to objects, properties, and propositions. And this will illustrate the role of ontological questions in the associated metaphysical problems more generally.

As we will see, the relevant issues for understanding talk about numbers, objects, properties, and propositions will be quite different in these different cases. Although talk about properties and propositions are closely related, they are significantly different from talk about numbers and objects. We will need to look at each of these cases separately, since we can't assume that they are alike, and since they in fact are not alike at all. We have already seen some results about talk about numbers in chapter 2, but they were restricted to a very limited set of examples. Now we need to look at number talk more generally, including the cases already discussed, but in particular giving emphasis on talk about numbers in arithmetic. Some overlap with the issues raised in chapter 2 is unavoidable here, but the problems are more general than were discussed there. What, then, do we do when we talk about natural numbers?

5

Talk About Natural Numbers

5.1 Frege's Other Puzzle

In his groundbreaking *Grundlagen*, Frege (1884), Frege pointed out that number words like "four" occur in natural language in two quite different ways, and that this gives rise to a philosophical puzzle.[1] On the one hand, "four" occurs as an adjective, which is to say that it occurs grammatically in sentences in a position that is commonly occupied by adjectives. Frege's example was

(J4m)   Jupiter has four moons.

where the occurrence of "four" seems to be similar to "green" in

(178)   Jupiter has green moons.

On the other hand, "four" occurs as a singular term, which is to say that it occurs in a position that is commonly occupied by paradigmatic cases of singular terms, like proper names:

(Nis4)   The number of moons of Jupiter is four.

Here "four" seems to be just like "Wagner" in

(CTisW)   The composer of *Tannhäuser* is Wagner.

and both of these statements seem to be identity statements, ones with which we claim that what two singular terms stand for is identical.

But that number words can occur both as singular terms and as adjectives is puzzling. Usually adjectives cannot occur in a position occupied by a singular term, and the other way round, without resulting in ungrammaticality and nonsense. To give just one example, it would be ungrammatical to replace "four" with "the number of moons of Jupiter" in (J4m):

(179)   *Jupiter has the number of moons of Jupiter moons.

This ungrammaticality results even though "four" and "the number of moons of Jupiter" are both apparently singular terms standing for the same object in (Nis4). So,

---

[1] This chapter is based on Hofweber (2005a). The overlapping material has been edited and updated, and this chapter contains new material, in particular towards the end.

how can it be that number words can occur both as singular terms and as adjectives, while other adjectives cannot occur as singular terms, and other singular terms cannot occur as adjectives?

Even though Frege raised this question more than 100 years ago, I dare say that no satisfactory answer has ever been given to it. Some attempts to answer it are lacking in a number of ways, and in this chapter I hope to make some progress towards an answer to this puzzle. Since Frege first raised the puzzle it might be called Frege's Puzzle, but that term is already reserved for the puzzle that Frege also raised about identity statements and belief ascriptions, which is unrelated to our puzzle here.[2] I will thus call the puzzle about the different uses of number words *Frege's other Puzzle*. Frege's other Puzzle is strictly speaking only a puzzle about natural language, but its importance goes beyond that. I described it as a puzzle about grammar and syntax, but it quickly turns into a puzzle about the semantic function of number words as well. Singular terms paradigmatically have the semantic function of standing for an object, whereas adjectives paradigmatically modify a noun and do not by themselves stand for objects. If number words fall into one or the other of these categories, then this will be of great interest for the philosophy of mathematics. If number words are singular terms that stand for objects, then arithmetic presumably is a discipline about these objects. But if number words are adjectives that do not stand for objects, then arithmetic will have to be understood along different lines. Whether or not arithmetic is a discipline that aims to describe a domain of objects, or does something else, is a question that can be closely associated with the question of what the semantic function of number words is. Frege's other Puzzle is thus not only a puzzle about the syntax and semantics of natural language, but also of great interest for the philosophy of mathematics.

In this chapter I will discuss Frege's other Puzzle in general, looking at a variety of different occurrences of number words, in particular ones that relate to their use in arithmetic. It is now time to go beyond our discussion of the relationship between the innocent statements and their metaphysically loaded counterparts from chapter 2, which, for the case of number words, is just a special case of Frege's other Puzzle. Solving Frege's other Puzzle in general promises to be essential for settling the question of whether internalism or externalism is true for talk about natural numbers, as discussed in chapter 4. And from this we should expect significant consequences for the philosophy of arithmetic. The next chapter will be devoted to working out just what these consequences are using the results we will hopefully achieve in this chapter. We should thus try to solve Frege's other Puzzle.

To be more precise we should distinguish the *simple version* of Frege's other Puzzle, which is the puzzle about the different uses of number words in natural language, from the *extended version* of Frege's other Puzzle. The extended version covers not only number words in natural language, but also symbolic numerals, like "4," which are pronounced just the same way as number words in natural language. These

---

[2] See, for example, Salmon (1986).

symbolic numerals are the ones used in mathematics proper.[3] How they relate to number words in ordinary everyday language is a question that leaves room for some debate. One simple, though not implausible, view is that symbolic numerals are merely abbreviations of natural language number words. But if so, which uses of number words do they abbreviate? Let's call a *uniform solution* to the simple version of Frege's other Puzzle a solution according to which in natural language number words are the very same word when they occur as singular terms and as adjectives. Such a solution will have to explain how one and the same word can occur in these two different ways. And let's call a *non-uniform solution* one where they are not one and the same word. Such a solution will have to explain how these different words relate to each other. If we have a uniform solution to the simple version of Frege's other Puzzle, then it will plausibly extend to a solution to the extended version of Frege's other Puzzle. If in natural language number words are one and the same word in both kinds of occurrences, then it is reasonable to think that symbolic numerals are abbreviations of natural language number words. This is plausible, but not guaranteed to be so. After all, it could be that mathematical uses are different from ordinary natural language uses. But that the symbols and number words are pronounced the same way doesn't seem to be a mere coincidence. Thus it is not unlikely that symbolic uses of numerals are derivative on either the singular term or the adjectival use of number words, and that they abbreviate one or the other in symbolic notation. But the interaction between these words and symbols could also be more complex. It could be that the symbolic uses of numerals have an effect on the uses of number words in natural language. In fact, we will closely explore this possibility below.

We can thus distinguish at least three different uses of number words: the *singular term* use, as in (Nis4); the *adjectival*, or as we will also call it from now on for reasons to be explained shortly, the *determiner* use, as in (J4m); and the *symbolic* use, as in "4." The main question for the following is how they relate to each other.

5.2 How the Puzzle Can't Be Solved

One might think that one can avoid all this difficulty quite easily by simply claiming that "four" is ambiguous and thus there is one word that is an adjective and another that is a singular term, both happening to be spelled and pronounced the same way. This would avoid the difficulty of how one and the same word can occupy two different syntactic positions and have two different semantic functions. But this isn't correct. Ambiguity occurs when two different and unrelated words happen to be written and

---

[3] Symbolic uses of number words here are ordinary mathematical uses of them. These have to be distinguished from what could be called *formal* uses. The latter are expressions in an artificial language, expressions that are also pronounced "four," like a term in a formal language used to state, say, Peano Arithmetic. It is a further question how formal uses relate to symbolic uses. This issue is addressed further below, as well as in Hofweber (2009b).

spoken the same way. "Bank" is ambiguous in this way. But "four" is not ambiguous in this way. There are two good reasons for this. First, every number word would have to be ambiguous in the same way. It is not just "four" but also "two" and "three" and all the others. But such a systematic ambiguity can't really be an ambiguity any more, at least not in the same sense in which "bank" is ambiguous. Rather it would mean that number words systematically appear in these two different ways, for some reason or other. But that is just the puzzle that we hope to solve. We want to know why number words systematically appear in these two ways. Second, there clearly are connections between the two appearances of number words. That (J4m) and (Nis4) seem to be quite clearly truth conditionally equivalent is no accident. But if there is a connection between these two occurrences of number words, then it can't just be ambiguity analogous to "bank." Solving Frege's other Puzzle should shed light on what this connection is. To simply hold that number words are ambiguous either is mistaken, if ambiguity is understood as two unrelated words being pronounced the same way, or it is merely a restatement of the puzzle, if ambiguity is understood as that number words can appear in these different ways. What we need to understand is why and how number words have this feature.

There is a long-standing tendency in the philosophy of mathematics to discard the adjectival or determiner uses of number words. There are two main lines to justify this; one is very widespread, the other one less so. The less important line sees adjectival uses of number words as merely a curious feature of natural language, but not something to be taken too seriously, in particular by those who are mainly concerned with science and mathematics. This goes back at least to Frege. In fact, this is what he suggests shortly after pointing out in the *Grundlagen* that number words can occur with apparently two different syntactic and semantic functions. Frege says:

Now our concern here is to arrive at a conception of number usable for the purposes of science; we should not, therefore, be deterred by the fact that in the language of everyday life number appears also in attributive constructions. That can always be gotten round. For example, the proposition "Jupiter has four moons" can be converted into "The number of moons of Jupiter is four." Frege (1950: §57), (translation J. L. Austin)[4]

Of course, it is not completely clear what Frege's considered judgment is on these issues, even if we only look at the *Grundlagen*, and this chapter is not the place to settle questions about Frege. But Frege quite clearly gives primacy to the singular term uses of number words, and he holds that numbers are objects. The above passage seems to suggest that the adjectival or attributive uses of number words can be put aside for a more serious investigation, since they are merely an avoidable feature of everyday

---

[4] The original is:

"Da es uns hier darauf ankommt, den Zahlbegriff so zu fassen, wie er für die Wissenschaft brauchbar ist, so darf es uns nicht stören, dass im Sprachgebrauch des Lebens die Zahl auch attributive erscheint. Dies lässt sich immer vermeiden. Z.B. kann man den Satz 'Jupiter hat vier Monde' umsetzen in 'die Zahl der Jupitermonde ist vier.'" Frege (1884: §57)

language, and not to be expected in the language that will be suitable for science. And this seems to suggest that adjectival uses of number words are a feature of natural language that would not be found in a more ideal language suitable for science. The attributive uses of number words are thus cases where an ideal language and natural language come apart.

Whatever Frege's considered judgment is on these issues, we should not be satisfied with the answer outlined above. First of all, it is not clear why scientific language should not also partake in determiner uses of number words. Why is such use not to be taken at face value? Even if it can be avoided, why should we avoid it? And why should we avoid the adjectival use and not the singular term use? The singular term use often can't be paraphrased away with just the adjectival use, but we will see cases below where the opposite is true as well, i.e. adjectival uses that can't be paraphrased as singular term uses. In addition, this attitude does not help us to solve Frege's other Puzzle, since the latter is after all about natural language. Putting aside natural language won't help us here, even if we could paraphrase all determiner uses away.

A further, more important, and more widely attempted way to deal with Frege's other Puzzle, or at least to get around it, is what I'll call the *syncategorematic account*. According to this proposal, determiner uses of number words are to be understood as syncategorematic, they disappear upon analysis. This proposal is inspired by a proposal that Russell made about the word "the" in his theory of descriptions, and it can be motivated as follows: A number word can be combined with a noun to form a numerical quantifier. Such quantifiers, the syncategorematic analysis goes, can be understood as complexes of the quantifiers ∃ and ∀. These quantifiers are part of the first-order predicate calculus, and this calculus, in turn, is part of logic, and thus unproblematic. Take for example

(180)   A man entered.

which contains a quantifier, and which could semi-formally be written as

(181) $\exists x : x$ is a man and x entered.

Similarly,

(182)   Two men entered.

can be understood as involving more than one quantifier. Semi-formally it could be written as

(183) $\exists x \exists y : x \neq y$ and x is a man and y is a man and x entered and y entered.[5]

And so on for other number words. Thus at the level of semantic representation the number words disappear when they are used as determiners. What are left are blocks

---

[5] This gives the reading of "two men" as "at least two men." The "exactly two men" reading is, of course, also expressible the usual way in a similar manner.

of quantifiers that also occur in the first-order predicate calculus, and which are thus part of logic.[6] And in this way the proposal is like Russell's proposal about descriptions. According to Russell, as he is usually understood, the word "the" in a description is syncategorematic and disappears upon analysis.[7] It doesn't make an isolatable, single contribution to the truth conditions. Rather, it is analyzed away in context. According to Russell, the underlying form of

(184)   The man entered.[8]

can be revealingly spelled out with the semi-formal

(185) $\exists x$: x is a man and x entered and $\forall y$ if y is a man then $y = x$.

The syncategorematic analysis thus attempts to solve Frege's other Puzzle by proposing that determiner uses of number words disappear in the semantic analysis, whereas singular term uses do not. Semantically, number words can thus be understood as having the function of standing for objects. That is to say, all number words that are still left at the level of "logical form" have the function to stand for objects. Number words in their determiner use only appear in that position in the surface syntax. They will have disappeared into blocks of $\forall$ and $\exists$ at the level of logical form.

This proposal to solve Frege's other Puzzle has a variety of problems. First, it at most works semantically, but not syntactically. There is no explanation of why the word "four" syntactically occurs both as a determiner and also as a singular term. After all, the word "the" never occurs as a singular term. According to Russell's proposal the word "the" is of a fixed syntactic category, but it disappears at the level of semantics into blocks of $\forall$ and $\exists$. The word "four," according to the syncategorematic account, disappears at the semantic level in its determiner uses, but not in its singular term uses. The syntactic issue of how it can be that apparently one and the same word can occur in these different syntactic positions is not answered by this.

Second, the puzzle that we are concerned with isn't answered by this proposal. We want to know how it can be that apparently one and the same word can do these different things: being a singular term and also being a part of quantifiers, or more generally how these different uses of "four" relate to each other. Whether or not number words as determiners disappear at the level of logical form, the puzzle remains how they can both be singular terms and determiners. How can they sometimes stand

---

[6] The view that number words are syncategorematic when they are part of quantifiers, and that these quantifiers are unproblematic, since they come down to first-order predicate calculus quantifiers, is rarely explicitly discussed at any length, though widely assumed. This position, and the corresponding dismissive stance towards Frege's other Puzzle, can be found in, for example, Hodes (1984), Field (1989a), and many others.

[7] See Russell (1905).

[8] Of course, "the man" by itself is hard to parse as a description and is at best an incomplete description, which needs a contextually provided restrictor. We could just as well use "the man with the red hat" or the like. But we can ignore all these complexities for our main point here.

for objects, but also sometimes syntactically occur as determiners or adjectives, and then semantically disappear with only blocks of quantifiers as a trace?

Finally, the view that words like "the" or number words as determiners should be understood as syncategorematic should be rejected for completely different reasons. These reasons are widely discussed, and accepted, in the case of the syncategorematic treatment of "the" that Russell proposed. But they are not as widely appreciated in the philosophy of mathematics literature, where the syncategorematic treatment of number determiners is often endorsed, even though they apply there equally well.

Words like "the," "four," "some," "many," "most," etc. are rather similar. They combine with nouns to make full noun phrases, and they all can form similar sentences, like

(186)   The F is G.

(187)   Some F is G.

(188)   Most F are G.

However, according to the syncategorematic analysis, some of them are syncategorematic, others are not syncategorematic, and others again are not treated by this analysis at all. "The" is supposed to be syncategorematic, whereas "some" is not. It directly gets represented as a first-order predicate calculus quantifier at the level of logical form. "Most" on the other hand, can't be treated syncategorematically along the lines discussed above, since it is known not to be first-order expressible. By "first-order expressible" I mean "expressible in the first-order predicate calculus with the quantifiers $\exists$ and $\forall$." The syncategorematic analysis only gets off the ground since some determiners are first-order expressible. The quantifiers that can be formed with these determiners can be expressed as complexes of $\forall$, $\exists$, variables, and predicates. Other determiners, like "most," can't be expressed this way.

But what is the relationship between a natural language sentence and its predicate calculus counterpart, even in cases where the truth conditions can be expressed this way? Does the sentence in the first-order predicate calculus reveal the underlying logical form, or is it merely another way to express the same truth conditions? As we will see in just a minute, a unified semantics of all determiners can be given. It takes to heart that the first-order predicate calculus is not enough to properly treat natural language quantification, and it will assign the same underlying structure to all of our above examples (ignoring issues about plural and singular for now). The syncategorematic treatment of number determiners would assign completely different underlying logical forms even to these two sentences:

(189)   Four F are G.

(190)   Two hundred F are G.

The latter would have 200 quantifiers in its logical form. Since a semantics for determiners can be given that gets the truth conditions right and doesn't have any of the flaws of the syncategorematic analysis we will have to conclude that the syn-categorematic analysis gives merely another way to express the same truth conditions with a sentence in the first-order predicate calculus, but not one that is semantically revealing and that brings out the underlying logical form.[9]

The syncategorematic analysis thus has to be rejected. A closer look at the semantics of number determiners in the semantic theory that replaced the syncategorematic one is, however, most useful for getting closer to solving Frege's other Puzzle. Thus we will look at it in some more detail next.

5.3 Number Determiners

5.3.1 Determiners and Quantifiers

As we already saw in chapter 3, there are several limitations to the view that natural language quantification is closely tied to quantification in the predicate calculus with the basic quantifiers $\forall$ and $\exists$. First, these quantifiers take a full noun phrase to be the basic unit of analysis. Thus "something" is the smallest unit of a quantifier, analyzed as "$\exists$." However, natural language quantifiers are often complex and built up from smaller units. In fact, even "something" seems to be built up from "some" and "thing." Natural language quantifiers contain phrases such as "most men," "every tall child," etc. It would be nice to have a view that specifies the semantics of a complex quantifier on the basis of the semantics of its simpler parts. That is, it would be nice to have a compositional semantics of complex quantified NPs, which includes a semantic treatment of the smaller parts. Second, many natural language quantifiers are not first-order expressible, as spelled out above. These quantifiers are provably not expressible in the first-order predicate calculus, for example "most men." But a unified account of natural language quantification should treat all quantifiers, not just the ones that are first-order expressible.

These requirements have been met in generalized quantifier theory (GQT), a very successful and widely accepted theory of natural language quantification. It is based on work of Mostowski and Montague[10] and has been developed by various logicians and semanticists.[11] We have seen some simple aspects of this approach to quantification in natural language in chapter 3, but we will very briefly repeat some of it again, this time from a somewhat different angle.

---

[9] We use the term "logical form" here as a level of representation that reveals certain semantic facts about sentences, and these semantic facts can go beyond simply what the truth conditions are. They can include issues of scope, what the relevant semantic parts are, and more. Thus two sentences with the same truth conditions can have different logical forms. In addition, the logical form should be free of syncategorematic expressions, if there are any. Other notions of "logical form" exist in the literature, but we do not have to investigate this issue further as long as the above remarks are kept in mind.

[10] See Mostowski (1957) and Montague (1974).

[11] See Keenan and Westerstahl (1997) for a survey of work that has been done in this area.

Natural language quantifiers have an internal structure whose basic parts are a noun and a determiner. Focusing on the simplest case, nouns are words like "man," "thing," etc. and determiners are words like "some," "all," "many," "most," "the," and many more, including number words in their determiner use. The semantics of these quantifiers is compositionally determined by the semantics of the noun and the determiner. The appropriate semantic values for these expressions are "higher-order objects," usually understood as follows. The semantic value of a sentence is a truth value. A sentence splits up into an NP and a VP. The semantic value of a VP is a property, and the semantic value of an NP is a function from properties to truth values. The semantic value of a sentence is thus determined by the semantic values of its immediate parts: simply apply the function that is the semantic value of the NP to the semantic value of the VP. We will ignore the internal structure of VPs here, but we will look at the internal structure of a quantified NP. It splits up into a determiner and a noun. The semantic value of a noun can be a property as well, and the semantic value of a determiner is thus a function from properties to functions from properties to truth values. Thus the semantic value of a determiner is a function from semantic values of nouns to semantic values of full NPs. In a useful notation to specify these higher-type objects: an object of the type of truth value is of type t, an object of the type of an entity is type e, and a function from one type t_1 to another t_2 is written as (t_1, t_2). If we understand properties to be functions from entities to truth values[12] determiners are thus of type $((e, t), ((e, t), t))$. These types can be understood semantically as well as syntactically. Semantically they specify what kind of a semantic value an expression has. Syntactically they specify how an expression combines with others to ultimately form a sentence, which is of type t.

As it turns out, this theory works perfectly well, at least for the cases we are considering here, and it is widely accepted. It gives us a unified compositional semantics for quantified noun phrases and in fact other noun phrases as well. Different determiners just get assigned different functions, but all get a semantic value of the same kind.

For our discussion in this chapter there are two important points to note here. First, according to GQT the word "two" (in its determiner use) is not syncategorematic. It makes a discernible contribution to the truth conditions and is not explained away at the level of logical form. Second, number determiners, i.e. number words in their determiner use, have a semantic value, just like all determiners, but they are not referring expressions or singular terms. Having a semantic value and having a referent have to be distinguished. Every expression has a semantic value, and might have different ones in different semantic theories. But not every expression is a referring expression. Names and some other singular terms refer, but "some" and "many"

---

[12] This is merely for convenience, and does not involve a metaphysical view of properties. We can just directly say "the semantic value of X is a function of type (e, t)" instead of "the semantic value of X is a property." A more complete account will use intensional types, but this complexity is not important for us here.

don't. In addition, even referring expressions can have a semantic value other than their referent. In Montague's treatment of proper names, for example, names get sets of properties (or, equivalently, functions from properties to truth values) as their semantic value, but they don't refer to such functions. They refer to people, dogs, and the like. Thus even though determiners like "some," "many," and "two," have certain semantic values in GQT, they do not refer[13] to these semantic values.[14]

That number determiners are meaningful expressions that are not singular terms and that are not syncategorematic must be beyond question given the success and the wide acceptance that generalized quantifier theory enjoys and the obvious advantages it has over the alternatives we have discussed. It was not beyond question when Frege wrote the *Grundlagen*, and maybe not when early neo-Fregeans remarked that number

---

[13] The use of "refer" in the above comment contrasts with that of some philosophers who simply use it for the relationship that holds between an expression and its semantic value. On such a use almost all words and phrases are referring expressions, and even parts of words can be referring expressions. But on our use we mark the intuitive difference between some words that have the function to contribute objects to the content of an utterance, paradigmatically names, and others that don't have this function, paradigmatically words like "by" or "very." "reference," as a technical philosophical term, is used in different ways, and the present use is one common use of this word, but not the only one. We will discuss reference and its relationship to semantic values in much more detail in chapter 8.

[14] See Barwise and Cooper (1981) or Gamut (1991) for more on generalized quantifiers. Further issues arise about plural. See, for example, van der Does (1995) for both a survey and an account of various issues about plural quantifiers. We will ignore many of the real complications about plural in this chapter, but van der Does (1995) provides an account that in certain ways is congenial to the view taken below. We in particular ignore the issue of different types for plural and singular quantifiers. This is only for reasons of simplicity in presentation.

There is some controversy about whether number words in the relevant uses are determiners, modifiers, or adjectives. This is also an issue which is insignificant for us here. All arguments below will be motivated by examples which are valid no matter which one of these groups number words belong to in the end. All that matters for us here is that number words in these uses belong to one of these categories. I will continue to primarily call them "determiners," but they can occur as modifiers as well (and we will see more on this just below). Proponents of the view that number words are adjectives or modifiers, but not determiners, usually point to some disanalogies between number words and other determiners. For example, "three men" can combine with "the," which many other determiners can't:

(191)   The three men who entered the bar got drunk.

(192)   *The some men who entered the bar got drunk.

However, some other determiners exhibit the same behavior, for example

(193)   The many children who died in the war will not be forgotten.

In addition, not all determiners behave syntactically the same way. See Barwise and Cooper (1981) for cases of this and a well-known attempt to explain the different behavior semantically. Number words also behave differently than classic adjectives. For example, they fail the seem-test which can be used to demarcate adjectives (thanks to Randall Hendrick for pointing this out to me):

(194)   They seem green.

(195)   *They seem four.

However, the classification issue is of no real importance for the overall debate here. What ultimately matters for our discussion is that number words in their determiner use can form complexes as will be discussed shortly, and that they are not themselves referring expressions in this use. Whether they are in the end adjectives, determiners, or form a separate class of their own, is secondary.

words in such uses are ultimately singular terms.[15] But today this cannot be accepted and it at least has to be taken as a radical proposal about natural language quantifiers. I stress this to make vivid that we then have no solution available to Frege's other Puzzle, either in its simple or extended version. Generalized quantifier theory does not cover number words when they occur as singular terms, but only when they are determiners. We thus have theories that deal with one or the other of the two occurrences, but none that gives a unified account of both.

In the following I will propose a further account, one that starts with ordinary uses of number determiners, and that together with some empirical considerations attempts to provide a unified account of all uses of number words. Not all aspects of this proposal are fully contained in this chapter and some are subject to empirical confirmation or refutation. But the proposal to follow gives us a new solution to Frege's other Puzzle. Given the failure of the proposals outlined above we badly need to try something else. The key to a different account of the various uses of number words is to take the determiner uses more seriously, and not to brush them aside as syncategorematic. In fact, the determiner uses are quite a bit more interesting and complex than one might think.

5.3.2 Bare Determiners

As a determiner, "two" takes a noun argument to make a full noun phrase. Determiners can also occur in sentences without their argument being made explicit. Consider:

(196)   Three men entered the bar, and two stayed until sunrise.

(197)   Every man entered the bar, but only some stayed until sunrise.

(198)   Three eggs for breakfast is too many, two is about right.

In all these examples, the last occurrence of a determiner in these sentences is without an argument, at least without it showing up explicitly in the sentence. This is not too puzzling and easily explained as ellipsis. The argument was just mentioned half a sentence ago and isn't repeated again. There are, however, a number of examples where the determiner also occurs without the argument being explicit, but where it does not seem to be a case of ellipsis. Consider:

(199)   Some are more than none.

(200)   Many are called upon, but few are chosen.

(201)   None is not very many.

(202)   All is better than most.

---

[15] See Wright (1983) for similar remarks.

Such examples (except maybe (200)) can be used either elliptically or to express generalizations. In the use in which (199), for example, expresses a generalization it basically means that, of whatever, some are more than none. In the elliptical use a contextually salient kind of thing will be such that one claims that some of them are more than none of them. I am using the term "elliptical" broadly here to include the case of a contextually salient, but not explicitly mentioned kind. This phenomenon also occurs with number determiners. They can be used elliptically, as above, or to express generalizations, as in standard utterances of

(203)   Five are more than three.

(204)   Two are at least some.

Let's call (an occurrence of) a determiner without explicit argument or noun a *bare determiner*. As we saw, we have to distinguish between two kinds of bare determiners: the ones that are like the elliptical ones and the ones that are used to express generalizations. Let's call the elliptical ones *elliptically bare determiners*, keeping in mind our broad use of the term "elliptical," and the others *semantically bare determiners*. The distinction is supposed to be exclusive and applies to occurrences of determiners in particular utterances. An example of the first kind are also utterances of

(205)   After dinner I will have one, too.

The kind of thing of which the speaker said he will have one will be fixed by the context of the utterance. So, the context will fix a certain kind of thing, X, such that what the speaker said is true just in case the speaker has one X after dinner. So, "one" is an elliptically bare determiner here. An example of semantically bare determiners, again, is an ordinary use of

(206)   Two are more than none.

where the speaker is not intending to talk about any particular kind of thing.

5.3.3 Complex Determiners and Quantifiers

Before we can go to issues more directly related to Frege's other Puzzle we should also look at operations on determiners and on quantifiers. Natural language allows us to build more complex determiners out of simpler ones. The best examples of this are Boolean combinations of determiners as in

(207)   Two or three men entered the bar.

(208)   Some, but not all, women smoked.

And we can build more complex quantifier phrases out of simpler ones. Boolean combinations are again a good example:

(209)   Some men and every woman smoked.

(210)   Two men and three children saw the movie.

Such operations can also occur on bare determiners:

(211)   Two or three are a lot better than none.

(212)   Few or many, I don't care, as long as there are some.

Consider now:

(213)   Two apples and two bananas make a real meal.

(214)   Two apples and two bananas are only a few pieces of fruit.

(215)   Two apples and two bananas are four pieces of fruit.

Here "Two apples and two bananas" forms a complex quantified NP that consists of two quantified NPs that are joined together by "and." The semantic function of "and" in these examples is worthy of a few words here. It can't be seen as merely the familiar sentential conjunction. In these uses "and" combines quantifiers to make more complex quantifiers, and it thus conjoins quantifiers, not sentences. In particular, we are dealing with plural here and operations on plural NPs usually allow both so-called distributive and collective readings, just like plural NPs in general. To illustrate this, consider

(216)   Three men carried a piano.

This could either be read collectively, where three men together, as a group, carry a piano, or distributively, where three men each carry a piano. Similarly

(217)   Three men and two women carried a piano.

could be read in a variety of different ways: a group of three men carried one piano, a group of two women carried another, or all five carried one piano, or each one of them carried one, and maybe others as well. This distinction is important when it comes to understanding cases like

(218)   Pizza and bad beer makes me sick.

which might or might not imply

(219)   Pizza makes me sick and bad beer makes me sick.

depending on whether or not "and" is read collectively. In particular, if it is read collectively then the "and" involved can't be understood as a truth functional sentential connective.

Just as there are sentences that involve in essence only simple bare determiners, like our examples (199), (201), and (202) above, there are also other sentences that involve in essence only bare determiners, but these determiners are complex. For example, take

(220)   Two or three are at least some.

All this is pretty straightforward. But this gets us closer to something that almost sounds like a bit of arithmetic.

5.3.4 Arithmetic-Like Bare Determiner Statements

Some statements that crucially involve bare determiners are apparently very close to arithmetic statements, though, as we will see below there are also important differences among standard expressions of arithmetical statements. Consider:

(are4)   Two and two are four.

Here we have three bare determiners at once in the sentence, the first two combined by "and" (in its collective reading). In most uses, these determiners will be semantically bare. Note also that (are4) is plural, as the bare determiner use would seem to require. These statements not only sound a lot like arithmetic, they in fact are quite close to it as we will see in the following. However, before we can consider arithmetic we should look at these bare determiner statements in some more detail.

   (are4) is a general statement involving semantically bare determiners. What it says can be spelled out as: two X and two (more) X are four X, for whatever X. In fact, there is another option which is equally good for our discussion to follow, and which we in the following won't carefully distinguish from the reading just mentioned. It involves understanding "and" as an operation on determiners, not NPs. So, according to this reading (are4) can be understood as: two and two (more) X are four X, whatever X is. For our discussion in the following either one of these two readings will do. Note also that the qualification "more" does not usually have to be made explicit. "and" is mostly read as "and in addition," not just in the above examples, but in many similar cases. Consider:

(221)   She only had an apple and dessert.

A usual utterance of this wouldn't be true if she just had an apple, even though fruit is perfectly fine dessert. There are a variety of mechanisms that can guarantee for this to be so. It could simply be ellipsis, or pragmatic, or something else, but which one is the correct one in cases like this does not matter for our discussion here.

   How should we think about statements like (are4) with larger-scale philosophical issues in mind? First of all, (are4) contains no referring or denoting expressions. The words "two" and "four" occur in their determiner use, and as determiners they are not referring expressions. They are part of NPs, but not by themselves NPs. In particular, these words do not refer to determiners, just as "some" or "none" do not refer to determiners in (199). In both of these statements no reference occurs. The truth of these statements does thus not depend on the existence of any objects that are referred to, since none are referred to. In fact it seems that these statements are true no matter what exists. Whatever there might be, some are more than none, and two and two are four. I will argue for these claims in more detail in chapter 6.

5.4 Numbers and Arithmetic

We have seen that statements formulated with bare determiners can be a lot like arithmetical statements, and that they apparently can be true no matter what objects exist. But it would be quite premature to think that this can be easily extended into an account of arithmetic and talk about numbers in general. It is now time to have a closer look at how what we have seen so far relates to arithmetic and the theory of numbers. This will bring us closer to a new solution of Frege's other Puzzle. It might not be completely clear why looking at arithmetic and number symbols is the natural thing to do if we want to solve Frege's other Puzzle, since it, in its simple version, was about number words in natural language. But as mentioned above, it might well be that arithmetic and symbolic numerals affect the uses of number words in natural language. A look at how the bare determiner statements relate to arithmetic will clarify this and lead to a solution of Frege's other Puzzle that can't be seen by looking at natural language in isolation. Thus we will now discuss basic, simple arithmetical statements.

5.4.1 Basic Arithmetical Statements

Basic, quantifier-free arithmetic is a trivial part of mathematics, but it is philosophically important for at least two reasons. First, understanding it is an important step towards understanding arithmetic. Second, it is in many ways the first and most basic part of mathematics. It is the first mathematics a child learns, and it is most closely connected to ordinary non-mathematical discourse. Besides that, it is there where mathematical symbols are first introduced. A good starting point for understanding mathematics is thus trying to understand basic arithmetical statements. Among these is

(=4) $2 + 2 = 4$

I use the mathematical notation here, since it is itself an important issue what these symbols really mean. What does (=4) mean? How and why does this notation get introduced? A first step to answering the first question is simply to read (=4) out loud. But, as it turns out, different people say different things (at different times) when asked to do so. When one asks different mathematically competent people to read (=4) out loud one gets at least the following answers (I report here on an informal survey):

(are4)   Two and two are four.

(is4)   Two and two is four.

(222)   Two and two equal four.

(223)   Two and two equals four.

and all the above with "plus" instead of "and."[16] It seems that there are basically two ways to read "2 + 2 = 4," in the plural and in the singular. And on reflection, these two ways of looking at it seem to be quite different. As we have seen, when we speak in the plural and say

(are4)   Two and two are four.

we are using bare determiners and we are not talking about particular objects. But when we speak in the singular and say

(is4)   Two and two is four.

then it seems that we are saying something about particular objects. First of all, it seems that in (is4) "is" is the "is" of identity. With it we are claiming that what one singular term, "two and two," stands for is identical with what another singular term, "four," stands for. And surely, for this to be true the things these terms stand for have to exist. In addition, (is4) interacts with quantifiers in exactly the way that seems to be required to establish that the terms "two and two" and "four" are referring expressions. In particular, we can infer that there is something which is four, namely two and two. So, it seems that (is4) is an identity statement involving two singular terms that stand for objects, whereas (are4) is a statement involving no referring terms, but only bare determiners.

The plural and singular ways of reading symbolic arithmetical equations are a striking case of Frege's other Puzzle. The two sentences are as close to each other as possible, besides that in one number words are used as singular terms, and in the other they are used as determiners. These pairs of examples provide an interesting addition to Frege's pair (J4m) and (Nis4). In both the puzzle is about the relation of the different uses of the number words. In Frege's original pair he claimed, and many but not all agreed, that they were equivalent in truth conditions. This seems less clearly so in our pair. As we have seen, on reflection the truth conditions seem to be quite different, and thus only one of them could be the correct way to read the mathematical equation, assuming it is unambiguous.

But is there really only one correct way to read the symbolic equation? To understand how the plural and the singular ways of reading the symbolic arithmetical equations relate to each other would be to solve a special case of Frege's other Puzzle. To understand this we should not only try to find out which one of these is the correct reading of the symbolic equation, if there is only one correct reading, but also why there doesn't seem to be such a clear-cut answer to this. Why do many mathematically

---

[16] There are of course a number of further possibilities that also seem to be right, like

(224)   Two and two make/s four.

competent people say one or the other? We will have to look at how the arithmetical symbols are first introduced to us, and what meaning is given to them then, to make progress on this. This can't be decided by looking at natural language or mathematics by themselves. We will have to look at how this actually happens when we learn arithmetic.

5.4.2 Learning Basic Arithmetic

How a child learns arithmetic is of considerable interest from many points of view. On the one hand, it is of interest to developmental psychologists, who take children's mastering of counting, arithmetic operations, and the number concepts as an important case study for large-scale issues in developmental psychology. On the other hand, it is of interest to educators who would like to understand better how children learn in order to improve their teaching. For this reason there is quite a bit of literature on what happens when a child learns arithmetic. A good part of this literature is orthogonal to our main concerns here, but some of the issues discussed there are quite relevant for us now. I can't possibly hope to thoroughly discuss the many fascinating things that have been discovered about number concepts and the learning of arithmetic, but fortunately much of that isn't directly relevant for our question, yet some of the more basic facts are.

Among the things that a child needs to learn are at least the following:

1. They have to learn, or memorize, how to continue the sequence: *one, two, three, four.* . . .
2. They have to learn how to count down some collection of objects by associating with each object one of the first n numerals.
3. They have to be able to determine size, or answer "How many?" questions.
4. They have to master change of size, adding things to a collection or taking them away.
5. They have to master the mathematical formalism, like "2," "+," etc.
6. They have to learn how to solve arithmetical problems purely within the formalism, i.e. give the right answer to $2789 + 9867 - 34 =$?

How this learning works temporally, i.e. what needs to be learned before what, is not completely straightforward, and one obviously doesn't have to master one area to move on to the next. We will mostly be concerned with what happens towards the end of this list. There is much debate in the psychological literature on how and at what age a child acquires the concept of a number in the first place. Here the classic proposal by Piaget that children do not have any numerical competence at birth and only acquire the relevant concepts through abstraction at a fairly old age has received much criticism.[17] Instead it turned out that babies already have basic competence with

---

[17] See Piaget (1952) for the proposal and Wynn (1992) for a well-known experiment that seems to show the opposite. Much of this debate is surveyed in Carey (2009), which contains a discussion and many references to further literature.

small numbers and they have an innate ability to form judgments about, or at least properly react to, small magnitudes without counting. There is much discussion and controversy about how these magnitude judgments get extended to learning of the full number series, whether or not counting itself is innate, or whether it is acquired from scratch via simply memorizing a sequence of words. Susan Carey gives a very detailed overview of this debate in Carey (2009) and defends a view like the latter. Another book length survey of the psychological literature is in Stanislas Dehaene's Dehaene (1997). But neither Carey nor Dehaene discusses the puzzle about talking, and thinking, about numbers which we labeled Frege's other Puzzle above. Although they discuss the cardinal and ordinal aspect of thinking about numbers, this aspect is shared by number words as names for objects and number words as determiners. Much of this discussion of how we acquire our concept of number, and which part of it is innate or learned, strikes me as quite orthogonal to our concern here. In fact, neither Carey nor Dehaene is very explicit about what kind of concept the concept of natural numbers is supposed to be on their views. The debate is mostly about how we get our concept of number, not what it is a concept about. Is it a concept that stands for an object, or one that corresponds to a determiner, a quantifier, or an adjective? These questions are not given any prominence in their debates, and much of what they discuss does not seem to settle the issue one way or another, as far as I can tell at least.

The issues most relevant for us are ones that arise later in the developmental process, when we carry out arithmetic with symbolic numerals. Here it is well-known that children face difficulty in doing this. I want to suggest one proposal about what happens in light of this difficulty that directly addresses Frege's other Puzzle. This proposal is compatible with other aspects that make learning arithmetic difficult, for example, those discussed in chapters 4 and 5 of Dehaene (1997): difficulties tied to complex natural language number words ("two hundred and forty seven"), difficulties with memorizing multiplication tables, and so on. To see how Frege's other Puzzle is connected to the difficulties in learning arithmetic it is helpful to remind ourselves how the symbolic numerals and arithmetical operations are first introduced in a classroom setting.

After a child has learned the number sequence "one, two, three, etc." they will have to be able to answer questions about how many things there are in a particular situation, and they will learn to count down the objects in the situation with the number words they have memorized. And they will master judgments about changes of size, both in actual collections as well as in imagined ones. There are a variety of exercises to do this, and a standard textbook on teaching mathematics to first graders will contain a collection of a good mixture of them.[18] Examples are

---

[18] See, for example, Verschaffel and Corte (1996: 112f.) for a classification of such exercises and also Becker and Selter (1996).

(225)   Here we have three marbles on the floor. Now I put two others there. How many marbles are now on the floor?

(226)   Suppose Johnny has two marbles, and Susie has three more than Johnny, how many does Susie have?

During this learning process, which takes quite some time, the teacher will introduce the mathematical symbols. The students will learn the decimal system, that "2" is read "two," they will learn to count in symbols, i.e. continue the sequence *1, 2, 3, 4, ...* And the student will learn to represent what was learned in exercises like (225) and (226) in symbols. After doing exercise (225) a classroom situation might continue:

(227)   That's right, Pat, three marbles and two more make one, two, three, four, five marbles. (*the teacher will write on the blackboard:*) $3 + 2 = 5$. (*and say out loud any of the following:*) Three and/plus two is/are/make five.

After such exercises are mastered to a reasonable degree the education will continue in teaching the children to add, subtract, and multiply using the symbols alone. At this stage children will learn tricks for adding that are based on the use of the decimal system, like carrying over ones, or multiplying with the tens first and the ones later, and the like. The child is then supposed to solve simple arithmetical problems abstractly, without imagining a collection of marbles that gets increased or diminished. The child is supposed to be able to solve problems like

(228) $26789 - 789 + (2 \times 23) = ?$

In this last case, for example, it is supposed to see more or less directly and without much calculation that "789" are the last three digits of "26789" and thus subtracting the former from the latter gives "26000."

All this sounds easy, if not trivial, to us. But in fact it is extremely hard for a child to learn this. It will take several years before children make the transition from being able to count and use natural number words to being able to complete simple arithmetical calculations. By the time children have learned to do basic arithmetic they will already have mastered the most complicated things, like making up a story, or finding their way home, or knowing what they shouldn't do. I think it should be quite surprising that learning arithmetic is so hard. It literally takes years of hard work and repeated training for children to be able to solve simple arithmetical equations. But why is arithmetic so hard, and why does it seem so easy to us now? I believe there are a variety of different reasons for this, many of which are discussed in the psychological literature and reviewed, for example, in Dehaene (1997). But one further reason is not discussed in that literature, and it is most relevant for our attempt to understand Frege's other Puzzle. That reason for the difficulty is the following.

5.4.3 Cognitive Type Coercion

The contexts in which arithmetical symbols are introduced, and the examples with which arithmetical equations are illustrated, suggest that arithmetical equations at first express bare determiner statements. After all, teachers introduce the symbolism with explanations about sizes of collections, about there being so many marbles or cats. Number words in such examples are determiners, and since arithmetical equations are introduced in a context where they are illustrated with such examples it is natural to take them, at first, to be based on bare determiner statements. To give a primacy to the bare determiner statements in the beginnings of the introduction of the mathematical symbolism is not really to decide the question of the nature of arithmetic in any way. The crucial question is how arithmetic develops from this starting point, and what it ends up being in the end. We will have to see how the singular arithmetical equations arise from a starting point of plural, bare determiner statements.

Numerical equations like

(229) $3 + 2 = 5$

and basic arithmetical truths more generally are first learned in the context of thinking about the sizes of collections, and so they might in these contexts be appropriately expressed as

(230)   Three and two are five.

However, thinking about arithmetic in this way, involving bare number determiners, has its cognitive obstacles, in particular when the numbers get larger. Once we try to make calculations that are not obvious any more, and once we try to solve arithmetical problems of a somewhat greater complexity, we run into cognitive difficulties. Thoughts that are expressed with bare determiners and that involve operations on determiners are quite unusual. In ordinary thinking there are only very few cases of this besides the ones involving number determiners. "Some but not all" and a few more come to mind, but their complexity is rather limited. Number determiners are special in this respect, that they allow for the expression of complicated thoughts that involve essentially only bare number determiners and operations on them. Our minds, in particular when we are young children, are not very well suited to reason with such thoughts. As we have seen above, learning even fairly basic arithmetical truths and doing simple arithmetical calculations is a substantial task for a small child and takes years to accomplish. Anything that helps in solving arithmetical problems will be gladly adopted.

Our minds mainly reason about objects. Most cognitive problems we are faced with deal with particular objects, whether they are people or other material things. Reasoning about them is what our mind is good at. And this is no surprise. We are material creatures in a material world of objects, and the things that matter the most for our survival and well-being are material objects. But what precisely is the difference between reasoning with thoughts that are paradigmatically about objects, and reasoning with thoughts that are expressed with bare determiners? Why is our reasoning generally better with one rather than the other? This can be nicely illustrated by adopting a certain widely held picture of reasoning and the role of mental representations in reasoning. According to this view reasoning is a process of going from one mental state to another such that what facilitates the transition between mental states does not directly operate on the contents of the mental states, but rather on representations that have these contents.[19] This process has to track certain properties of the contents of the mental states for it to be good reasoning, but since the contents themselves are not directly accessible, reasoning is a process that directly operates on the representations that have content, and only indirectly on the contents, via these representations. The reasoning process thus primarily gets a grip on the representations not by their representational features, but by their non-representational features. To put this neutrally, reasoning primarily operates on the *form* of a representation, and these operations on the form of a representation have to properly mirror operations on the contents of the representations. The hypothesis that mental representations form a language of thought is one way to spell this out.[20] In this formulation reasoning can be seen as operating on the syntax of the language of thought, and not directly on its semantic features. Representations that are about objects will have a particular form, or "syntax," and representations that can be expressed with bare determiners will have a different form, all things being equal. Our reasoning ability corresponds to our ability to make the transition from certain mental representations to others, thereby preserving certain representational properties like truth. Thus the observation made above, that our minds are better at reasoning about objects than at reasoning with thoughts that can be expressed using bare determiners, can be reformulated as follows: In reasoning our minds favor representations that have a certain form, the one paradigmatically had by representations about objects, over others that have a different, more complex form. In particular, our reasoning is more efficient when it operates on representations that have the form that is paradigmatically had by representations which are about objects. This, of course, is a simplistic picture of how reasoning works. There are many other aspects to it, and many other sources of roadblocks. But this nonetheless is one such roadblock, and it means, in essence, that when things get hard it gets easier again when we use our reasoning-about-objects abilities.

Now, consider again the difference between the plural and the singular basic arithmetic statements:

(are4) Two and two are four.

(is4) Two and two is four.

[19] This view, often called the representational theory of mind, is widely discussed by various authors, accepted by many, but, of course, it is controversial. See, for example, Fodor (1987) and Fodor (1990) for a discussion and defense, with many further references.
[20] Again, see Fodor (1990).

We have seen above that we can understand determiners to belong to a particular type of expression, namely ((e,t),((e,t),t)). The type of "and" in the plural statement correspondingly is the rather high type mapping two determiners onto a determiner. Thus if we abbreviate the type of determiners as "d," then the type of "and" in the plural uses in our type notation is (d,(d,d)), which is the type of a function that maps two determiners onto another determiner.[21] In the singular statement the number words are of a low type, the type of objects is e, and "and" corresponds to an operation on the type of objects which is correspondingly of the low type (e,(e,e)). These features, *mutatis mutandis* will carry over to the mental representations that have the same content. In fact, we can see that the individual expressions in the singular and plural statements pairwise correspond to a "type raising" or "type lowering" of each other on certain ways of treating plural in a type framework. The plural determiner "two" corresponds to the singular "two" via a type lowering, and the corresponding type lowering holds between the two uses of "and" and the other number words. In fact, the difference between singular and plural can quite generally be associated with a lowering or raising of certain types, and this feature has been used in the semantics of plural in natural language.[22] That there is a systematic correspondence between the singular and the plural statements in terms of type change together with the above story about the role of the form of a representation in reasoning suggests the following account of the relationship between the singular and plural uses of number words in the basic arithmetic equations.

When we encounter arithmetical problems and we attempt to reason about them with representations involving high types we quickly run into cognitive difficulties. Our reasoning is not very good at working with representations having this form. On the other hand, we have very powerful resources available for reasoning, those that operate on representations that have the syntactic form of representations which are paradigmatically about objects. These operations on representations of this kind are well developed in creatures like us, but we can't use them to reason with thoughts that would be expressed with bare determiners and that involve higher types. We thus have a mismatch between the form of the representations that we want to reason with and the form of a representation that is required for our powerful reasoning mechanisms to be employed. But this mismatch can be overcome quite simply. We can force the representation to take on a form that fits our reasoning mechanism. The representation will have to change its syntactic form by systematically

---

[21] Or more explicitly, it maps a determiner onto a function from a determiner to a determiner. Taken together, it is a function that gives you a determiner when you put in two determiners, thus a function from two determiners to a determiner. Since we don't directly have two-place functions in our setup we have to take one argument at a time.

[22] See van Benthem (1991: 67ff.) and van der Does (1995). Not everyone agrees with this. For example, approaches congenial to Link (1983) enrich the domain of type (e) to contain plural as well as singular objects and associate the difference between singular and plural to what kind of object the semantic value of a phrase is. How and whether these two approaches really differ is a further question we will leave aside for now.

lowering the type of the operation on determiners and of the bare determiners to the lower type of operations on objects and objects, respectively. Once this is done the reasoning mechanisms we have can get a grip. This type lowering corresponds exactly to the difference between the plural and singular arithmetical statements. I will call the process of changing the type of the form of a representation to facilitate cognition *cognitive type coercion*. It is a special case of the more general phenomenon of type coercion, which can occur for a variety of reasons, not necessarily to facilitate cognition, and which can have a variety of other results, not necessarily focused on the form of a representation. We will discuss other kinds of type coercion shortly.

The process of cognitive type coercion forces a representation to take on a certain form so that a certain cognitive process can operate with this representation. Systematically lowering the type of all expressions (or the mental analogue thereof) is a way of doing this, and the difference between our ability to reason with representations involving low types rather than high types explains why this type lowering occurs in the case of arithmetic. It will occur in the process of learning arithmetic, once the arithmetical problems that we are asked to solve become complicated enough that thinking about them with thoughts involving higher types becomes a cognitive burden. Such type coercion will at first occur for mental representations in the language of thought, the primary objects of our reasoning mechanism, but then also affect the linguistic expression of these thoughts in a public language. The change of the form of the mental representation affects the public language with which we communicate our thoughts. On this way of understanding the difference between the singular and plural readings of (=4) the difference is not explained solely by considerations about natural language, but rather in part by considerations about cognition.

Note that according to the cognitive type coercion account we merely change the form of the representation. We do not replace one representation with another one that has a different content. We take the same representation and change its syntactic form so that our reasoning mechanism can operate on it. The content of what is represented remains untouched by this. To put it in terms of the language of thought, we change the syntax of a representation so that our reasoning mechanism can get a grip on these representations. Other than that we leave it the same. And what holds good for mental representations will hold good, *mutatis mutandis*, for their linguistic expression in language. The singular arithmetical statements are the linguistic expression of thoughts involving type lowered mental representations.

5.4.4 Contrast: Other Kinds of Type Coercion

Type coercion is the general phenomenon of something of one type being forced to take on a different type. This phenomenon is widely discussed in computer science, and it is familiar from semantics as well. In computer science, for example, an expression in a programming language can be coerced or forced to take on a certain type so that it can be interpreted in a certain situation. In semantics sometimes an expression can be forced to be of a certain type so that a sentence as a whole becomes interpretable. Let's look at some cases of type coercion, and how they differ from cognitive type coercion.

One surprisingly neglected approach to solving Frege's other Puzzle is one that uses the idea of semantic type shifting that has been developed in natural language semantics.[23] Semantic type shifting has been proposed to solve problems in natural language semantics that arise from expressions apparently having different types on different occasions. The multiple uses of "and" are a good example of this. "And" can conjoin expressions of many different types: sentences, verbs, determiners, and others. Because of this it is hard to say what type should be assigned to "and" itself. What seems to be required is that "and" is of different types on different occasions. But it would be a mistake to think that these cases involve different words "and" all pronounced the same way. A better way to go is to think of "and" as having variable type: it can take on different types on different occasions, but all the types it can take on are related in a certain way. Such a more flexible approach to assigning types to expressions will make these assignments simpler and more systematic. What type a particular occurrence of "and" will take is left open by this so far. The semantics of the word as such does not determine that it takes a particular type, it only specifies a range of possible types, and what contribution to the truth conditions it makes in what type. It is rather the occurrence of "and" in a particular sentence that determines what type it takes. For example, in

(231)   John sang and Mary danced.

"and" is conjoining sentences, whereas in

(232)   John and Mary danced together.

it is conjoining noun phrases. Such an account of "and" thus takes it to be of variable type. There is a range of types that "and" can take, in each type which it can take the contribution to the truth conditions is of a particular kind, and what type it takes on an occasion is determined by its occurrence in a sentence.

In general, the situation is a little more complicated, since it might be that there are more expressions than just "and" which are of variable type, and thus that there are different ways to specify these types to make the sentence as a whole come out meaningful. In these cases there will be different readings of the sentence corresponding to different ways to specify the types. Once we have multiple possible readings corresponding to multiple type assignments of various expressions the question arises whether any of the readings have priority. In the type-shifting tradition priority is generally given to the lowest possible types. For example, Partee and Rooth in Partee

---

[23] See Partee and Rooth (1983) and Partee (1987) for two of the classic papers in this tradition.

and Rooth (1983: 340) work with the principle that sentences are at first interpreted with the lowest possible type assignments that make them coherent, and higher types are invoked only when necessary. Thus the preferred reading of a sentence with multiple readings due to a variety of permissible type assignments should be the one where the assignments are of the lowest type. In any case, all expressions of variable type must take on some type or other for the sentence to be well-formed and semantically interpretable.

Semantic type coercion is thus the phenomenon where an expression of variable type is forced to take a particular type on a particular occasion so that the sentence as a whole in which it occurs is semantically interpretable. The case of conjunction, also called "generalized conjunction," is one example that illustrates this. This is a case of type coercion since the semantic type of a particular occurrence of a word of variable type is determined by or coerced by the types of the other phrases in the sentence in which it occurs. It is coerced to be the type (or one of the types) which makes the sentence as a whole meaningful.

Type-shifting principles are principles that tell us what type an expression can take and how these types relate to each other. Such principles are widely used and discussed in linguistics, in particular semantics, but to my knowledge they have not been used to attempt to show how number words can occur both as singular terms as well as determiners.[24] To do this one would have to specify what types these expressions can take, how they relate to each other, and what contribution they make to the truth conditions in what type. One could then attempt to solve Frege's other Puzzle along these lines, using only semantic type shifting and semantic type coercion to explain the different occurrences of number words. It is somewhat involved to spell out the details of such a proposal and to account for the examples that need to be accounted for. I will not try to do this here, since I am not advocating this proposal. I do, however, take it to be the second best option for solving Frege's other Puzzle, and for a while I took it to be the first best option. However, I am by now persuaded that semantic type shifting alone is not what is going on in our use of number words, and that the correct account instead must rely on considerations from linguistics as well as psychology. Here is, in part, why.

Without having a detailed type-shifting account formulated, we can nonetheless appreciate some of the more problematic examples for such an account. On any version of such an account number words must be able to take at least the type of a noun phrase (which is either (e) or ((e,t),t)) and the type of a determiner: ((e,t),((e,t),t)). Note that the type of a determiner is higher than that of a noun phrase, and that by the standard interpretations strategy widely assumed in the literature on type-shifting principles and cited above, the lowest type should produce the standard reading when both types are available. But this just does not seem to be the case. When both types

---

[24] Work related to such attempts are Geurts (2006) and an in progress dissertation by Eric Snyder.

are available the determiner reading is more prominent, and the noun phrase reading is harder to get. Examples that makes this vivid are examples of this kind:

(233)   I want two or three beers.

This sentence on such a type-shifting proposal would have both of these readings:

(234)   a. I want either the number two or else three beers.
          b. I want either two beers or three beers.[25]

But the first reading (234a) is barely possible, while the second one (234b) is almost mandatory. But on standard semantic type-shifting principles it should be the other way round. On the first reading (234a), the number word has lower type. It is of the type of a noun phrase, $((e,t),t)$, in this case. This should be the preferred reading, but it clearly is not. It is tempting, but not in the end satisfactory, to try to explain away the unavailability of the reading (234a) on more pragmatic grounds. After all, who wants a number, and who doesn't want beer? But this unavailability of the reading where the disjunction combines noun phrases rather than determiners applies throughout. To consider another example:

(235)   I'd like to investigate two or three proposals.

And it is certainly possible to investigate the number two, but I suspect no ordinary speaker will find that reading in that sentence without quite a bit of setup of the context. Not that such a context can't be specified to make that reading available. Imagine a student who has to make a choice between small numbers, or accounting, etc., etc. The point is not that this reading is impossible, but that it is far fetched, while the other one is the natural and default one. On standard proposals about type shifting it should be the other way round. Furthermore, it seems impossible to get such readings with other determiners. Other determiners can not be understood as having the type of a full noun phrase, although they, too, can be bare determiners. Consider:

(236)   I'd like to investigate many but not all proposals.

There is no reading where you would like to investigate many (not many proposals, or many X, just many). But semantic type-shifting principles usually work for the whole category of expression, in our case determiners, not just for some of them. On the other hand, cognitive type shifting does not apply for a whole category of expressions. It applies only to particular expressions where we have the cognitive need to change the form of expressions to help us in cognition. This applies to number determiners, since there are complex cognitive problems to be overcome with them, but not to other determiners. Reasoning with representations involving other

---

[25] For our purposes here we can ignore the reading where my desire has a content indeterminate between 2 or 3 beers, thus I neither desire 2 nor do I desire 3, but something indeterminate between them.

determiners is simple enough to do it without cognitive type shifting, reasoning with number determiners is not.

Semantic type shifting is in many ways compatible with cognitive type coercion. I have no objections to number words being of variable type in general. For example, it might well be that "two dogs" has a different type when it occurs in subject position than when it occurs in predicate position. Systematic type-shifting principles might tell us how their types relate to each other in such cases. What I do not believe, however, is that type-shifting principles alone will give us a solution to Frege's other Puzzle. More than purely general linguistic mechanisms are involved in the explanation of why number words have the features they have, in particular as they occur in arithmetic. The proposal of cognitive type coercion hopes to make that clear.

As we will discuss later in more detail in chapter 8, even if semantic type shifting were correct and number words would take semantic type (e) on certain occasions, this would not guarantee that number words are referential on such occasions. It would only show what the appropriate category of a semantic value is for these expressions on those occasions. But semantic values have to be distinguished from reference, as we will discuss in detail in particular in section 8.3. We will see the details of this issue there, but this highlights another aspect where the debate about type-shifting principles is orthogonal to one of our concerns: whether number words are referential on at least some of their uses. Expressions of type (e) don't have to be referential. On the other hand, if we grant that determiners are not referential and if the cognitive type coercion account is correct, then number words in their singular use would not be referential either, as cognitive type coercion does not affect their semantic function. We will see more on this and its significance below.

I think that semantic type coercion is the second best attempt to solve Frege's other Puzzle, but once we look at the details we can see that it can't quite be the answer. I favor cognitive type coercion instead. All this does not mean that number words might not have more than one semantic type. In fact, it is plausible that number words sometimes occur as modifiers and sometimes as determiners, and that these two occurrences are indeed of different type. And it is plausible that noun phrases in general can have different type, for example when they occur in predicative position. I don't deny such flexibility in type in any way. But I do hold that semantic type shifting isn't the solution to Frege's other Puzzle. Not because there can't be type shifting, but because semantic type shifting doesn't account for the data. The puzzle isn't solved by semantic considerations alone. What is needed in addition is the proper lesson from our difficulty with reasoning with information represented in a particular way. I take the cognitive type coercion proposal to be the proper answer to that difficulty.

Another proposal that could use a similar type lowering idea, and that could motivate it in a similar way, is a fictionalist or pretense proposal. According to it we do not change the representation involving the mental analogues of bare determiners to one that has the form of a representation about objects, while leaving the content the same, as the cognitive type coercion proposal has it. Nor does this proposal hold that number words are semantically of variable type. Rather it holds that we use a different representation, with a different content, and we use pretense to connect them. One way to spell this out is that we pretend that there are numbers and operations on them that exactly correspond to the operations on bare determiners. The number words that refer to these objects on such a proposal will, however, not be determiners, but rather new and different words: names for pretended-to-be-there numbers, which are pronounced the same way as the number determiners. This proposal is again distinctly different from the two proposals discussed above. The above two do not involve pretense. In the case of generalized conjunction we do not pretend that "and" conjoins verbs as well as sentences, it does conjoin both kinds. And in cognitive type coercion we do not pretend that the representations have a different form, they do have a different form. The present proposal thus is not a fictionalist proposal.[26]

A fictionalist proposal that is in other ways congenial to our concerns here is in the work of Harold Hodes, in particular Hodes (1984) and Hodes (1990). For Hodes the relationship between a number and the corresponding number quantifier is that of encoding or representing. Hodes, in particular in his Hodes (1984), likens numbers to fictional objects that are posited as representers or encoders of number quantifiers, and his position there was labeled "coding fictionalism." Hodes's work is more focused on the relationship between different formal languages than natural language number words. This and his fictionalism mark a clear difference between his view and the present proposal. However, the emphasis of the connection between number terms and number quantifiers makes them congenial in spirit.

To sum up, we can say that the three proposals to solve Frege's other Puzzle, cognitive type coercion, semantic type coercion, and fictionalism, make the following three different suggestions about how number words on their different uses relate to each other: a) cognitive type coercion holds that number words are the same word on their different uses, with a fairly determined semantic function, but with a different syntactic occurrence that reflects a cognitive need; (b) semantic type coercion holds that number words are the same word, but that they are semantically and syntactically of variable type according to general type-shifting principles; and finally (c) the fictionalist proposal holds that number words are different words in their determiner and singular term uses, but there is a relationship between them based on some form of pretense.

5.4.5 Other Arithmetical Operations

Although the above examples all focused on various expressions of addition, a similar story carries over to the other main arithmetical operations. Some such operations, in particular arithmetically more advanced ones, are clearly defined in terms of simpler

---

[26] Unfortunately, contemporary fictionalists usually do not consider Frege's other Puzzle, as a puzzle about natural language. They generally tend to understand number determiners as syncategorematic. Field (1989a) and Yablo (2005) are two examples of this.

ones, if only implicitly. But several of the most basic ones have a root in ordinary discourse just as addition has with the Boolean operation "and" on bare determiners. Here we will briefly consider two others: subtraction and multiplication.

Although subtraction, or "minus," is less directly present in natural language than addition, the same general story of how to understand number words in subtraction statements like

(237) $4 - 2 = 2$

carries over from our account of addition statements. Equations like (237) have a plural and a singular reading as well as their addition counterparts:

(238)   Four minus two are two.

(239)   Four minus two is two.

And "minus" can naturally appear between non-bare determiner phrases:

(240)   Four cats minus two cats are two cats.

Here the only crucial difference with addition is that addition has a natural expression in "and" as an operation on bare determiners, whereas "minus" is a less natural word, but it expresses an equally good operation on bare determiners.

The situation is slightly different with multiplication. Here multiplication is naturally expressed with "n times," but this phrase is more complicated than simply an operation on determiners. Just to illustrate, take

(241)   Two men fished the lake four times.

This naturally is read as either that the same two men came to the lake to fish four times or that four times there were two fishermen at the lake, possibly different ones each time. On the former reading it is natural to see "four times" as modifying the verb phrase, but not affecting the noun phrase "two men." On the latter reading it is natural to think of it as saying that a certain type of event, two men fishing the lake, occurred four times. These extra complexities, and clear differences with natural expressions of addition, are significant, but they are not central for my main point here. In the central aspects multiplication is like addition. We have singular and plural readings of multiplication statements:

(242) $2 \times 2 = 4$

(243)   Two times two are / is four.

And we have natural expression of multiplication with full determiner phrases:

(244)   Two times two beers are four beers.

The same general problems that motivated cognitive type coercion in the case of addition carry over to the cases of multiplication and subtraction. Although there are differences between multiplication, subtraction, and addition, in the relevant ways that motivated the cognitive type coercion story they are alike. The cognitive type coercion proposal, and the corresponding solution to Frege's other Puzzle, will apply to them equally. Once the basic arithmetical operations are understood with the cognitive type coercion proposal, there is no different account that needs to be given for clearly defined arithmetical notions.

The cognitive type coercion proposal gives us an account of the relationship between the singular and plural arithmetical equations, and it explains, at least in outline, why thinking about numbers is a lot like thinking about objects. According to this account it is only like thinking about objects when it comes to the form of the representation involved in thoughts about numbers, but these thoughts are not about any objects since their content is the one expressed with the bare determiner statements. This so far is not an account of all uses of number words, but only one step towards such an account. It does not deal with all uses of number words, but only with a certain class of them. It is now time to have a closer look at what we can conclude from all this.

5.4.6 "Two" vs. "The Number Two" vs. "The Number of Fs"

Much of our discussion in this was focused on number words which are simple words like "two." Our discussion of "the number of moons" in chapter 2 was focused on their occurrence in the loaded counterparts as they are used in the trivial inferences. This leaves a number of cases of number talk open, and some of them are clearly interestingly different from the cases we focused on. For example

(245)   A number of students failed the exam.

(246)   The number of students who failed is shocking.

are not dealt with by what we have said so far. However, these examples are not as pressing for our main topic now as the ones we have focused on so far. First, it is clear what the truth conditions of these sentences are and also that it is implausible that these truth conditions are achieved by reference to numbers. In the above cases the truth conditions are more or less the same, respectively, as those of

(247)   Many students failed the exam.

(248)   How many students failed is shocking.

These examples are in contrast to our main examples where a referential picture of number talk seems to suggest itself, mistakenly as it turns out. Second, examples like these are not central for understanding mathematics, in particular arithmetic, which will be our next topic. However, there are other examples that so far remain open and that are more pressing. First there are examples where number words appear directly in subject position of a subject-predicate sentence, as in

(249)   Two is a number.

Here there is a natural way to extend our account so far to cover cases like (249). After cognitive type coercion happens for bare determiner statements number determiners become available to be subjects in singular subject–predicate sentences. They still are not referential expressions on such uses, but determiners. Correspondingly, the predicate "is a number" in (249) is not predicated off the referent of "two," but forms a meaningful and true subject–predicate sentence nonetheless. To properly understand this way of thinking of sentences like (249) we need to rethink a certain "referential paradigm" of how language in general works, and we will discuss this issue in a more general setting a bit later, in section 8.2.

A second remaining issue is the difference between number words like "two" and phrases like "the number two." Here one might be tempted to hold that they are fundamentally different. After all, it is not true that one can always be replaced by the other. Most clearly, "the number two" can not occur in determiner position, whereas "two" frequently does occur in that position:

(250)   I ate two bagels.

(251)   *I ate the number two bagels.

One possible explanation is that "the number two" is a referring or denoting expression after all, even if "two" is not. In fact, a mixed view like this is defended by Friederike Moltmann in Moltmann (2013b). There she holds that number words like "two" are not referential, but other number phrases are. For example, "the number of moons" refers not to a number, but to a number trope, the particular "numerosity" of the several moons. Number tropes have to be considered to be the particular "fourness" of certain four things, which is different from the particular "fourness" of four different things. Furthermore, "the number two" also refers, according to Moltmann, but not to a trope, but to a fictional object. Such a view of the semantic function of number phrases can be called an *impure* view, since it mixes several different functions for different number phrases.[27] On the other hand, in the present chapter we argued for a *pure* view: all number phrases are non-referential, although there are important other differences between them.

Moltmann's view faces several serious obstacles. First, it is hard to see what number trope we might be referring to when we say that

(252)   The number of unicorns is zero.

---

[27] In section 5.1 we introduced the terminology of a uniform and non-uniform solution to Frege's other Puzzle. It is slightly different from a pure account of the function of number phrases, since the former was, by definition, only concerned with bare number words, not all number phrases.

There are no unicorns whose "zeroness" could be the object of reference. Furthermore, if "the number of men" is a referring phrase, then it can be used to form identity statements, claiming that the objects referred to are one and the same. However, on Moltmann's view the following statement taken as an identity statement can never be true:

(253)   The number of men is identical to the number of women.

no matter how many men and women there are. All that can be true is that the tropes referred to are similar, but not that they are the same.[28] Furthermore, for Moltmann sentences like

(254)   The number of moons is the number four.

always are false, since what the two terms refer to are neither identical nor similar: one is a trope, the other a fictional object.[29]

Impure views are thus dubious. If number phrases are referential then they better all refer to numbers, and if some of them are not referential then they better all be non-referential. On the pure, non-referential view defended here, what then is the difference between "two" and "the number two"? I think "the number" in "the number two" is an apposition, a side remark like "the philosopher" in "the philosopher Aristotle." Here "the philosopher" is a comment on who Aristotle is. It could be stated just as well as "Aristotle, the philosopher" or "Aristotle, who is a philosopher." In that example "Aristotle" is a referring expression, and "the philosopher" is an apposition to it. However, "the philosopher Aristotle" as a whole is not a single referring phrase, but rather contains one: "Aristotle." In our main example "the number two," "two" is a non-referring expression and "the number" is an apposition to it. The whole phrase "the number two" is equally not a referring expression. We will see more on apposition in a related case later on, in section 8.4.

Overall we can conclude that the variety of different uses of number words and phrases nicely fit into a unified account that takes all of them to be non-referring. We have not discussed all cases of this, but we did discuss the ones that are central for arithmetic and that are the most promising for the opposite view: the ones that appear to be referring expressions. Overall then, we can conclude that number words and other number phrases are not referential and that talk about numbers is not in the business of reference.

---

[28] Moltmann discusses this example, although not as a problem, but as a datum. She says: "Also the predicate 'is identical to' can apply to qualitatively exactly similar but numerically distinct tropes" and then cites (253). But this is not the right reaction. To be sure, (253) can clearly be true, but the question is how that can be given that to the left and right of "is identical to" are referring expressions that refer to different things. That these things are similar doesn't help, in particular since "is identical" here can hardly only be read as "is qualitatively identical" or "is similar."

[29] Moltmann takes such sentences to be of dubious grammaticality, but I disagree with her on this.

5.5 The Solution to Frege's Other Puzzle

The proposal made in this chapter as well as in chapter 2 solves Frege's other Puzzle. It explains the connection between the plural and singular uses of arithmetical statements, and how they relate to the arithmetical symbolism. It gives a unified account of many, though not all, uses of natural number words. According to this proposal, plural arithmetical statements are formulated with (semantically) bare determiners. These determiners are an ordinary part of natural language. The use of these determiners plays a central role in the introduction of the mathematical symbolism when basic arithmetic is introduced and learned. However, there are cognitive obstacles to arithmetic reasoning that relate to the unusually high type of these determiners. To overcome these difficulties a child who is faced with more and more complex arithmetical problems will adopt a systematic type lowering at the level of mental representation to employ reasoning mechanisms that require representations of this lower type. This type lowering we called *cognitive type coercion* because representations of a certain type are forced into a different type for cognitive reasons. This type coercion is at the level of the form of the representation that represents basic arithmetical truth, not at the level of the content of what is represented. The content remains untouched and is the same as the content of the plural basic arithmetical statements which are formulated with bare determiners. Thus both the singular and the plural readings of symbolic equations are correct. The plural one is the natural expression of the bare determiner statement, the singular one is the linguistic expression of the cognitively type coerced representation. And this account solves a special case of Frege's other Puzzle. It explains how number words, in this limited set of examples, can syntactically occur both as determiners and as singular terms. According to this account, number words are one and the same word in both occasions, they are determiners, but, for the reasons spelled out, these words also occur syntactically as singular terms.

This account of the use of number words in the above cases is congenial to the account from chapter 2 of why number words occur in singular term position in sentences like

(Nis4)   The number of moons of Jupiter is four.

when they occur in the trivial arguments, and how they relate to the occurrence of number words as determiners, as in

(J4m)   Jupiter has four moons.

According to this account number words in examples like (Nis4) are displaced for the purpose of achieving a focus effect. This displacement is the result of a syntactic process that does not affect the status of number words as non-referring expressions. Thus even in (Nis4) the word "four" is not a referring expression, and the sentence, despite appearances to the contrary, is not an identity statement. We discussed these claims in detail in chapter 2.

This account of what "four" is doing in (Nis4) is congenial to, but independent of, the account given in this chapter of number words in arithmetical statements in that they nicely go together to support a uniform solution to Frege's other Puzzle. According to this solution, number words are not referring expressions, but they appear syntactically in singular term position for a variety of different reasons. We have not established this uniform solution here, since we have not looked at all cases of number words, but the cases discussed here nicely fit into such a uniform solution, and since they are the most compelling cases for referential uses of number words they strongly suggest, but don't firmly establish, such a solution.

It is important to note that even though in both of the cases we discussed where number words appear in singular term position, the case of arithmetical equations and the case of sentences like (Nis4), these number words are non-referring expressions, the reasons why they appear in singular term position are completely different and independent. Number words appear as singular terms in arithmetical equations as a result of cognitive type coercion. Number words appear as singular terms in sentences like (Nis4) because such sentences are focus constructions. These are completely independent accounts. One could be true even if the other isn't.

The proposed solution of Frege's other Puzzle is a uniform solution. It takes number words to be non-referring expressions throughout. Number words are non-referring not in the sense that they aim to refer, but fail, but in the sense that they do something other than referring. On the present account they are determiners that occur for various reasons in singular term position. This solution of Frege's other Puzzle thus supports internalism about talk about numbers. At least the part of internalism that is concerned with what the semantic function of number words is when they occur as singular terms. What is left is to look at quantification over natural numbers.

5.6 Quantifying over Natural Numbers

That natural number words are not referring expressions suggests that internalism is true about talk about natural numbers. But for internalism to be true not only do number words in general have to be used non-referentially, quantifiers over natural numbers in general have to be used in their internal reading as well. If quantifiers over numbers in general are used in their internal reading, then we would have a stable account of our talk about numbers, as discussed in section 4.1. The question thus remains whether it is reasonable to think that quantifiers over natural numbers are in general used that way.

Internal quantifiers generalize over the instances in our own language. One worry about conceiving of our uses of quantifiers in particular cases as being used in their internal reading is that there might not be the right instances in our language. And one way that can happen is if there might not be enough instances.[30] We will discuss

---

[30] This is also a standard objection to a substitutional interpretation of quantifiers that was widely discussed in the 1970s and 1980s. See Bonevac (1985) for a discussion and further references.

this worry about internal readings of quantifiers over properties and propositions in great detail in chapter 9. It doesn't arise, however, in the case of internal readings of quantifiers over natural numbers. Here we do have an instance, a number term, for every natural number: one, two, three, etc., and we presently have a nice simple symbolic notation for numbers in the Arabic numerals: 1, 2, 3, etc. This, however, is not universal for all human languages and it wasn't always so. Before the Arabic numerals were in use, writing down larger numbers was difficult. In addition, not all natural languages have simple and systematic number words.[31] It might thus seem that even if quantification over natural numbers in English can be understood along internalist lines, it is implausible that this is so in general. In particular, it is implausible that quantification over numbers in, say, the Latin of Ancient Rome can be understood this way, given the limitations of Roman numerals, despite the fact that the Romans engaged in arithmetic. Similarly, it can seem that quantification in more impoverished languages, when it comes to number expressions, like the now famous Pirahã language of the Amazon region in Brazil that allegedly has no number words, is in conflict with an internalist picture of talk about numbers, in particular quantification over numbers.[32] But none of this is in fact in conflict with internalism. First, even if speakers of Pirahã do not engage in arithmetic, since they are supposed to have no words for numbers or quantities other than those for one or many, they nonetheless might have all the resources to do so. It is not clear that they do not have the expressive resources to say "and one more," which would really be enough to say what needs to be said. The case of the Ancient Romans is quite different. They clearly did engage in arithmetic, but their Roman numerals are not as limited as they are made out to be. There is a Roman numeral for every number, but the Roman numbers are not as nice and convenient as the Arabic numerals. Once the numbers get larger, there are no simple abbreviations any more, one has to repeat M (for 1000) again and again to get a notation for a large number. So, 30,000 can be notated in Roman numbers with thirty consecutive Ms. The Romans apparently also had various abbreviations for larger numbers such as a bar over a number to indicate that number times a thousand, or special parenthesis.[33] So, even the Roman numerals have a numeral for every number. In general, any language in which we can carry out arithmetic will need to give us at least the ability to articulate "and one more" or the like, or have a number notation system where iterations give us larger and larger number symbols. And this is all we need to find instances for internal quantifiers.

Number quantifiers can thus naturally be used in their internal reading without there being a danger that we run out of the relevant instances. A statement like

(255)   Some natural number is Φ.

is guaranteed to have a true instance if it is true at all, i.e. some sentence

---

[31] A nice survey of differences among number representations in human languages is in Dehaene (1997).

[32] See Frank et al. (2008) for a recent discussion of Pirahã and further references.

[33] For a historical overview of different number notion systems, including Roman numerals, see Menninger (1969). The number 0 is an exception to this, as apparently there was no Roman numeral for it.

(256) n is Φ

is true, and "n" is a number term in our own language. Similarly for more complex quantified statements like

(257)   Every number has a successor.

Here we have for every instance of the first quantifier an instance of the second one which is true. This sentence is truth conditionally equivalent, on its internal reading, to an infinitary conjunction where each conjunct contains an infinitary disjunction. Other quantified statements like

(258)   There are infinitely many primes.

can be seen to be true with our analysis of internal generalized quantifiers like "infinitely many" given in the appendix to chapter 3. None of this carries over to talk about other kinds of numbers, like real numbers. It might well be that

(259)   Some real number is Φ.

is true while every real number for which we have a term in our language is not Φ. After all, there are uncountably many real numbers, but only countably many terms in our language. We will return to this general issue in chapter 9, and to talk about the real numbers in chapter 6, but it is clear that this worry does not apply to talk about natural numbers. We have an instance for every natural number, and so we are not in danger of having too few instances in this particular case.

But there is a flip side to this coin: maybe we have too many instances. And this could be turned into an objection against holding that quantification over natural numbers in general is used internally.[34] Internal quantifiers are often trivial. It is usually trivial to conclude that there is something, on the internal reading. When Fred admires Sherlock then there is something which he admires, namely Sherlock. And thus it is trivial to conclude that there is something which is Sherlock, in this innocent, internal reading. Similarly, Sherlock is Sherlock, we can grant, even if Sherlock doesn't exist, and so there is something which is Sherlock, namely Sherlock. But if quantification over natural numbers is just internal quantification then won't it be too easy to get results about what numbers there are, in particular, won't it be possible to get the results that there are some numbers which we know from arithmetic there aren't? Consider the largest prime. There is no largest prime, as arithmetic has shown. But John might search for the largest prime. And the largest prime is the largest prime, we can again grant. So, on the internal reading, there is something which is the largest prime, namely the largest prime. But how can this be, given that arithmetic has shown that there is no largest prime, and quantification in arithmetic is just internal quantification? It would

---

[34] A version of this objection was first presented to me by Warren Whipple, and, independently, Tristan Haze.

seem that "there is a largest prime" should be true on the internal reading, contrary to what arithmetic has shown. This is the threat for internalism from, intuitively, too many instances.

What this shows is that quantification in arithmetic can't just be *unrestricted* internal quantification. With unrestricted internal quantification it is always possible to show that there is something which is *t*, for any term *t*: just *t* itself. But this doesn't mean that quantification in arithmetic isn't internal quantification. Internal quantification, just as external quantification, can be, and usually is, restricted. In the case of the external, domain conditions, reading, the restriction is to a subset of the domain. In case of the internal reading, it is just a part of the instances over which one generalizes. And in arithmetic we, in general, generalize over the instances that are formed with the natural number terms: 1, 2, 3, and so on. That the largest prime is the largest prime might be trivial, on an innocent reading of it, and thus that there is something which is the largest prime (namely the largest prime). But whether or not one of the numbers is the largest prime is so far left open. And that is the question of whether or not 1 is the largest prime, or 2 is the largest prime, and so on. And here the answer is no. So, on this, restricted, but internal, reading of the quantifier we get that there is no largest prime.

The same issue arises in essence for those who hold that quantification over numbers is used in the external reading. Arithmetic has shown that 17 is prime, and thus no numbers (other than 1 and 17) have 17 as their product. But, of course, $\frac{17}{2}$ and 2 have 17 as their product. This does not contradict the arithmetical result, since "no number" in the above statement was a restricted quantifier. It was restricted to natural numbers and thus didn't include rational numbers like $\frac{17}{2}$. When we said that there are no numbers other than 1 and 17 whose product is 17 we intended this quantifier to be restricted to natural numbers, and on this reading it is true. On the unrestricted reading, however, it is false. The results of arithmetic have to be understood as involving implicit restriction of the quantifier. And such restriction has to happen whether the quantifiers are used on their internal or external readings.

As with quantifier domain restriction in general, such a restriction can be overcome. So, when I say that

(260)   John admires some number.

it is natural to take this quantifier not to be restricted to the paradigmatic number words, but to include all kinds of other number phrases. Here "the largest prime" can well be a legitimate instance of the quantifier. But in arithmetic it won't be a legitimate instance. Here we try to find out about the numbers 1, 2, 3, etc., not about the number of women, the largest prime, and so on. Although in general unrestricted quantification is available, in arithmetic quantification over numbers is restricted, and one should hold this whether or not one holds that such quantification is in the internal or external reading.

For the case of arithmetic we can thus propose a stable solution to the problem of how number words and number quantifiers are used. Number words are not referring expressions. Even when they appear in singular term position they are not referring expressions, and we have seen two quite independent reasons why number words appear in this position. Number words instead are determiners, modifiers, or adjectives. Number quantifiers generalize over the instances, although in certain cases not over all possible instances, but over the paradigmatic instances exemplified by the number words. Putting all those together we can see that internalism about talk about natural numbers is correct.

Internalism is a striking and powerful thesis. It entails that in the whole domain of discourse, talk about natural numbers, no reference is, in general, even attempted. Even in subject–predicate sentences number phrases do not aim to refer to anything. Nonetheless arithmetical statements can be true. This fact about our talk about natural numbers is bound to be significant for the philosophy of arithmetic, and in the next chapter we will see how. In this chapter I argued that internalism is true for talk about natural numbers. It is now time to see what follows from this for the philosophy of arithmetic.

6

The Philosophy of Arithmetic

Arithmetic is philosophically problematic in a number of ways, and the problems it gives rise to are closely connected to problems in ontology. In fact, the philosophy of arithmetic is a paradigm case of a discipline where large-scale philosophical and metaphysical questions are closely tied to ontological questions. Many, but not all, of the following considerations that are focused on arithmetic also carry over to other parts of mathematics. But we will at first focus just on arithmetic and discuss the larger philosophy of mathematics only at the end, for reasons I hope to make clear then. To start, I will briefly review why arithmetic is philosophically problematic, how these problems are tied to ontology, and what a philosophy of arithmetic is supposed to do. After that we will see that the results from the previous chapters about ontology and talk about natural numbers give us a philosophy of arithmetic that solves these philosophical problems.

6.1 Philosophical Problems About Arithmetic

6.1.1 The Central Tension

Arithmetic gives rise to philosophical puzzles, since there is a clear prima facie tension among some of the apparently obvious facts about arithmetic. I'll call this tension *the central tension*. It arises from the following apparent facts:

1. *Objectivity* Arithmetic is fully objective. No matter what one's cultural background, personal preferences, or opinions, it is simply a fact for all equally that there are infinitely many primes, or that $2 + 2 = 4$.
2. *Thinking alone* You can find out about these facts by thinking alone. Having eyes and ears is useful for doing arithmetic. They can help you read textbooks and listen to others who have done arithmetic. But in principle you could do it all with your eyes and ears closed, by yourself, by just thinking. To prove something, to find the answer to a question, all you have to do is think.
3. *Subject matter* Arithmetic has a subject matter: it is about the natural numbers. What makes arithmetic distinct from other parts of mathematics and other parts of inquiry is that it is about the natural numbers and what is true about them. The natural numbers are the subject matter and domain of arithmetical investigation.

These three facts are in a well-known tension, at least prima facie. It is easy enough to reconcile two of them, but hard to fit all three into a coherent picture. Some pairs of these facts nicely go together. For example, the subject matter of arithmetic can be used to support its objectivity, but it remains unclear how these two can be combined with the "thinking alone" aspect of arithmetic. It would be no wonder that arithmetic is so objective if it has as its subject matter a domain of objects which exist independently of us and our opinions. But if there is such a domain of objects that arithmetic is about, and which are the source of its objectivity, then why are we able to find out about these objects by thinking alone? In general it seems that we are not able to find out about objects, with the possible exception of ourselves, by thinking alone.

On the other hand, if the source of arithmetical objectivity is tied to our thinking, maybe via the objectivity of reason, or some other source of objectivity that can be connected to thought, then it would be no surprise both that arithmetic is objective and that arithmetical discovery can happen by thinking alone. But it is unclear how this picture can be tied to the apparent fact that arithmetic has as its subject matter the natural numbers. In particular, the natural numbers themselves don't seem to be mental things. If one holds that numbers do exist, and are what arithmetic is about, then it seems hard to also hold that these numbers are mental objects. Are they the objects of my thoughts or yours? Do only those exist that are, or have been, thought about? Aren't there too many numbers and too few minds? All these considerations have led those who hold that numbers exist at all to hold that they are not mental objects. But if they are non-mental objects, then the problem again arises how they can be the subject matter of arithmetic while arithmetic can be figured out by thinking alone.

A number of philosophers have proposed answers to this dilemma. Although this is not the place to survey the literature in the philosophy of mathematics, let's nonetheless mention a few options on how one can deal with this situation. First, and maybe foremost, there is Frege's proposal in his Frege (1884). Frege ties the objectivity of arithmetic to the objectivity of logic, which makes sense of the fact that we can find out about arithmetic by thinking alone. The natural numbers are not the source of arithmetic objectivity, but instead secondary to the objective arithmetical truths, in particular the singular terms that occur in expressions of these truths. Numbers are, in some elusive sense, derivative on true arithmetical sentences and the singular terms in them. In particular, they are not the source of arithmetical objectivity.[1] Frege's view depends on the so-called *syntactic priority thesis*, a thesis that closely ties being an object to singular terms. In particular, what singular terms occur in true sentences is prior, in some once more elusive sense, to the objects that are referred to.[2] The syntactic priority thesis takes syntactically singular terms to be paradigmatically referential terms. I have criticized this thesis earlier, in chapter 2. It assumes a semantic uniformity among singular terms which simply isn't there. And it is hard to see how the numbers

---

[1] Of course, philosophers disagree what Frege's considered position was. This is how I read him.
[2] See Wright (1983), which is a main source of a revival of Frege's position in the contemporary debate.

can be real objects, ones that exist independently of us, but nonetheless they are secondary to the arithmetical truths about them. A version of this possibility has recently been developed in an intriguing way by Stephen Schiffer, in particular in his Schiffer (2003). I have expressed my doubts about his position in Hofweber (2016c).

A second attempt to resolve the tension is to adopt a version of fictionalism.[3] The tension might be resolved since the things which form the subject matter of arithmetic, the numbers, are only pretended to be there. The source of arithmetical objectivity is not the numbers themselves, but rather the stipulations that are made in the pretense that there are numbers in the first place and what they are like. If one has these stipulations, somehow, in mind, then thinking alone might be enough to figure out what features these numbers must have, given the stipulations. But whether such stipulations are in fact ever made is an empirical hypothesis. Although this has some plausibility for certain parts of mathematics, as we will discuss below, it is especially dubious for the case of arithmetic. Fictionalism outlines how the tension might be resolved, but that by itself does not give us any reason to think that it in fact is resolved that way.

Finally, there is the option to take numbers to exist, and to exist independently of us, and to be the source of mathematical objectivity, but nonetheless to argue that somehow we can find out about them by thinking alone. Here there are several different lines pursued by proponents of this idea. First there is the mostly negative strategy of arguing that those who think there is a tension between what mathematics is about and how we know about it are in the grip of a mistaken epistemological picture. For example, one might argue that we should not expect there to be a tension between arithmetic being about causally inert abstract numbers and our finding out about what they are like unless we held on to a dubious causal theory of warranted belief. But this strategy alone isn't sufficient since, as Field has argued in Field (1989b), it leaves us with the task of explaining how we come to have true beliefs about these things more often than false ones, and more generally, explaining that we are reliably forming true beliefs about arithmetic. Here the proponent of this way of trying to relieve the tension could, with John Burgess in Burgess (2008), maintain that if you want to see how we find out about what the numbers are like you have to look no further than seeing how we do arithmetic. After all, arithmetic is the discipline in charge of doing just that, and it does it really well. One should not think that philosophy can step back from the practice of finding out about numbers as it is best carried out in this specialized part of inquiry and assess the validity of it from its distinguished and external standpoint.

---

[3] Prominent versions of fictionalism are defended by Hartry Field, in Field (1989b), and Stephen Yablo, in Yablo (2005). A more recent version is developed by Mary Leng in Leng (2010). Fictionalism in the philosophy of mathematics is a widely discussed topic. See Balaguer (2013) for a detailed overview, as well as Burgess and Rosen (1997) and Burgess (2008) for criticisms. The latter two references contain many criticisms of non-fictionalist forms of nominalism as well, as for example that of Geoffrey Hellman in Hellman (1989). We won't delve into this large literature in more detail here, but rather focus on the positive proposal supported by what we have seen in earlier chapters. A survey of nominalism in the philosophy of mathematics can be found in Bueno (2013).

And there is certainly something to this. It would be absurd for philosophy to try to find out about numbers in its own way, but this shouldn't stop philosophy from trying to understand how the practice works, what its source of objectivity is, and so on. We don't hope to change the practice, but we hope to understand it. And so far we don't understand it, and simply engaging in more of the practice doesn't help us here.

A separate version of this final strategy to resolve the tension is a holistic one, following Quine, at least in spirit: arithmetic is part of our best account of the world and this account, as a whole, is well supported by empirical evidence. As it turns out, we find out about arithmetic by thinking, but it is tied into a larger theory which is empirically confirmed. Arithmetic gets its support from being a part, and a central part at that, of this larger theory. This, again, is especially implausible for arithmetic. Arithmetic is so obvious and compelling that it seems a real stretch to hold that it gets its support from being a central part of our best account of what the world is like. It seems to be able to stand on its own.[4]

It is tempting to identify the source of the tension in our conception of arithmetic in a conflict between *ontologism* and *rationalism*. On the one hand, a rationalist conception of arithmetic is compelling. Arithmetic is a discipline carried out by thinking alone, it needs nothing more than reason to flourish. On the other hand, there seems to be a domain of things, the numbers, that arithmetic is about, and which exist independently of us as the source of arithmetical objectivity, a view which I will call "ontologism."[5] How both of these can be true is mysterious. There are several not completely satisfactory ways to try to make that tension more precise, but it being there prima facie should be beyond dispute. To have a clean resolution of the tension either rationalism or ontologism will have to go, or so it seems.[6]

Both rationalism and ontologism are problematic on their own, and thus even if we can dispense with one of them, problems will remain. For ontologism we have seen a variety of problems in the opening chapters of this book. For rationalism the problem remains how the objectivity of arithmetic can be traced back to the objectivity of reason, or wherever it is supposed to come from, and whether there is enough of the latter to be the source of the former. We will discuss this some more below.

Besides the general tension in our conception of what arithmetic is like, there are at least two further problems that need to be addressed for a satisfactory philosophy of arithmetic. One might suspect that a solution of these two well-known problems and an easing of the central tension will go hand in hand. The two problems are, first, the problem of identifying numbers with ordinary objects, and, second, the application of arithmetic to counting and collections of objects. Let's look at them in turn.

---

[4] See Parsons (1980a).

[5] I am indebted to Andreas Kemmerling for suggesting this terminology.

[6] Of course, not everyone agrees that this tension is a real one in the end. A defense of harmony between them is given, for example, by Jerrold Katz in Katz (1997).

6.1.2 The Caesar Problem

Suppose that natural numbers do exist. Then these numbers have some mathematical properties, like being even or being prime. That leaves open what non-mathematical properties they have besides their mathematical properties. For example, are they at a particular place? Do they have mass? Are they abstract, outside of space and time? These are all good questions. What is striking about them is that some answers to them seem obviously absurd, and it is not so clear why they should be absurd. We can say that an answer is absurd if it is properly dismissed as being beyond taking seriously as the right answer. In particular, it seems absurd to hold that numbers are ordinary material objects. Could my cup be the number 7? Clearly not. But why not? One consideration is this: why would my cup be so special to be a number, but not yours? And why would it be that number, and not a different one? Surely it would be arbitrary to think it is my cup that is the number 7, not the number 8. That is clearly right, but it only shows that it would be arbitrary to think my cup is the number 7, not that my cup isn't the number 7. Maybe numbers are ordinary objects then, but we can't figure out which ones are which? More importantly, it seems that my cup can't be the number 7, since I can break, and thus destroy, the cup. But I can't destroy the number 7 and thereby make it true that there is no number between 6 and 8, and that the successor of 6 is 8, and so on. This is a little question-begging, of course, but even if we grant it one could still hold that if numbers are material objects then they must be indestructible ones. But then, why not think that there are infinitely many indestructible material objects somewhere on a faraway star which are the numbers? Should we ask NASA or the NSF for a grant to support a search for these arithmetical objects? I wonder how big they are! All this is quite obviously absurd. But it is not so clear why it is absurd.

A version of this problem is well known as the Julius Caesar problem for Frege's philosophy of arithmetic. Frege had all the resources to make clear why different numbers are different from each other, but no resources to make clear why the number 4 is different from Julius Caesar. But the problem doesn't just apply to Frege's view. It is more generally the problem of why and on what grounds it is universally judged, by philosophers and non-philosophers alike, to be false and even absurd to think that Caesar is the number 4. We will use the term "Caesar problem" for this general problem. This problem isn't a problem for a particular philosophy of arithmetic, like Frege's, but a problem for anyone. Different philosophers mean slightly different things by the Caesar problem, but here we will take this problem to be the following two-part problem:[7]

---

[7] A version of the problem of identifying numbers with objects described in other terms is discussed by Paul Benacerraf in Benacerraf (1965).

(261)   *The Caesar problem:*
    a. What reason do we have to think that identity statements which identify numbers with regular objects, e.g. "the number 4 is identical to Caesar," are always false?
    b. Why are such identifications universally judged to be absurd?

The two parts of this problem are likely related. To identify numbers with regular objects gets something badly wrong, or so it seems, and whatever it is, we might all be picking up on it, and that's why we judge such identifications to be absurd, that is, beyond taking seriously as a hypothesis. But it is also important to keep the two parts of the problem separate. The issue is not just why such identifications are wrong, and what reasons we have for holding this. The issue also is that such identifications are generally, by everyone, judged to be beyond taking seriously. Why that is so is a worthwhile question all by itself. On the face of it, it is not at all clear why this should be so. If numbers are objects, why is it so absurd to hold that they are certain material objects? After all, numbers being regular material objects does not have to be in conflict with any of their mathematical properties. This hypothesis simply holds that besides their mathematical properties they also have certain other material properties. To properly evaluate ontologism in the philosophy of arithmetic, and correspondingly what the subject matter of arithmetic is, we must understand and solve this problem.

6.1.3 The Application Problem

Our second problem in the philosophy of arithmetic concerns the application of arithmetic. This problem also comes in two parts. The first concerns the question of why numbers, and not something else, are used for counting; the second concerns the question of why the results of arithmetic can be relied upon in counting. We will discuss them in order.

The first of these problems is almost silly, but I think a substantial and even somewhat deep problem. Why do we give a number as the answer to a question about how many things of a kind there are? It is natural to think that, of course, the answer to a question "how many?" is a number. Numbers are tailor-made for counting, and they alone are suitable as answers to "how many?" questions. But why is that so? To bring out the puzzle, assume for the moment that numbers are abstract objects, as many believe. Why do we use these abstract objects rather than some other things as the answer to a "how many?" question? To be sure, numbers as abstract objects would be suitable as answers to these questions. Their ordering as an ω-sequence would make them suitable as measures of sizes of (finite) collections. But many other things are ordered in an ω-sequence as well. They would in principle appear to be just as suitable to be the measures of sizes of collections. But clearly and obviously, the answer to a "how many?" question has to be a number, not any other kind of object, no matter how similar to numbers they might be. Numbers alone are suitable to answer questions about how many things there are. But why are numbers the uniquely suitable measures of sizes of collections?

A related, but different, concern is about the application of the results of arithmetic to questions about counting. Although this application of arithmetic is slightly more complicated in detail, its basic problem is fairly straightforward. Suppose I have n apples, and I get m more apples. How many apples do I have? The answer is, of course, $n + m$. But why am I justified in using the arithmetical function of addition to calculate how many apples I end up with? One worry is that some apples might somehow disappear when more apples are added to the collection. But that is not the worry I would like to focus on, and so I am leaving that aside for now. Let's thus assume that none disappear or appear. There is still the worry about what justifies the use of addition. If I conjoin a group of n apples with a group of m apples, what justifies me in holding that value of the addition function on n and m gives the size of the resulting group of apples? Why is it that function? This might be a well-supported empirical hypothesis. The addition function ended up being confirmed over a number of alternative hypotheses. If it indeed is a well-confirmed empirical hypothesis, then this would answer the question of why we are justified in using addition when we try to find out how many things we end up with when we conjoin two groups of things. But that feels all wrong. It seems to be a priori that if n apples are combined with m apples and none disappear or appear, then the result is the sum, in the sense of addition, of the numbers n and m. But how can this be a priori? How can it be a priori that addition on natural numbers is the right function for calculating the answer to "how many" questions of this kind? After all, addition is merely one of infinitely many possible two-place functions on these objects, the numbers.

There are, of course, various strategies for trying to solve this problem. One could try to argue that by relying on the recursive definition of addition in terms of the successor function we are a priori entitled to use addition. But as usually with recursive definitions, it just pushes the issue back to the functions in terms of which addition is recursively defined. Why am I entitled to hold that the successor function is the right one to use in the relevant clause of the definition?

For the general problem of the application of arithmetic, and the connection of natural numbers to counting, Frege in his Frege (1884) again was groundbreaking, and neo-Fregean approaches to the philosophy of mathematics are trying to defend in some form or other Frege's main idea. The neo-Fregeans in particular focus on the alleged connection between our concept of a number and Hume's Principle, which connects sameness of number with sameness of the sizes of collections. But the neo-Fregean program, and this application of it, suffers from a major flaw: it is more impressed by a technical result, namely that Peano Arithmetic can be interpreted in second-order logic plus Hume's Principle, than by actual empirical considerations about how our concept of number comes about and how we come to think of numbers just in the way we think of objects. The latter was the focus of chapter 5, where we reached non-Fregean conclusions. What matters for our in fact being entitled to apply arithmetic is not some interpretation of one formal theory in another, but what actually goes on in our heads. We thus should hope for a better solution than the neo-Fregean one.[8]

The problem of the application of arithmetic is thus a two-part problem as well:

(262)   *The application problem:*
     a. Why do we use the natural numbers for counting?
     b. What justifies us to use the results of arithmetic in the application to "how many" questions?

The application problem is in a sense the philosophy of mathematics equivalent of Frege's other Puzzle in the philosophy of language, the puzzle about number words in natural language from chapter 5. Number words as singular terms seem to refer to objects; number words in determiner or adjectival position are tied to counting. How these objects are related to counting is what the application problem is about. How these two uses of number words function in language is what Frege's other Puzzle was about. We will see below how our solution to Frege's other Puzzle connects to the solution of the application problem to be proposed.

Any respectable philosophy of arithmetic will have to try to solve the Caesar problem and the application problem. The key to a solution to these problems is to see how the tension between our three apparent facts about arithmetic—subject matter, objectivity, and thinking alone—can be overcome, which will likely require taking sides with either ontologism or rationalism. What we have seen in chapter 5 allows us to do all this as follows.

6.2 Internalism and Ontological Independence

To briefly review, in chapter 5 we have seen an account of the function of talk about numbers in both ordinary discourse as well as arithmetic, both formal as well as informal. On this account number words are not referring expressions. They are determiners that are, for a variety of reasons, sometimes displaced into singular term position. In the case of

(Nis4)   The number of moons of Jupiter is four.

the number word "four" appears in this position to achieve a focus effect, as spelled out in chapter 2. In arithmetical equations like

(=4) $2 + 2 = 4$

---

[8] For the neo-Fregean program in general, see Hale and Wright (2001). For the issue of the connection between numbers and counting, see Dummett (1991), Wright (2000), and Heck (2000). A paradigm example of trying to solve the problem technically is Rayo (2002).

the number words are symbolic abbreviations of determiners, and the semantic content of this statement is the same as that of the bare determiner statement

(are4)   Two and two are four.

The singular reading of (=4), namely

(is4)   Two plus two is four.

is the expression of the result of cognitive type coercion of the bare determiner statement. Cognitive type coercion in turn is performed to simplify arithmetical calculations, and to mesh up what our minds are good at thinking about with what needs to be figured out. The details were given in chapter 5. Furthermore, quantification over numbers is commonly employed on the internal reading of the quantifiers, as discussed in chapter 3. This account of what we do when we talk about numbers thus supports internalism: number words are not in the business of referring, and quantifiers over numbers are commonly used in their internal reading. Nonetheless, number talk is often literally true.

Internalism in the form defended here could be called *bare determiner internalism*, since it goes beyond internalism in general in that it not only holds that number words are non-referential, but also that basic arithmetical statements have the truth conditions of bare determiner statements.[9] This is the key to breaking the stalemate in the philosophy of arithmetic, and to relieving the tension among the apparently obvious facts about arithmetic. It resolves the ultimate tension between ontologism and rationalism in favor of rationalism, and preserves all three of the apparent facts about arithmetic: thinking alone, objectivity, and subject matter. In particular, we can see that ontologism is mistaken: the subject matter of arithmetic is not some domain of entities. Furthermore, the truth of arithmetic does not depend on what exists in general. We can now see why that is so.

First, we can note that in arithmetical statements like (=4) no referring expressions occur at all. The number words in (=4) do not refer to anything in these statements in the sense that they are not even in the business of referring. They are determiners, and thus non-referring expressions. Thus no reference is even attempted in arithmetical statements. And thus no objects that are being referred to are required to exist for these statements to be true. Arithmetical equations have the contents of bare determiner statements. And such statements make no claims about any particular objects. This is one step towards *ontological independence*: the position that the truths of arithmetic

---

[9] As noted in chapter 5, it is inessential for the present view what the precise classification of number words in their "adjectival" use comes down to, in particular whether number words should be classified as modifiers rather than determiners. For our purposes this wouldn't make any difference, and so I take the classification of being a determiner to be broad enough to include number words in their "adjectival" use, as well as "many," "few," and other expressions that might give rise to such a classification challenge.

are independent of what exists.[10] But it is only a first step towards ontological independence. For arithmetic to enjoy the privilege of ontological independence it is not enough that the truth of arithmetic is independent of which particular objects exist. We have to distinguish two notions of ontological independence. First, a class C of statements enjoys *weak ontological independence* just in case the truth of the statements in C is independent of which particular objects exist. Second, a class C of statements enjoys *strong ontological independence* just in case the truth of members of C is independent of what exists in general. The two can come apart for cases like

(263)   There are at least two things.

This sentence has weak, but not strong ontological independence, on its external reading. It doesn't matter for its truth which two things exist, but it does matter what exists: there have to be at least two. And similarly it can be objected that on the present account of the content of arithmetical statements they have only weak, and not strong, ontological independence. Since no particular objects are referred to in arithmetical statements weak ontological independence seems to be guaranteed. But strong ontological independence is what is required to relieve the tension between rationalism and ontologism. If a certain number of objects is required to exist for the truth of arithmetic, then the tension will remain as long it isn't clear that we can find out by thinking alone that there are that many objects. And just this is a standard objection to a congenial, but different, approach to the philosophy of arithmetic. The objection is that an account like the present one has to assume that there are infinitely many objects. It is important to see how the present approach differs from the one against which this objection is commonly raised, and why it does apply to it, but doesn't apply to the present position.

6.3 The Adjectival Strategy and Second-Order Logic

Applying the proposal about basic arithmetical statements defended in chapter 5 to the philosophy of arithmetic is related to what Michael Dummett labeled as "the adjectival strategy" in Dummett (1991: 99ff.). The adjectival strategy claims that number words in arithmetic are derivative on their adjectival use in natural language, and it is contrasted with "the substantival strategy," which claims that their use in arithmetic is derivative on their singular term use. Frege was a defender of the substantival strategy; the proposal of this chapter and the last is in the ballpark of the adjectival strategy.

---

10   The relevant notions of "dependence" and "independence" should be seen as broadly counterfactual ones: even if certain things wouldn't exist, the arithmetical truth would still be true. I take these counter-factuals broadly here in that they might well concern objects which exist necessarily, say certain sets or positions in an ω-sequence or what have you. Here the counterfactual has to be taken as a counterpossible: even in the impossible situation where these things do not exist (i.e. where there are no sets or positions), the arithmetic truth would still be true. How such counterpossible conditionals are to be understood more precisely is not completely clear, but see footnote 35 in section 13.4.3 for more.

The adjectival strategy, as it is commonly carried out, however has a well-known problem. It might seem that this problem carries over to the view defended here, and thus threatens strong ontological independence. In this section we will contrast this common way to spell out a version of the adjectival strategy with the position defended here, and argue that the objection to the former does not carry over to the latter.

In the philosophy of mathematics literature the preferred way to spell out the adjectival strategy is to map symbolic arithmetical sentences like

(264) $5 + 7 = 12$

onto sentences in second-order logic, like

(265) $\forall F \forall G (\exists_5 x F(x) \wedge \exists_7 y G(y) \wedge \neg \exists z (F(z) \wedge G(z)) \rightarrow \exists_{12} w (F(w) \vee G(w)))$

whereby the number quantifiers, like $\exists_5 x$, are abbreviations of blocks of first-order quantifiers in the usual way. Let's call this the *second-order logic strategy*, or *SOL strategy*, for short. How precisely the relationship between (264) and (265) is supposed to be understood deserves further discussion. Does (265) make the underlying logical form of (264) explicit? Is one just a replacement for the other with the same truth conditions? No matter what one says here, any claim that (265) spells out the truth conditions of (264) faces the following serious objection, which I will call *the objection from finite domains*. The second-order sentences that are supposed to correspond to the arithmetic ones will not even get the truth values right if there are only finitely many objects. If there are only n many objects and the arithmetical statements involve numbers larger than n, then the antecedent of the second-order statement is false and the whole statement thus is vacuously true. Thus if there are only finitely many objects, then arithmetical equations with numbers larger than the number of objects will all be vacuously true, no matter what they say about addition, an absurd consequence. In particular, the SOL strategy does not achieve strong ontological independence and thus does not overcome the tension between rationalism and ontologism. One can try to do better, but it won't go all the way. For example, one can try to modalize the second-order logic statement, adding a necessity operator out front.[11] This helps, since it doesn't require that infinitely many things exist, only that more and more things could exist. But this doesn't achieve ontological independence in the sense that it achieves harmony with rationalism, unless, of course, one holds that we can find out by thinking alone that more and more things can exist. I won't hope to pass judgment on what one might be able to find out by thinking alone about what could be. I will leave it to those who pursue this strategy to defend this. I instead would like to contrast it with the view defended here.

---

[11] See Hodes (1984) for such a proposal. For his later view, see Hodes (1990), which instead requires infinite models for the relevant arithmetical languages. This view thus holds that arithmetic requires the existence of infinitely many objects. See also Hellman (1989).

It is important to note that and how the SOL strategy is different from the present proposal. This difference will make clear that the objection from finite domains, which is a very good objection at least against the unmodalized SOL strategy, does not carry over to our proposal. The objection against the SOL strategy is a good objection, as long as the SOL strategy holds that the truth conditions of the arithmetical equations are captured by the second-order statements. Under certain conditions, when there are only finitely many objects, the truth values of some arithmetical statements and their corresponding second-order statements come apart, and thus their truth conditions can't be the same. "$5 + 7 = 10$" is always false, but the corresponding second-order statement is true under the condition that there are only four objects. Thus the objection from finite domains is a good objection against any view that holds that the arithmetical statements are truth conditionally equivalent to such second-order logic statements. It would carry over to our proposal if it were true that the truth conditions of the bare determiner statements, in this case

(266)   Five and seven are twelve.

are the same as the truth conditions of the second-order logic statements, in this case (265). Thus to extend the objection against the SOL strategy to the present proposal one would have to hold that the bare determiner statements are truth conditionally equivalent to the second-order logic statements. But I think we can see quite clearly that this is not so. Not only is there no good reason to think that these are equivalent, it seems clear upon reflection on some examples to be discussed shortly that they are not equivalent, and that no objection similar to the objection from finite domains will carry over to our proposal.

One way in which one might argue that (265) has the same truth conditions as (266) is to argue that the former spells out the underlying logical form of the latter. I take logical form here in a sense in which the logical form of a sentence is semantically revealing. That is to say that the logical form of a sentence makes certain semantic features of the sentence explicit, ones that might have been left implicit in the natural language expression of this sentence. Given this conception of logical form, the second-order statements do not articulate the logical form of the bare determiner statements. Many important semantic features of them are not captured by the second-order logic statements, and the second-order logic statements bring in several things that are not in the bare determiner statements. For example, the "and" in (266) expresses a collective operation on plural bare determiners. In standard second-order logic there is no room for plural, there are no collective operations, no bare determiners, and conjunction is restricted to formulas, that is sentences and predicates, but it isn't defined for terms. All this semantic structure is misrepresented in (265). (265) is of conditional form, but (266) appears not to be. (265) treats number quantifiers as syncategorematic, but we have seen in chapter 5 that this is incorrect as a proposal about the logical form of number determiners. (265) thus does not make the underlying logical form of (266) explicit. (265) is as good as it gets *in second-order logic*, but that just means that the second-order logic is not up to the task of spelling out the logical form of (266).

That (265) does not capture the logical form of (266) does not establish that they do not have the same truth conditions. But we can see quite independently that the truth conditions of these statements are different. As we have seen, the second-order logic statements depend on the size of the domain for their truth value, in particular, if this size is finite then many of them become vacuously true, even though they correspond to false arithmetical equations. To see that the truth value of the bare determiner statements does not likewise depend on how many things there are we should look at some examples. Here I am simply asking you to judge the truth value of ordinary English sentences. Consider:

(267)   Two dogs are more than one.

That sentence is quite clearly true. Does its truth depend on the existence of dogs? To see that it doesn't, consider

(268)   Two unicorns are more than one.

It is also true, even though there are no unicorns. And

(269)   Two unicorns are more than three.

is false, even though there are no unicorns. More generally, the ordinary English sentence

(270)   Two are more than one.

is true, but its truth is not dependent on how many things there are, and

(271)   Two are more than three.

is false, and its falsity also is not dependent on how many things there are. Similarly, the English sentence

(272)   Five unicorns and seven more are twelve unicorns.

is true and

(273)   Five unicorns and seven more are ten unicorns.

is false, even though there are no unicorns. In fact, these sentences are true and false as stated even if there can be no unicorns, that is, even if unicorns are impossible. To illustrate this, consider examples where the relevant objects are impossible:

(274)   Three round squares are more than two.

and similarly for the other examples. More generally,

(266)   Five and seven are twelve.

is true no matter how many objects there are, and

(275)   Five and seven are ten.

is false no matter how many objects there are. I base these claims on judgments about the truth value of the ordinary English statements. As such I find them hard to dispute. Thus the truth conditions of the bare determiner statements are not captured in the second-order statements, modalized or unmodalized, since the truth values of the latter depend on how many things there are or could be, but those of the former do not.

All this is not to say that we cannot capture the truth conditions and logical form of the bare determiner statements in a formal language. Some of the things that are lacking in standard second-order logic, namely plural, collective operations on determiners, etc., can be found in other formal languages, for example in a type theoretic framework with a proper modeling of plural.[12] But any argument about which one of these models captures the truth conditions of (266) will have to be carried out at the level of natural language. We have to understand what the truth conditions of (266) are first, and then we have to see which formal language is expressive enough to capture it, and how to capture it in that formal language. Standard second-order logic alone won't do. The objection from finite domains, which is a serious problem for the SOL strategy, thus does not carry over to our proposal. To the contrary, the truth and falsity of the bare determiner statements does not depend on what there is, nor on how many things there are.

By similar considerations we can also see that (266) is not equivalent to a counter-factual conditional like the following:

(276)   If there were five Fs and seven more Fs then there would be twelve Fs.

On a standard understanding of counterfactual conditionals along the lines of Lewis in Lewis (2001a) this gets the wrong truth conditions. If the Fs are impossible things, like unicorns or round squares, then the antecedent of the conditional becomes impossible to satisfy and thus the conditional becomes vacuous, no matter what the numbers involved are. But the bare determiner statements don't have those truth conditions, as we saw above. Any such counterfactual that aims to approximate the truth conditions of the bare determiner statements needs to accommodate impossible antecedents in the counterfactuals, and how closely such an analysis approximates the truth conditions of the bare determiner statements will depend on how it goes.[13] But the straightforward and traditional analysis won't be good enough.

A different problem with a counterfactual analysis is simply that it might not be true that if there were five and seven more, then there would be twelve. It might well

---

[12] See, for example, van der Does (1995) for some of the details. To spell this out properly we will have to consider some of the issues that we mainly ignored here, like the proper representation of collective readings of plural, the semantic type raising that van der Does holds this comes with, and other issues.

[13] See Nolan (1997) for a proposal of dealing with impossible antecedents and Williamson (2008) for some criticism.

be that if you add seven more to the five there are, then you get an explosion, not twelve things. Balls of uranium are an example, in a world that is small enough that ten or more goes beyond the critical mass for a nuclear explosion. The counterfactual statement seems to be more sensitive to worldly matters, whereas the bare determiner statements are not. This is an old objection to understanding arithmetic as being about ordinary things in some way or other. I am not sure who first made this objection, but I think it makes a good point. The content of the bare determiner statements, that is, the proper content of the arithmetical equations, is not the same as that of some generalized counterfactual conditional.

How then should we give the semantics for bare determiner statements? This is a great question that I don't know the answer to. It is one question under what conditions such sentences are true and false, and here we have clear judgments for many cases. It is another question what formal semantics can capture this data. The standard way to give an analysis in terms of generalized quantifier theory won't do, since it concerns considerations about the subsets of the domain. But if the domain is understood more or less as it is commonly understood, then this won't be good enough to capture the truth conditions, since some bare determiner statements don't seem to properly connect with the domain, as when we wonder about unicorns and round squares, none of which are in the domain. The situation here might in general be no different than with the semantics of generics:[14] the standard tools of semantics are not good enough to capture the truth conditions of generic sentences. What is needed is a further, primitive, generic quantifier. Bare determiner statements might be analogous. We might need further resources to properly capture their truth conditions, but I won't be able to pursue this here. None of the considerations about ontological independence were motivated via a particular semantic analysis of bare determiner statements, but via fairly strong judgments of the truth conditions and truth value of particular examples. What formal semantics can capture these facts is a further question I have not answered.

We can conclude from all these considerations that on the present account the truth of arithmetic enjoys ontological independence in the strong form. It does not matter for the truth or falsity of bare determiner statements which particular things exist, nor does it matter how many things exist. Since the content of basic arithmetical statements is that of bare determiner statements this carries over to basic arithmetical statements. Since quantification over numbers is internal quantification this also carries over to quantified arithmetical statements. Thus arithmetic as a whole enjoys strong ontological independence. This allows us now to see how the second puzzle about ontology, discussed in chapter 1, can be solved for the case of talk about numbers, and after that how this solves the philosophical problems about arithmetic specified above.

---

[14] See Carlson and Pelletier (1995).

6.4 The Existence of Numbers and the Second Puzzle

Our considerations above solve the second puzzle about ontology for the case of natural numbers.[15] The second puzzle was the puzzle about how important ontological questions are. On the one hand, it seems everything depends on them; on the other hand they seem to be only of philosophical interest. On the present internalist account of number talk and its resulting ontological independence, we can see that nothing in arithmetic depends on ontology. The truths of arithmetic enjoy strong ontological independence: they would hold no matter what exists or how many things exist. Although it appears that everything in arithmetic depends on the ontological question about numbers to have an affirmative answer, this turns out to be false, and we have seen over the last couple of chapters why that is so. Even though the ontological question about (natural) numbers just is the question "Are there (natural) numbers?," and almost every arithmetical statement implies that there are numbers, nonetheless the results of arithmetic do not imply an answer to the ontological question about numbers. Since internalism is true about talk about numbers, all that is implied is "There are numbers" on the internal reading. The ontological question about numbers is "Are there numbers?" on the external reading, and it is not answered by "There are numbers" in the internal reading. Internalism about number talk and its ontological independence guarantee that what we say in arithmetic does not imply an answer to the ontological question about numbers.

What then is the answer to the external, ontological question about numbers? Even though the results of arithmetic do not imply such an answer, the fact that internalism is true about talk about natural numbers does guarantee that the answer is "no." We have seen in chapter 4 why internalism guarantees a negative answer to the ontological question in the relevant domain, and it is now the time to revisit this argument for the case of numbers and the particular version of internalism about number talk defended in chapter 5. The basic argument is this: On the internalist account number words are non-referring expressions, and on our particular version of internalism, bare determiner internalism, this is so since they are determiners that, for various reasons, can nonetheless appear in singular term position. They are non-referring expressions not in the sense that they aim to refer, but fail to succeed in this as empty names do, but in the sense that they are not even in the business of referring. They are non-referential expressions in the sense that they semantically do something other than referring, and so, in particular, they do not refer to anything. So, none of the objects in our domain, none of the existing things, are referred to by expressions like "2" or "the number 2." But that means that none of these things are the number 2, assuming internalism is true and I am presently employing number talk in the standard internalist way. It must be true that nothing (used externally) is

---

[15] See section 9.3 below for a more general discussion of the second puzzle.

the number 2, since the latter phrase "the number 2" does not denote any object in the domain that the external quantifier ranges over. So there is no number 2. Among all the objects that exist, none is the number 2. And similarly for all the other natural numbers: 0, 1, 3, 4, and so on. None of these exist either. The natural numbers are just those numbers, and thus it follows that there are no natural numbers. The ontological question about natural numbers has a negative answer, and internalism makes clear that and why this is so.

Although this is the real, objective answer to the ontological question about natural numbers, it is important to make clear what is not implied by this. None of this is support for nominalism more generally, i.e. the view that there are no abstract objects. But it does mean that whatever abstract objects there might be, none of them are natural numbers. For example, nothing I have said rules out that the von Neumann ordinals exist.[16] But even if the von Neumann ordinals exist, none of them are natural numbers. And similarly for all other entities that some philosophers have proposed are the natural numbers. Maybe there are positions in an abstract ω-sequence.[17] But even if there are, none of them is a natural number. If internalism is true then our number words do not refer to von Neumann ordinals, nor positions in patterns. They don't refer at all. And thus whatever there might be, none of it is a natural number.

The above argument invites a forceful objection: what if we had done things differently? Maybe the above is right about how we contingently did things, but we could have done it differently. We could be Neumannians, an imaginary group of people who are like us as much as possible, but who use number words with the intention to refer to von Neumann ordinals. And since nothing I said above rules out that there are von Neumann ordinals, we can imagine that, somehow, Neumannians succeed when they intend to refer to von Neumann ordinals with their use of words, which they also write "one, two, three, ... " Neumannians could end up with an arithmetic just like ours. They, too, would accept "every number has a successor," "there are infinitely many primes," and so on. But for them their arithmetic would be about objects. But how can it depend on contingent features of speakers whether arithmetic is about objects? And why does it matter what we happen to do with number words if we could just as well have done it differently to apparently the same overall effect? And won't all the same puzzles arise for Neumannians as they did for us? Shouldn't the solution to these puzzles be more uniform, and not rely on contingent features that distinguish us from Neumannians?

This objection gets some things exactly right, and some things completely wrong. It is important to see which is which to make clear what the view defended here is

---

[16] The von Neumann ordinals are the standard way to identify ordinal numbers with certain sets. See Hrbacek and Jech (1999).

[17] See Resnik (1997) or Shapiro (1997).

and what it isn't. Neumannians can use words like "one, two, . . ." with the intention to refer to von Neumann ordinals, and be successful at doing it for all we have seen. We could have done this, too, although we didn't. But Neumannians would not be referring to natural numbers when they do this. Their use of "two" or "the number 2" would not refer to the number 2. There is no number 2, and the number 2 is not any von Neumann ordinal. Similarly, what Neumannians call "arithmetic" is not about natural numbers. There are no natural numbers. They use words that do not refer in our language referentially in a different language.

Neumannians might or might not face the same puzzles we faced here. This will depend in part on how the story is fleshed out more. If they also use number words in determiner position then the question will arise how a word that is referring to a von Neumann ordinal can also be a determiner, i.e. they will face their own version of Frege's other Puzzle. But the solution to the puzzle will be different than it is for us. They won't have focus effects in statements about what the number of moons is that we encountered in chapter 2. For such a focus effect to arise it was crucial that the number word is not a referring expression in "the number of Fs is n." Neumannians also won't undergo cognitive type coercion in learning arithmetic, assuming they start out with number words as names for von Neumann ordinals. If so then they must learn it differently than we did. Their solution to the application problem and the Caesar problem must be different than the one to be proposed momentarily. The difference between us and Neumannians makes all the difference when it comes to the philosophy of arithmetic. Philosophically they are very different, technically they are in a sense similar. The sentences of the von Neumannian's "arithmetic" could be mapped onto our arithmetical ones in a way that preserves truth and falsity (assuming there are finite von Neumann ordinals). But this similarity is only on the surface. It maps one sentence in one language onto another one in a different language that is written the same, but they mean different things and are also completely different in several other ways.

What we hope to understand here is arithmetic, that is the theory of the natural numbers, that is the numbers 0, 1, 2, and so on. This is how the subject matter of arithmetic, as well as the philosophy of arithmetic, should be specified. And so understood it matters for what the philosophy of arithmetic is supposed to do whether in our talk about numbers we talk about objects or not. Others could have done it differently in that they could have used symbols and sentences that look the same and are equally true as our arithmetical ones, but that mean completely different things. But they couldn't have done it differently in that they refer to numbers when they do arithmetic while we don't.

Internalism, as defended in chapter 5, the solution to the second puzzle defended above, and the corresponding answer to the ontological question about numbers now put us in a position to resolve the tension among the apparent facts about arithmetic, solve the Caesar problem as well as the application problem, and propose a philosophy of arithmetic.

6.5 The Internalist Solutions to the Problems

6.5.1 Resolving the Tension

The philosophy of arithmetic has to deal with the tension between three apparent truths: subject matter, objectivity, and thinking alone. On the present view we can resolve this tension as follows. Arithmetic has the natural numbers as its subject matter, but it was a mistake to give this an ontological reading. On the internalist conception of talk about numbers, arithmetic is about natural numbers. But it is "about" them not in the referential sense of aboutness, but in the topical sense, as discussed in section 4.1. Natural numbers are the topic of arithmetic, and thus what arithmetic is about. But they are not what phrases in arithmetical statements refer to. Understood this way the subject matter of arithmetic is in no conflict with the thinking alone aspect of it. As argued above, the truths of arithmetic enjoy ontological independence in the strong sense. This ontological independence is compatible with arithmetic having the natural numbers as their subject matter, in accordance with topical aboutness, and thus resolves the tension between subject matter and thinking alone. The objectivity of arithmetic is thus on this account not tied to a domain of objects that exist independently of us. Arithmetical objectivity can instead be tied to reason and thinking. Not that we would know so far how it is so tied, but arithmetic having natural numbers as its subject matter leaves this connection now open. This gives us the promise of a viable form of rationalism. We do not yet know if the objectivity of arithmetic can indeed be derived from reason, but the central tension is relieved and that makes this line a promising approach to rationalism.

6.5.2 The Solution to the Caesar Problem

And we can see how the internalist conception of talk about numbers resolves the Caesar problem. No identity statement of the form

(277)   The number 4 is identical to x.

can be true when "x" is replaced by a referring or denoting expression. Since the number term "the number 4" on the internalist account is not a referring or denoting phrase it can't be a term in a true identity statement when the other term is a referring or denoting phrase. For such a statement to be true it has to be that what the terms on either side of "is identical to" denote is the same thing. And that can't be so when only one of the terms denotes anything at all. But non-referring expressions can nonetheless occur in identity statements as discussed in chapter 2. Such statements are not identity statements in the sense that they claim that what two referring expressions stand for is one and the same thing. Instead such statements simply assert that two things are the same, and those two can be numbers, ways of making cakes, how we both like it, and so on. "identity" is used more broadly than merely in the connection with referential expressions. But when we combine a referential with a non-referential term in an identity statement, the result can never be a true statement. Nonetheless, the sentence uttered is grammatically well-formed. This is why it is false to claim that the number 4 is identical to Julius Caesar, although it appears to be a coherent identification.

Tying the mistake in a Caesar style identification of a number with a regular object to a fact about language, in particular a kind of general mistake in a mismatch of referentiality in the identifying terms, nicely explains why in general such identifications are judged to be absurd. We are not explicitly aware of many facts of our language in that if we were asked to judge whether this fact obtains, we would not know the answer. But such facts nonetheless can manifest themselves in our reactions when exposed to certain assertions. Judgments of when a sentence is grammatical are a well-known example of this. We often don't know what is wrong with a sentence grammatically, while we nonetheless have a clear sense that it is off. Syntactic focus effects, discussed in chapter 2, are another good example of this. We are not explicitly aware of when a focus effect arises, but we have strong reactions to question–answer incongruences that are tied to focus effects. That same kind of explanation applies to our case of identifying numbers with objects. We are not explicitly aware of the non-referentiality of number words. But we do have strong reactions tied to it. These reactions are not judgments about what has gone wrong in a sentence, but the feeling that something has gone wrong badly somewhere, even if we can't point to what it is. And this happens when someone proposes that Caesar is the number 4. It is this connection to general facts about our shared language that explains why a sense of absurdity is shared by us, no matter what our other philosophical views are. I find it hard to explain our shared feeling of the absurdity in any other way, but easy to explain it this way. A full solution to the Caesar problem must explain not only why the identification of numbers with regular objects isn't correct, but why we in general take such identifications to be absurd proposals. The internalist solution solves the Caesar problem in both of its parts.

6.5.3 The Solution to the Application Problem

The internalist view defended here also solves the application problem. The application problem came in two parts. First, why do we give numbers as answers to "how many" questions? Second, why are we entitled to use the results of arithmetic to answer more complex "how many" questions? The first part is answered straightforwardly. The paradigm answer to a question like

(278)   How many students failed?

is a statement like

(279)   Four students failed.

We can give the same answer, exploiting the phenomenon of background deletion, discussed in section 2.4.3, alternatively as:

(280)   Four.

Here "students failed" is elided, since it is the background of the question, and only "four" is articulated, since it is the new part of the information. This works equally with other determiners. Instead the answer could have been either one of the following pairs:

(281)   a. Almost all students failed.
        b. Almost all.

(282)   a. Few students failed.
        b. Few.

When we give a number as the answer to a "how many" question, we thereby articulate a number determiner, and elide the rest of the answer. That this is so can be seen quite easily by the fact that we can do the same thing with other determiners. We do not refer to an object when we do this, and thus we don't have to face the question of why we name that object and not some other, equally suitable one in our answer.

In a sense everyone can accept this solution to the first application problem, unless they hold that number words aren't properly determiners, but referential expressions, even when they appear in determiner position.[18] Anyone who holds that this is the answer to the first part of the application problem, but also holds that at other times number words are referring expressions will have to explain how these two parts go together, and thus will have to solve Frege's other Puzzle in a non-uniform way, as discussed in section 5.4.2. However, unless one believes, as I do, that number words are determiners even in arithmetical equations, one won't be able to extend the solution to the first part of the application problem to a solution to the second part, to which we will now turn.

The second part of the application problem, the problem of why we are entitled, apparently a priori, to use arithmetic to figure out the answer to "how many" questions, is directly solved by the version of internalism defended here. To repeat, the puzzle was why we are, apparently a priori, entitled to use arithmetical results like

(=4) $2 + 2 = 4$

to answer questions like

(283)   How many are two dogs and two more dogs?

The answer according to bare determiner internalism is simply this: (=4) is the symbolic abbreviation of

(is4)   Two and two is four.

---

[18] Although it is natural to take this to be Frege's view in Frege (1884), I take it to be untenable nowadays in light of what we have learned about natural language since 1884. It is not relevant for this issue whether number words are determiners or some other kind of higher type expressions, like modifiers. That is a good question, but even if they are modifiers and not determiners they are nonetheless neither referring expressions nor names of objects.

(is4) has the same content, although expressed in a different way as a result of cognitive type coercion, as

(are4)   Two and two are four.

And (are4) is a bare determiner statement that leaves the noun position following the determiner open. Its content could also be specified as

(284)   Two X and two (more) X are four X.

Here "X" could be instantiated with anything, and the bare determiner statement holds for whatever X may be. Keeping in mind the bare determiner character of arithmetical equations it becomes clear why they are immediately relevant for answering "how many" questions. If I want to find the answer to our "how many" question (283), I can use the arithmetical result (=4) and simply instantiate it with whatever noun I like, in particular "dog." Application on this account is simply instantiation. In particular, the arithmetical operation of addition and the operation expressed by "and" as it occurs in the question (283) are one and the same operation. Addition, expressed by the "+" in (=4), is merely a symbolic notation for "and" when it operates on bare determiners, as it does in (are4). Although sometimes it is pronounced "plus" instead of "and" these are merely different words for the same thing, as we saw in chapter 5. It is thus no surprise that addition, the arithmetical operation, is used, a priori, to answer "how many" questions involving "and." Addition and the operation expressed by "and" as it occurs in (283) are one and the same operation.

   Although the solution to the first part of the application problem can be modified to suit rival positions to the one defended here, the solution to the second part won't carry over to other views. It requires that arithmetical equations really are bare determiner statements, and thus it requires a view at least very close to the one defended here. With that view, however, the application problem can be solved quite directly in both of its parts.

6.6 Logicism Without Logic

Logicism is, first and foremost, a grand vision of mathematics. It is the vision of mathematics having a number of grand philosophical features: that we can make progress by thinking alone, without empirical help, that mathematics is fully objective, that it is objectively true, and maybe others as well. The traditional way to defend that grand vision is to tie it directly to logic. If mathematics just is logic then mathematics would have these grand features, since logic, it was widely assumed, has these grand philosophical features. And the standard way to try to show that mathematics is logic is to try to reduce it to logic. To do that one specifies what one takes to be logic and what one takes to be mathematics in precise formal languages, and then one hopes to find an effective mapping from the mathematical truths to the logical truths. But this way to try to establish logicism has two problems. First, but not even foremost, one can't reduce mathematics to logic in this sense. If one means by logic and mathematics what is usually meant, then for technical reasons alone there can be no effective mapping of mathematical truths to logical truths. There are some attempts by contemporary logicists to get around this problem by either counting more as logic, or less as mathematics, or something else as reduction.[19] But this doesn't help with the next, second, and more serious problem: No such technical result, even if it could be achieved, would show that mathematics has the grand philosophical features that logic is assumed to have. All we would get is that there is a certain function which effectively maps one set of true sentences onto another. But that there is such a function doesn't mean anything about the contents of these sentences, or the philosophical features that members of these sets have. It just shows that each mathematical sentence can be assigned a logical sentence such that they are either both true or both false. But what other philosophical features these sentences have is not addressed by this. One might try to avoid this by simply introducing new artificial mathematical sentences that by stipulation have the content of the logical ones that they get assigned to, but this won't show us anything about actual mathematical sentences which have their content already, and not by stipulation.[20] A mere technical result won't be enough to show that logicism is true for mathematics as it is. What would need to be established in addition is that mathematical sentence and the logical sentence either are something like synonymous, or at least have the same grand philosophical features. The former is quite clearly false, and the latter leaves all the real work to be done. The traditional way of trying to establish logicism is thus hopeless, or so it seems to me.

But there is an alternative, and more promising, way of thinking of it. A logicist should concentrate directly on the grand vision of what mathematics was supposed to be like. Instead of reducing mathematics to logic, the new logicist should attempt to show directly that mathematics has the grand philosophical features that logic is believed to have. To do this one wouldn't have to show that mathematics is logic, or is reducible to logic, but that mathematics is in large-scale philosophical ways just as logic is believed to be. And it wouldn't even matter for this whether in the end it turns out that logic indeed has these grand philosophical features outside of what is used in mathematics itself. This way of establishing logicism is not a technical project. It won't depend on a mathematical result, a proof of an effective mapping of one class of formulas onto another. Logicism so understood doesn't depend on formal logic, since it does not involve a technical formal result. And it doesn't depend on logic in that it is not required that logic has the grand philosophical features that it is believed to have. It is logicism without logic, but still logicism in that it aims to defend the grand philosophical vision of mathematics that goes by the name "logicism," and goes by

---

[19] See, for example, Zalta (2000).
[20] An approach not unrelated to this is developed by Agustín Rayo in Rayo (2013). I critically discuss his proposal in more detail in Hofweber (2014b).

that name with some justification. It is simply based on a different way of realizing that vision.

In fact, tying logicism to logic, in particular formal logic, is one of the main obstacles for achieving it. A paradigm case of this is the adjectival strategy as discussed by Dummett and others. On one version of the adjectival strategy, the second-order logic (or SOL) version we discussed above in section 6.3, one maps arithmetical sentences onto sentences in second-order logic. But the second-order logic sentences do not capture the logical form of the arithmetical sentences, nor do they capture their truth conditions. They are simply as close as one could get within second-order logic. Once the truth conditions of the arithmetical sentences are spelled out properly we can see that the arithmetical sentences enjoy strong ontological independence, while the second-order sentences they get mapped onto do not. Trying to map arithmetical sentences to sentences in a formal language is a mistake and not the route to logicism. Reflecting on the arithmetical sentences and their content directly is the way to go. Similar remarks apply to the neo-Fregean program in the philosophy of arithmetic.[21] The technical result that Peano Arithmetic can be interpreted into second-order logic together with Hume's Principle is a bad start for a philosophy of arithmetic, since the question is not what can be interpreted in what, but what we in fact do with our use of number words and our thinking about number. The technical result is certainly of interest, but it doesn't answer the questions we are trying to answer, instead it distracts from them.

The internalist account of arithmetic defended here is a plausible candidate for a logicist account of arithmetic. It avoids the biggest obstacle to other logicist accounts, in particular Frege's, of having to reconcile the thinking alone aspect of arithmetic with the dependence of arithmetical truth on ontological issues. Since on the present account arithmetic truth enjoys strong ontological independence it avoids the tension between how thinking alone can be good enough for figuring out which statements are true, on the one hand, and that these statements appear to be about mind-independent objects, on the other hand. This removes the main stumbling block of logicist approaches to arithmetic, and opens the door to understanding why thinking alone is good enough in arithmetic. The bare determiner statements are paradigmatic cases of statements that we can know by thinking alone if we can know them at all. Thinking alone is sufficient to figure out that

(285)   Some are more than none.

or that

(286)   Three are more than two.

And in a similar vein we will be able to determine the truth of basic arithmetical equations, and the rest of arithmetic. None of these claims about what we can

---

[21] See Hale and Wright (2001).

know by thinking alone are based on a particular view in epistemology, nor do they provide a view in epistemology or anything like an account of knowledge or a priori knowledge. They rely on the intuitive difference between some claims that you can know by thinking alone, and some that you can't. On that intuitive difference the bare determiner statements are clearly in the thinking-alone camp. And on the flip side, arithmetic as conceived as having as its subject matter a domain of independently existing objects is clearly in a prima facie tension with the thinking-alone aspect. Of course, more will have to be said about the epistemology of such bare determiner statements, but the general line looks rather promising, since bare determiner internalism can overcome this tension. It makes rationalism about arithmetic a plausible and reasonable position to take. This rationalist view in the philosophy of arithmetic is not a consequence of tying arithmetic to logic, but based on what we do when we talk about numbers in the first place.

6.7 The Philosophy of Mathematics

Although the internalist conception of arithmetic defended here is a plausible candidate for logicism, conceived as a rationalist view of arithmetic, it is not, and isn't intended to be, a logicist view of all of mathematics. The considerations brought forward to defend internalism about talk about natural numbers were particular to talk about natural numbers. Number words as determiners, the solution to Frege's other Puzzle, the internalist truth conditions of quantified statements, all these considerations are particular to arithmetic, and thus do not carry over to mathematical discourse in general. We should thus not expect from the considerations given here that internalism is true for all of mathematics. And we should not expect the resulting logicist view defended above to carry over from arithmetic to the rest of mathematics. The question thus remains what we should say about the rest of mathematics, and how the logicist picture of arithmetic goes together with what might be true about the rest. After all, some of the puzzles and problems we started with apply to other parts of mathematics as well. These or similar puzzles will thus need to be solved differently, since our solution for the arithmetical ones was particular to arithmetic. But which puzzles and problems carry over to which parts of mathematics will differ, and what their solution is might differ as well. It might be tempting to demand that solving these problems for just one part of mathematics is no solution unless it carries over to all other parts as well, but we should resist this temptation. It is based on the assumption of a certain unity in mathematics that might not be there. In this section I want to outline some of the issues that will need to be addressed about how the proposed philosophy of arithmetic fits into a larger philosophy of all of mathematics. I can only hope to outline some of the questions here and point in various directions for an answer. A more complete proposal about the philosophy of mathematics in general will have to be left for future work.

Although the reason why logicism is true about arithmetic won't carry over to other parts of mathematics, the logicist vision might apply elsewhere for different reasons. But, of course, a completely different vision of mathematics might be correct for other parts. It might well be that logicism is true for arithmetic, and something else is true for other parts of mathematics. On the one hand, this is appropriate, but, on the other hand, it is problematic. It is appropriate that arithmetic is philosophically different from other parts of mathematics, since it does seem to be a philosophically special part of mathematics. But it is problematic since it isn't clear how different parts of mathematics can be different in philosophically significant ways while still fitting into one overall philosophy of mathematics.

In the philosophy of mathematics there are two traditions on this issue. On the one hand, there is *monism*, the more popular view that when it comes to large-scale philosophical questions then all of mathematics is the same. On the monist view, platonism or fictionalism or structuralism or logicism or some other -isms apply to all of mathematics if they apply to any part of mathematics. Whichever fits one part of mathematics must fit all others. On the other hand, there is *pluralism*, the view that different large-scale philosophical accounts apply to different parts of mathematics. Historically there were a number of well-known pluralists. The most famous is probably David Hilbert, who held that the finitist part of mathematics has philosophically a completely different status than the rest of mathematics.[22] But Hilbert's pluralism is extreme in that it divides mathematics philosophically into just two parts, and in that it holds that the two parts are radically different. Pluralism does not have to come in this form. It might well be that there are more than two philosophically significantly different parts of mathematics, and that no one is distinguished with respect to the others.[23] However logicism about arithmetic is integrated into a larger philosophy of mathematics, it will likely be a pluralist account. Logicism, as defended, applies to arithmetic and closely similar parts of mathematics like the basic theory of the rational numbers, but not to all of mathematics. What then are the options for the rest, and is a coherent integration possible?

There are many different options for a coherent pluralist view in the philosophy of mathematics that contains the present logicism about arithmetic as one of its parts. To pick one example, it is coherent that logicism is true about arithmetic and fictionalism is true about the rest of mathematics. In this case, there would be a sharp separation between the two parts of mathematics with respect to their large-scale philosophical features. What such a bifurcated story would have to deal with in particular are cases of overlap between the two parts, for example the study of functions from numbers to

---

[22]   See Hilbert (1925). A broadly neo-Hilbertian position is defended by Solomon Feferman in Feferman (1998) where the crucial distinction is not between finitist and non-finitist mathematics, but between predicative and non-predicative mathematics.

[23]   Another way in which pluralism is historically common is by giving geometry a different status than the rest of mathematics, in particular via its connection to physical space. I will, however, sideline this possible source of pluralism.

something else. Here the fictionalist story about functions would be augmented with the non-fictional truths about natural numbers, just as truth in fiction in general is augmented with factual truth.[24]

Similarly, it might be that internalism is true for talk about natural numbers, while externalism is true for talk in other domains of mathematics, for example talk about real numbers. This, too, is perfectly coherent, although it might seem to be a bit strange. If such externalist talk about real numbers is true then we would have an ontology of real numbers, but not of natural numbers. This requires, for example, that the real number 1 is not identical to the natural number 1. Natural numbers, on such a view, can be embedded in, or assigned to, real numbers, but they are not identical to them. No object can be identical to them, not even certain real numbers, assuming they exist. Even for those who believe in the existence of natural numbers as well as real numbers it should be an open question whether the natural numbers are simply a subset of the real numbers, or whether they instead are different, but embeddable. Combining internalism about arithmetic with externalism about real number talk requires us to take one side, but again, that side is perfectly coherent.

Combining philosophically different parts of mathematics into one philosophy of mathematics is coherent, but that is not the real issue for determining whether a pluralist philosophy of mathematics is correct. The real issue is to find out which philosophical account is in fact correct for which part of mathematics. And here the issue is not what is coherent to hold but what is in fact the case. As we have seen for arithmetic, that internalism and logicism are true for it is a substantial, and partly empirical, claim. It depended on a proposal about what we in fact do when we talk about natural numbers. And the same general issue will arise for other parts of mathematics. For example, fictionalism, on the paradigmatic way of understanding it, requires that we do something different in the fictional part of discourse than in non-fictional parts of discourse. This difference could be engaging in a pretense, whether we are aware of it or not, but it could be something else, too.[25] But there must be a difference in what we do and whether there indeed is that difference is an empirical question. Thus whether this difference applies to some parts of mathematics is equally an empirical question. Similarly, platonism, as it is usually understood, requires that we don't do what fictionalism requires we do. Whether this is true is also an empirical question. As empirical questions go, they are usually complicated and their answers are not obvious even to those who engage in mathematical activity. We have seen above that what we do when we talk about natural numbers is not as straightforward as one might think. It won't be easy to find out what we do in other parts of mathematics as well. Here there are substantial questions that need to be addressed.

---

[24] See, for example, Lewis (1983b) or Walton (1990) or a discussion of how truth affects truth in fiction.
[25] For more, see Walton (1990), Walton (1993), and Crimmins (1998). For a more general survey, see Eklund (2011).

But besides the empirical questions about what we do in various parts of mathematics, there are also some more directly philosophical issues that give the present logicist account of arithmetic a special place in a pluralist philosophy of mathematics. This is tied to a difference about the status of axioms. Axioms are special statements among those that express a truth in a mathematical domain. In general we can think of the relationship between axioms and the domain of mathematical truths they axiomatize in two ways: as either *descriptive* or *constitutive*. On the descriptive conception of axioms they are special, but more in an organizing way and more for us than otherwise. They are statements that allow us to derive other statements in a systematic way. They form a small, but deductively powerful, subset of all the true statements. And each of them is especially compelling to us as being true. But axioms are otherwise just statements like the rest of the statements in that domain. In particular, there is no special connection to what is true in that domain of mathematics, and what the axioms are. What the axioms imply might be just a part of what is true, and truth in the domain in general is not tied to the axioms. On the constitutive conception of axioms things are different. Here the axioms are constitutive for what is true in that domain. Metaphorically speaking: all truth flows from the axioms. Truth in that domain of mathematics is determined by, and exhausted by, what the axioms guarantee to be the case. Whether or not axioms can ever be constitutive is not at all clear. To mention one worry: if the axioms are constitutive and whatever follows from them is thus true because of this, then don't the axioms guarantee their own truth since they follow from themselves?

On the logicist picture of arithmetic defended here, axiomatizations of arithmetic, like Peano's axioms, are descriptive axioms. The axioms of arithmetic play no role in determining what is true, and they are secondary to the facts in all relevant ways. The axioms were discovered long after arithmetic was a blooming discipline. The axioms don't exhaust what is true. Peano's Axioms are incomplete, like many other axiom systems, but some of the cases that are not settled by the axioms are clearly true, and not in some way indeterminate. But other parts of mathematics might not be like arithmetic in this respect. For them the axioms are like original stipulations, or postulates, that establish the domain in the first place and that determine all the facts about it. Expanding the number systems to realize more and more desirable features is a plausible candidate of this. The complex numbers, for example, are supposed to be numbers that have certain antecedently specified features. The fact that it is coherent to have these features then establishes the domain of the complex numbers. Here the original stipulations or axioms are constitutive of the domain, or so it is prima facie plausible to hold. Some, but not all, parts of mathematics appear to have constitutive axioms. Still, how to make sense more clearly of axioms being constitutive is largely an open question. A number of approaches in the end try to make sense of it, but all the ones I know of have their problems. Fictionalism is one way of trying to achieve it, but it must be seen as an empirical hypothesis, and it is in danger of replacing truth in mathematics with something else, pretended truth. Old style if-then-ism is another option, but it, too, is an empirical conjecture, and it faces the problem that it replaces truth of a mathematical statement that apparently is not of conditional form with something else, truth of a conditional that has the axioms as its antecedent. Hilbert's formalism for non-finitist mathematics is another attempt, but it is unclear how it gets from consistency to existence. Kit Fine's recent postulationism, in his Fine (2005a), is a new and original addition to this list, but I also can't sign on to his account of the connection between consistency and existence, and his relativism about unrestricted quantification. I thus regard the question of how axioms can be constitutive as an open question, despite the fact that they sometimes appear to be constitutive.

Even if some axioms are constitutive at all, which ones are can't be established by just looking at the axioms. That the axioms of arithmetic are not constitutive is not something that can be read off the axioms themselves. To see this one instead needs to look carefully at what we do when we talk about natural numbers. Similarly, to decide for any other domain whether the axioms are constitutive we need to look carefully at what we do when we talk about things in that domain. And this is a substantial task, but one that can be pursued with the relevant issues in mind.

But here is the rub: If any axioms are constitutive, then this is a perfect match with, and requires something like, the present logicist account of arithmetic. There is a crucial and special place that arithmetic has among other parts of mathematics that is tied to these considerations about the status of axioms and to the position in the philosophy of arithmetic defended here. Suppose that some axioms are constitutive. Then the fact that the axioms of arithmetic are descriptive and arithmetical truth is not tied to axioms is of crucial importance. Constitutive axioms are supposed to determine all truth in their domain, but what if these axioms are inconsistent? Then no coherent domain has been established at all. That the axioms are consistent is essential for any domain with constitutive axioms. But whether the axioms are consistent is itself a mathematical question. It is, in effect, a question in arithmetic. If arithmetic, too, had constitutive axioms, then there being a well-established domain of arithmetical facts would itself depend on an arithmetical fact, leaving the domain in limbo. The fact of consistence would only obtain if it is part of the range of facts established by the axioms. But if the axioms are not consistent then no domain is established. Thus if arithmetic had constitutive axioms then none of this could get off the ground: there being a domain of arithmetical facts at all would depend on one of the arithmetical facts obtaining. On the present logicist proposal, truth in arithmetic is not tied to axioms in any way. This limbo is thus avoided. If any part of mathematics has constitutive axioms, it requires that arithmetic doesn't have constitutive axioms for mathematics to be properly established. Thus a logicist philosophy of arithmetic will have a special place in almost any pluralist account of mathematics, at least any one that relies on constitutive axioms, as do most I know of. Thus not only is a pluralist philosophy of mathematics coherent with our logicism about arithmetic, any such pluralism should consider logicism about arithmetic as its dearest part. I have discussed how arithmetic plays a special role in a pluralist philosophy of mathematics in more detail in Hofweber (2000a) and Hofweber (2009b). But what part of mathematics deserves which philosophical treatment when we go beyond arithmetic is left open by all this, and will require substantial further work. For now I simply hope to make plausible that not having a philosophy of mathematics as a whole should not take away from the proposed philosophy of arithmetic.

7

Ordinary Objects

7.1 The Metaphysics of Ordinary Objects

Our usual conception of the world is that of a world full of objects. Among those objects are *ordinary objects*: the midsize objects that we commonly interact with like rocks, houses, bottles, people, and so on. Ordinary objects are contrasted with other things which would also be objects, but which are a bit less ordinary and more unusual: subatomic particles, mathematical objects, fictional objects, and so on. Ordinary objects, in contrast to some of the less ordinary ones, have a location, they are reasonably large, but not too large, and they have parts which together somehow make up the object. The world presents itself to us at first and foremost as a world of ordinary objects. But whether the picture of the world as one full of ordinary objects is in fact correct is widely debated in metaphysics. Common sense is very much attached to such objects, but metaphysics does not have to support common sense. Many metaphysicians hold that there is a substantial metaphysical project to find out whether there indeed are any ordinary objects. And several have argued that there are no such objects at all.[1] In this chapter we will investigate whether there are any ordinary objects at all, what part metaphysics has to play in answering this question, what work there is to do for a metaphysics of objects in general, and how the case of ordinary objects is similar to and different from the case of natural numbers discussed above.

The metaphysical debate about objects is centered around three questions. First, naturally, there is the question of whether there are any at all:

(I)   Are there any objects?

This question is associated with two different debates. First there is the debate about whether there are any objects at all, ordinary or otherwise. The term "object" is in this sense used widely, including all things, entities, and so forth. If there are no objects in this sense, then there are no atoms, no tables, no people, and no things more generally. If this position is correct, then there is nothing, and thus the view is generally labeled *ontological nihilism*. The view might seem absurd, but the idea is that whatever there is, it is different from objects. It is hard to express the view properly, since our language

---

[1] See in particular van Inwagen (1990) and Merricks (2001), who both allow only for living things, and Unger (1979), Goodman (2002), Dorr (2005), and Goodman (2005), who don't allow for any at all. Bennett (2009) argues that we have little reason to decide one way or another.

mostly talks about objects, but the idea at least is that whatever there is is not an object, it is of another kind, maybe just stuff, or a blob of being, or what have you. Some philosophers have explored this idea, and whether it is coherent.[2] Second there is a debate about complex objects, which include ordinary objects. This is a debate about whether there are things with parts, or whether all there is are just the simple things, which don't have any further proper parts. The negative answer is a different form of nihilism: *mereological nihilism*, also called *nihilism about composition*.[3] Ordinary objects, if they exist at all, have proper parts. If nothing has proper parts, then such objects simply do not exist. There are also other positions in this neighborhood which deny the existence of ordinary objects, but for slightly different reasons. One might hold that even if some things have parts, still, all there is is what microphysics tells us there is, and ordinary objects are not among those things. Or even though some things have parts, there are special problems about ordinary objects that rule out that there are any.

The first question is simply about whether there are any objects at all. It is not so far concerned with further, more loaded questions about whether objects are what in long tradition is called a substance or whether they are derivative in some sense on something else. These questions are related to the other two, following questions.

The second question is:

(II)   Assuming there are objects, how do they relate to their properties?

This question applies to objects more generally, but can also be debated specifically for ordinary objects. There is a static and a dynamic variant of the question. The former deals with the relationship between the object and the properties it has at a particular time. This problem is not unrelated to the problem of universals, except that here we focus on the objects, not the properties. It turns into the problem of whether the properties are located where the objects are, and if so, whether the objects just are the properties stuck together, so to speak, or whether they are something other than the properties, and if so what. The bundle theory of objects vs. a substance theory of objects are some of the main contenders in this debate. We will discuss this problem in more detail in the chapter on the problem of universals, chapter 11, since its solution will require a close look at properties and universals first.

The second, dynamic, version of the problem broadly deals with the persistence of an object through change of its properties. This version of the problem is closely tied to problems in the philosophy of time, and includes a substantial literature. It includes the so-called problem of change and the general debate about persistence.[4] This part

---

[2] See in particular Hawthorne and Cortens (1995) and Turner (2011).
[3] See the references in footnote 1 for more.
[4] See Haslanger (2003) and Hawley (2001) for more on these debates.

of the debate about objects is often seen as being closely tied to the philosophy of time, but to properly engage with it would take us too far afield.[5]

Finally, there is the question:

(III)   Assuming there are objects, how do they relate to other objects, in particular their parts?

This question encompasses another group of hotly debated metaphysical questions. It relates to the question of how the matter that constitutes an object relates to that object, as in the problem of the statue and the clay, and how the sum or fusion of the parts of an object relate to the object. It includes the question of when some things compose something, i.e. van Inwagen's special composition question,[6] and it includes the question of which parts an object can lose while remaining the same object, which is, of course, related to the problem of change mentioned above.

All three of these questions are often seen as distinctly metaphysical questions. Not only are they debated for the most part within metaphysics, they also seem to be distinctly metaphysical in the sense that they are not empirical questions, and thus not in principle in the domain of the empirical sciences. Take the question about composition, i.e. the question of when certain things compose something. It seems that this question can't be settled empirically, since there is no empirical way to distinguish the situation where there is a bunch of simple things, and nothing else, from the one where there are those simples, but also something composed of them. The world would look the same either way. To settle this question one needs metaphysical theorizing. And similarly for the other two questions.

I will argue in the following that this is mistaken. The question of whether ordinary objects exist and whether some things at least sometimes compose something is an empirical question, and we have lots of empirical evidence for an affirmative answer in both cases. In this chapter we will thus have a first look at the issue of how empirical considerations relate to metaphysical ones, a topic we will discuss at greater length in chapter 12. Furthermore, the second question about the relationship between objects and their properties is a misguided question. It has a false presupposition, and is thus a badly motivated question. We will see that this is so after our discussion of properties in chapters 8 and 11. The third question, on the other hand, does require real work, and we will see at the end of this chapter what it might be. But first, let's look at what we do when we talk about ordinary objects.

---

[5] I have discussed the alleged problem of change and in what sense it is a metaphysical problem in Hofweber (2009c), where I argued that there is no metaphysical problem of change. See also Rychter (2009) for a similar view. For a critical discussion, see Raven (2011b) and Einheuser (2012). The right view about how to understand the metaphysical debate about persistence seems to me to be the one defended in Hofweber and Velleman (2011). None of our discussion here will presuppose any of this.

[6] See van Inwagen (1990).

7.2 Talk About Ordinary Objects

In our discussion of the philosophy of arithmetic we have seen that the function of talk about numbers has important and direct consequences for the philosophy of arithmetic, in particular the question of whether there is a domain of numbers, and whether numbers exist at all. We will now first look at whether similar considerations answer the question about the existence of ordinary objects as well. And to do this we have to see whether internalism or externalism is true about talk about ordinary objects. Internalism about talk about ordinary objects would imply a negative answer to the external question about ordinary objects, just as was the case with talk about natural numbers which we discussed in section 6.4. We should thus first consider the question of how to understand talk about ordinary objects, with a particular eye on internalism and externalism, and then, second, look at what conclusions we can draw from this for the metaphysics of ordinary objects.

To settle the question of whether internalism or externalism is true about talk about ordinary objects we have to find out whether, in general, singular terms for ordinary objects are used with the intention to refer, and whether quantification over ordinary objects is generally used on the external reading of the quantifiers.

Contrary to settling the same question for talk about numbers, properties, or propositions, this question is much more straightforward for the case of ordinary objects. Our talk about ordinary objects is paradigmatically externalist. Regular objects, including people, are given names that aim to refer to these objects. In general we represent the world as containing a variety of objects, which exist in the world independently of our picking them out with a singular term. Correspondingly, when we quantify over ordinary objects we aim to make claims about this domain of objects.

The relationship between a regular object and a proper name for it is paradigmatically what reference is. But that doesn't mean that this isn't controversial among philosophers. Some philosophers have argued that appearances to the contrary, proper names are not referring expressions at all. Instead they are descriptions, and thus quantifiers, or predicates.[7] But it is important to be clear that much of the debate about whether proper names are descriptions or referential is orthogonal to our concerns here. What matters for the question of whether internalism or externalism is true for talk about objects is not whether our talk about them is referential in the narrow sense in which many believe names refer, or in the broad sense, which includes denoting expressions and thus descriptions.[8] Even those who hold on to a descriptive theory of names take them to be broadly referential, and that is the sense that matters for us. But if names are predicates, then things might be different. They might then not be referential even in the broad sense, since they would belong to a quite different kind. Here we should acknowledge that there are clearly predicative uses of proper names. For example, in

---

[7] There are other options as well, see in particular Dever (1998) and Cumming (2008). Although there are important differences in the semantics of names between these options, they won't matter for our purposes here, as I hope to make clear shortly.

[8] We discussed this difference and its relation to internalism in section 4.1.

(287)   Every Mike wishes he had a different name.

it is very plausible that "Mike" is not a referring expression, but instead predicative. Similarly in

(288)   I'd rather have a beer with a Mike than a Mortimer.

But this shouldn't take away from the fact that on other uses names appear to be referential. Although "Mike" might not be referential in the above examples, it seems to be referential, in the broad sense, in

(289)   Mike is tall.

where the speaker intends to talk about a particular person: Mike. What needs to be explained is how names can sometimes be predicative and sometimes be referential. This is reminiscent in certain ways of Frege's other Puzzle discussed in chapter 5. Names for objects, just as number words, appear to be in two apparently quite different positions. We need to explain how that can be, which one of them is more revealing of their true semantic function, and whether that function is to refer to objects. The traditional view holds that predicative occurrences of "Mike" in examples like (287) are meta-linguistic. They say in a short form that everyone called "Mike" wishes something. Names in argument position are instead not meta-linguistic, but referential in the broad sense. However, there is a recent movement to take names in predicate position to be more revealing. On that view, names are predicates. However, they can appear in argument position, since on these occasions they are accompanied by either an unpronounced determiner which turns them into a description, or else an unpronounced demonstrative which turns them into a complex demonstrative.[9] However, and this is what matters for us here, on these views names, together with their unpronounced accompaniments, are broadly referential when they are in argument position: they are either referential in the narrow sense, or they are disguised descriptions, and thus referential in the broad sense. All that matters for our questions here is whether names are referential in the broad sense in the cases where they are arguments, in particular in subject position. This is uncontroversial among these otherwise diverse views of names and what they do when they are predicates. On all of these views, names in argument position are used to refer broadly to objects, either by themselves or with unpronounced help.

A slightly more radical version of names has recently been defended by Paul Pietroski. For Pietroski, no expression in natural language is semantically referential. Natural language does not contain any expressions that have, semantically, the function to pick out objects.[10] Pietroski's view is motivated in part by a general picture of how composition in natural language is to be understood. Crucially, this is a view about

---

[9] For a discussion of names as predicates, see Burge (1973), Pietroski (2004), Pietroski (2007), Matushansky (2008), Fara (2011), Fara (2015), and Schoubye (2017).

[10] See Pietroski (2004) and Pietroski (2007). For a brief discussion, see Hawthorne and Manley (2012: 244ff.).

the semantic function of expressions in natural language. This view is compatible with
these expressions, which are semantically not referential, nonetheless being used by
speakers with the intention to refer. In particular, although the semantic content of
the sentence is not about any objects, the content of an utterance of such a sentence is
referentially about objects. As Pietroski himself holds, reference to objects occurs not
at the level of semantics, but at the level of thought or utterance. For our discussion
here this, too, is all that matters. Externalism concerns the question of whether or not
speakers use expressions that are topically about objects with the intention to refer, in
the broad sense, to objects. It doesn't matter if the expression used is semantically about
objects, as long as the content of the utterances of these sentences is about objects.

Thus although there is quite a bit of controversy in the semantics of names and
whether names are semantically referential, this does not undermine the picture of
talk about objects as involving attempts to refer to them, and thus undermine the
externalist picture of such talk. On all the diverse views we just briefly discussed,
speakers aim to refer to objects with utterances of names. Talk about ordinary objects
is the paradigm case of an externalist discourse, and despite the appearance that the
controversy about names as predicates undermines this, it is well supported in this
debate as well. But this by itself does of course not answer all relevant questions about
talk about objects.

First, it leaves open questions about the referential status of a variety of other phrases
that are about ordinary objects, at least in the topical sense of aboutness. There are the
questions about generics, about mass terms, and so on, but whatever one says about
them, it is clear that talk about ordinary objects in general is to be understood along
the externalist picture.

Second, and more importantly, even though we represent the world as containing a
domain of objects we can refer to and quantify over in the external reading, this does
not mean that this picture of the world is indeed correct. Even though we aim to refer
to such objects with proper names, there is no guarantee so far that these attempts
succeed. If there are ordinary objects, then we probably often succeed, but if there
are no ordinary objects, then such attempts might well systematically fail to pick out
anything. It is one question how we represent the world, and it is quite another how
accurate these representations are. We thus need to find out whether there are any
ordinary objects after all.

7.3 The Existence of Ordinary Objects

A common setup for making the existence of ordinary objects problematic is this:
ordinary objects, if there are any, have parts, and these parts again have smaller parts
until, hoping that such division will eventually come to an end, we reach the smallest
ones: the simples. The simples are arranged in a certain way, like that of a chair, or
a table, or a person. But why think there are the simples arranged in a certain way

and the table in addition, instead of just the simples arranged a certain way? If there is a table as well, then there must be at least one more thing than if there are just the simples. The table is not a simple, and so if there is a table besides the simples, then the number of things there are must be at least the number of simples plus one. To settle whether there is a table besides the simples is then a project for metaphysics to figure out. The (mereological) nihilist holds that the best overall metaphysical view of the material world is one of simples arranged certain ways, and nothing else. The nihilist opposition, which could come in different forms, holds that besides the simples there are also other things which have proper parts.

There is much to object to in this setup. It is not at all clear that we should take simples for granted, nor that we have any good reason that there are simples (as opposed to everything having smaller and smaller parts), nor that we have better reason to hold that there are simples than that there are ordinary objects.[11] But besides these worries, we will accept the setup in the following, for two reasons. First, even though it stacks the cards against ordinary objects, it nonetheless in the end leads to a support for ordinary objects, as I hope to make clear shortly. Second, it nicely allows us to think about what kind of a hypothesis it is to hold that ordinary objects exist over and above simples. What is the relationship between simples being arranged chair-wise, and there being a chair where those simples are arranged that way? One might hold that it is inconceivable that there are the simples arranged chair-wise, but there is no chair, and correspondingly that it is a conceptual or analytic truth that if simples are arranged that way, then there is a chair. Alternatively, one might hold that it is conceivable that there are the simples arranged chair-wise, but there is no chair. If the first holds, then the existence of the chair over and above the simples would be an analytic hypothesis: that the chair exists follows analytically from the distribution of simples. As such it seems hard to see this as a metaphysical hypothesis, since matters of meaning or conceptual matters already settle it. If, on the other hand, the existence of a chair does not analytically follow, then its existence or non-existence would be a synthetic hypothesis. Metaphysical hypotheses are generally conceived as synthetic hypotheses, and those who believe in a substantial metaphysical project of finding out about the existence of ordinary objects are generally in this camp. The question for those who think it is a synthetic hypothesis is then whether the existence of the chair over and above the simples is something settled empirically, or whether it is a distinctly different, non-empirical, problem, and how placing it in one camp or another here affects its status as a metaphysical problem.

Some philosophers, in particular Amie Thomasson in Thomasson (2007), have argued that the existence of ordinary objects follows analytically from the distribution of simples, assuming that there are any simples.[12] It is an analytic truth, Thomasson argues, that if there are simples arranged chair-wise, then there is a chair. She argues that requirements on a successful fixing of reference lead to analytical connections

---

[11] See Crawford Elder's discussion of this in Elder (2011).   [12] See also Hirsch (2011).

between the referring term and other more general terms. And more general terms, in turn, have application conditions that guarantee that they are properly applied to when simples are arranged in the right way. These views together support Thomasson's claim of an analytic connection between simples being arranged chair-wise and there being an object which is a chair.[13] I don't believe this position is correct, for at least the following reasons. First, requirements for reference fixing only constrain what the referent is, they do not get built into the meaning of the referring term. They at best constrain what kind of an object the referent is, but they don't establish an analytic connection between the referring term and a term for that kind. Second, it remains unclear why the application conditions of "chair" are met as long as there are simples arranged chair-wise, as opposed to a fusion of these simples. I don't want to dismiss Thomasson's sophisticated view, but I think we should reject it and I agree with others who have criticized this part of her view, for example Jonathan Schaffer in Schaffer (2009a). The connection between simples being arranged chair-wise and there being a chair is not an analytic connection. I take it to be conceptually coherent to hold either that there is a chair as well as that there isn't one, together with simples being arranged chair-wise. These are two competing synthetic hypotheses. The question remains, which one is correct, and how we are to find out.

Those who hold that the existence of a chair is a synthetic hypothesis over and above the chair-wise arranged simples generally also hold that it is not an empirical hypothesis.[14] Trenton Merricks's book Merricks (2001), which argues that inanimate composite objects do not exist, even has this as its first sentence: "Ontological discovery is not empirical." And one might think that there is good reason for this in our case. After all, it seems that you can't tell by looking whether composition occurs, nor does there seem to be a way to distinguish the hypothesis that it does or doesn't occur with scientific means. The world would look just the same, both for us ordinary observers as well as for science, whether or not the simples arranged chair-wise compose a chair. The problem about the existence of ordinary objects is thus, according to this line of reasoning, a paradigm metaphysical problem, one where science has to remain silent, and philosophy has to take over. To decide whether ordinary objects exist, and whether composition ever occurs, we need to formulate metaphysical hypotheses, and see which metaphysical view overall is the best. Most of the debate about the existence of ordinary objects is doing just that.

But this is mistaken. The existence of ordinary objects is not a distinctly metaphysical problem. Instead it is an empirical question whether such objects exist and whether composition ever occurs. Furthermore, the empirical evidence is strongly in favor of ordinary objects. In the next couple of sections I will defend these claims. After that we will look at what work there is for a metaphysics of ordinary objects. This will lead to a first discussion of how metaphysical questions relate to empirical ones, a topic we

---

[13] See Thomasson (2007) for the details.
[14] See, for example, Rosen and Dorr (2002: 7), van Inwagen (1990), Merricks (2001), and many others.

will return to in detail in chapter 12, and to the problem of what questions are left open once the empirical ones are settled.

7.3.1 Perceptual Evidence

It is tempting to hold that a world where composition occurs, and the simples combine into a chair, would look just the same to us as a world where it doesn't occur and all there is are the simples. If the world were one way rather than another things would look exactly the same to us. And thus looking at the world won't allow us to tell the difference. Therefore the question of whether composition occurs is not an empirical one, but a distinctly metaphysical one, or so the argument goes.

There is a sense in which this way of thinking is correct, but this does not answer the question of whether there is any empirical or perceptual evidence for one option rather than another. We can grant that we can't phenomenally distinguish the two cases, in the sense that the phenomenal character of our perceptual experience would be exactly the same in each case. In this sense it would look the same either way. But that is not the relevant question. The question is rather whether or not perception entitles us to believe that there are objects. This issue is not primarily about the phenomenology of perceptual experience. Instead the issue is what beliefs we are entitled to hold on the basis of perception. And here there is a crucial difference. Our perceptual apparatus is set up to produce in us perceptual beliefs. Perceptual beliefs are beliefs that are first in line at the level of belief as a result of perception, they are the immediate result at the level of belief from our perceptual experiences or whatever in fact produces these first beliefs in perception.[15] The issue is one about what we get evidence for in perception: is it just that the world looks, phenomenally, a certain way, or is it that the world is as our perceptual belief represents it to be? There is a long tradition in epistemology of trying to derive or infer what the world is like simply from the phenomenology of perception. But this project is at best seen as a project to refute a very radical form of skepticism, and even there as not a very promising one. We don't have to engage with skepticism here, as this is not what is at issue in our debate. We can grant here that skepticism is mistaken and that we are, in general, entitled to rely on perception. And what we learn in perception is not just what the world phenomenally looks like, but what our perceptual beliefs represent it to be like. We are prima facie and defeasibly entitled to the contents of our perceptual beliefs. The mere phenomenology of the perceptual experience might not have any propositional content at all, but the perceptual belief certainly has propositional content, and what we are entitled to believe in perception is the content of our perceptual beliefs. Any such entitlement is, of course, defeasible. We are, at first, entitled to our perceptual beliefs, but this entitlement can go away. To see what perception can do for us in the debate about ordinary objects, we should

---

[15] This qualification is necessary, since we can't simply assume that the perceptual experience is causally responsible for the perceptual belief. Maybe something causes both the experience and the belief, but the precise causal mechanism is not really relevant for the main issue in this section.

thus look both at what our perceptual beliefs defeasibly entitle us to believe, and at whether these entitlements are defeated.[16]

Our perceptual beliefs, as a matter of empirical fact, have contents about objects, not simples arranged object-wise, or the distribution of colors in phenomenal space, or what have you. This could have been different. Other perceptual systems might lead to perceptual beliefs with other contents. But ours leads to perceptual beliefs about objects. And as such we are defeasibly, and to a certain degree, entitled to believe these contents as a result of perception. Although the phenomenology of perception might not be able to distinguish a world where composition sometimes occurs from one where it never occurs, the contents of our perceptual beliefs do distinguish between these two cases. Some such beliefs represent the world as one where composition has occurred, since they represent the world as one containing ordinary objects. There is a real asymmetry here between the thought that there is a chair and the thought that there are simples arranged chair-wise. Perception entitles me only to hold a perceptual belief with the former content, not with the latter content. If I conclude that there are simples arranged chair-wise, then such a conclusion is derivative on the perceptual belief that there is a chair, and thus not a belief that is a frontline belief. The belief that there are simples arranged chair-wise is not a perceptual belief at all, and it can't be in our perceptual system. Our perceptual beliefs, as it turns out, are beliefs with a content about ordinary objects, not simples arranged object-wise. This is what perception defeasibly entitles us to believe.

Any such entitlement is defeasible and the question now is whether it is defeated, in particular whether the aspect concerning objects is defeated. It is common and useful to distinguish two kinds of defeat: rebutting and undercutting.[17] A belief can be rebutted by reasons for believing the opposite, and undercut if one has to realize that how it was formed is no longer to be trusted. To illustrate, if I believe that p since Fred told me so then this belief is rebutted if I get better reasons for believing not p, and undercut if I find out that Fred is notoriously unreliable. Individual perceptual beliefs are often rebutted by further perceptual evidence, but this doesn't speak against perceptual beliefs in general, only against particular instances of them. Our concern here is rather the more general one of whether we have reasons from perception to think that there are ordinary objects. In order to reject this one would have to hold that this whole aspect of what we believe on the basis of perception is either rebutted or undercut.

One serious possibility of a rebuttal of our perceptual beliefs about objects is that we find out in further inquiry, in particular in science, but also in metaphysics, that such beliefs are systematically mistaken. We will discuss the scientific and metaphysical

---

[16] For detailed accounts of our entitlement to perceptual beliefs along these lines see, for example, Pollock (1974) or Pryor (2000). Of course, this is not uncontroversial. For some criticism, see White (2006). Much of the controversy is tied to how this could be a response to skepticism, a topic we will touch on shortly, but which isn't our concern here.

[17] See Pollock (1986: 38ff.).

evidence for or against ordinary objects shortly. Another possibility, which we will focus on now, is that further considerations undercut our defeasible entitlement to our perceptual beliefs about objects. Here we should distinguish two levels of undercutting. The first and more radical one concerns considerations that we are not entitled to our perceptual beliefs at all. This is a largely skeptical consideration, affecting not just ordinary objects, but everything. A second and more modest one is that although in general perception leads to beliefs we are entitled to, nonetheless, our entitlement to our beliefs about ordinary objects in particular is defeated by being undercut. We are only concerned here with the second, not the refutation of skepticism. It is thus assumed that perception in general defeasibly entitles us to our perceptual beliefs, while it is put to debate what precisely it does entitle us to believe. The question here is thus not about whether we are entitled to our beliefs formed on the basis of perception, only about what beliefs we are so entitled to. In particular, are our beliefs about ordinary objects undercut by some philosophical or metaphysical considerations?

It is at first not implausible to think that it is enough to undercut our entitlement to our perceptual beliefs about objects to point to the fact that although the perceptual beliefs we happen to form are about objects, this could have been completely different had we had different belief forming mechanisms. However, this conclusion would be too quick. It is true that we could have had belief forming mechanisms that lead to beliefs about simples arranged a certain way, and then we would have been defeasibly entitled to those beliefs. But recognizing this doesn't make the entitlement we at present have go away. If it would, all of our entitlements would go away, since it is always true that we could have been different than we are, in perception, in reasoning, and so on. We could have had thoughts not about objects having properties, but about feature placing,[18] or phenomenal things,[19] or what have you. Recognizing this contingency might lower our confidence that we get it right, but it doesn't systematically undercut our entitlement.[20] To undercut it, we need more than that, and more has been proposed.

Trenton Merricks has argued that our entitlement to our perceptual beliefs in ordinary objects is systematically undercut by philosophical considerations. He holds that what causes us to have beliefs about ordinary objects is not the ordinary objects, even if there were any, but only the simples. Ordinary objects are causally inert, all the causal work in producing our beliefs is done by the simples and their arrangements. Thus even if there are fusions of these simples, and even if there are ordinary objects, they would play no role in the causation of our perceptual beliefs about them.

---

[18] See Turner (2011).   [19] See Carnap (1928).

[20] In Sider (2013), Ted Sider claims that taking nihilism seriously as a metaphysical hypothesis immediately defeats our entitlement to our perceptual beliefs about objects. He says: "To anyone who understands the challenge of nihilism and takes it seriously, any prior perceptual justification in favor of tables vanishes" (p. 260). But without any good reasons in favor of nihilism this strikes me as false. Simply entertaining a possibility as a real option doesn't undercut our entitlement.

Recognizing this, Merricks argues, undercuts our entitlement in our perceptual beliefs about ordinary objects, but preserves our entitlement coming from perception more generally, in particular concerning simples and their arrangement. It undercuts our entitlement to beliefs about ordinary objects, since the objects we form beliefs about are irrelevant in the production of those beliefs.[21] But this argument strikes me as being mistaken.

The argument is supposed to show that our original entitlement to our beliefs about objects is defeated by being undercut. In particular, it is not questioned that we at first have the entitlement, and furthermore the argument is supposed to work even if we assume that there are indeed objects, as Merricks does as well for the sake of the argument. The point simply is that even if there are objects, perception does not entitle us to belief in them once Merricks's considerations are pointed out. But in this imagined scenario it is true that our object beliefs track the objects and object-related facts. My belief that there is a cup in front of me does track the cup in the sense that if the cup didn't exist then I wouldn't have this belief. My belief counterfactually depends on the cup, and thus my cup beliefs track the cup facts. Merricks seems to doubt this, since he imagines that even if the cup wouldn't be there, the simples arranged cup-wise would still be there and would still cause me to have that belief. But this strikes me as the wrong reading of the relevant counterfactual. Assuming again that there in fact is a cup on the table in front of me, the counterfactual situation where the cup isn't there is one where it's completely gone in the sense that my table would be completely free of it: it together with the simples that make it up would be gone. The counterfactual situation relevant to evaluate the counterfactual conditional about my cup not being there isn't one where the simples are still there, but they somehow don't compose a cup, but rather one where both the simples and the cup are gone. But then nothing would cause me to believe in a cup in front of me, and so my cup beliefs track the cup facts: all things being equal, I wouldn't have that belief if there wasn't a cup. Thus nothing in this situation, once properly spelled out, should be taken to undercut our entitlement to our beliefs about ordinary objects.

Merricks focuses too much on the causal production of our beliefs, but this is not the crucial factor for assessing whether our defeasible entitlement goes away by being undercut. How our beliefs are causally produced in us is not the crucial question for epistemology. Whether they track the facts is relevant for this, and the answer is that they do, or so we have reason to think. Considerations about causal overdetermination do not undermine this, and thus do not undercut our entitlement. This is not essentially different than the fact that all the causal work in visual perception is done by the visible part of an object: the front and the surface. Recognizing this doesn't undercut our entitlement to believe that three-dimensional objects have backsides and internal parts, even though they are in this case not visible to the eye and are not causally relevant for the production of the perceptual belief. It isn't relevant

---

[21] See Merricks (2001: 74ff.) and Merricks (2003: 739ff.).

that not all of the object is causing the belief, but only that the beliefs track the facts. And they do, even if Merricks is right about the simples doing all the causal work.[22]

Assuming there are objects, our beliefs track them. The question, of course, remains whether our perceptual beliefs are undermined by our having to accept that they systematically mislead us. And that would be the case if we had reason to think that how perception makes us believe the world is is not how the world in fact is. This would defeat our entitlement to our perceptual beliefs about ordinary objects. However, it is not undercutting defeat, but rebutting defeat. This is the crucial question for us now. Although our defeasible entitlement to our perceptual beliefs is not undercut by considerations like Merricks's, is it nonetheless rebutted on further inquiry? If so, such rebuttal could either come from further empirical considerations, as in the sciences, or from further metaphysical considerations, as in metaphysical arguments directly against ordinary objects. We should look at them in turn.

Our preliminary conclusion so far simply is that we are defeasibly entitled to our perceptual beliefs and that perception does distinguish between there being a chair and there being simples arranged chair-wise. Perception makes this distinction not necessarily in the sense that the world would phenomenologically look different in these two cases,[23] but in the sense that we are defeasibly entitled to hold the belief that there is a chair on the basis of perception. No similar defeasible entitlement from perception supports the belief that there are simples arranged chair-wise. Even if these two scenarios would look the same, they are nonetheless different in that perception supports the perceptual belief in one, but not the other. This asymmetry is crucial for our being entitled empirically to hold that there are ordinary objects. It makes all the difference in the end.[24]

---

[22] See also Korman (2014) for a defense of similar undercutting arguments. Although a good part of Korman's discussion is aimed at a slightly different target, for the present case I take Korman to affirm that the causal chain is deviant enough to undercut our entitlement once the causal facts are recognized. I would like to disagree, for the reasons given above.

[23] It could be argued that the difference already comes in phenomenologically, and that the world would look differently if we would not represent it at all in terms of objects, but only in terms of simples arranged object-wise. This issue is not central for the present argument, though, since that goes through even if we assume that there is a sense in which the world would look the same in those two cases.

[24] There are some issues in the philosophy of perception that are controversial, but orthogonal to our issue here. All that matters for us is that we have some defeasible entitlement to our perceptual beliefs. It does not matter how this entitlement arises more precisely. Furthermore, the issue is one about perceptual belief, not the contents of perceptual experience. Everyone agrees that my perceptual belief that there is a chair in front of me has propositional content about a chair. What is controversial is what the content, if any, of my perceptual experience of the chair is. Perceptual experience is something phenomenal, but does the experience itself also have propositional or conceptual content? Does it have some other kind of content? None of this matters for us here. It is uncontroversial that our perceptual belief has propositional or conceptual content. Skepticism aside, we are entitled to that belief, defeasibly. It also is orthogonal to our issue whether or not the content of perceptual experience, if any, somehow directly justifies our perceptual belief, and if not what precisely the epistemic relation is between perceptual experience and perceptual belief. What matters for us is the in this debate uncontroversial fact that we are, somehow, defeasibly entitled to our perceptual beliefs. For a discussion of some of these further issues, see Byrne (2005) and the many references therein.

I can't stress enough that the above considerations are not simply a reliance on common sense. Common sense is often appealed to in the metaphysical debate about ordinary objects, with some philosophers explicitly drawing on common sense as a source of argument. But if common sense is simply understood as what we commonly hold to be true, then we should not rely on common sense. Many things we hold to be true might well be false, and some things might well be common sense even though we have no good reason whatsoever to hold them. Instead of trying to defend common sense we should critically question it, and retain or reject it where we find it to be appropriate or misguided. The view about our defeasible entitlement to our perceptual beliefs is one that defeasibly supports common sense. Not because a defense of common sense as such is a good thing in itself, but because perception gives us entitlement to our perceptual beliefs. These beliefs, unsurprisingly, are congenial to common sense, since we share a common perceptual setup. Perceptual beliefs defeasibly entitle us to hold these commonsensical beliefs. But nowhere did we assume here that these beliefs are to be accepted, since they are part of common sense. The question now becomes whether this entitlement to our perceptual beliefs is confirmed or disconfirmed upon further investigation. And the best way to look further is to turn to science.

7.3.2 Scientific Evidence

It is frequently said that the picture science draws of the world is one of atoms in the void, or worse, but not one with ordinary objects in it.[25] This is a gross overstatement, though. First, the story at best applies to physics, or better microphysics, but not at all to the rest of science. And, second, it is far from clear even for physics. Although microphysics doesn't mention ordinary objects, naturally, since it deals with the very small, it is not clear if it rules out ordinary objects. Although some have argued that there is a tension between quantum mechanics and ordinary objects, this tension usually relies on attributing ordinary objects features that they might simply not have.[26] For science more broadly the situation is really the opposite. Almost every branch of science supports that there are ordinary objects, or at least composite objects. To see whether microphysics is in conflict with ordinary objects will come down to seeing whether it is in conflict with the rest of science. Let's look at these issues one at a time.

Almost all sciences other than certain parts of physics imply that there are composite objects. And some sciences directly describe composite objects. Chemistry is

---

[25] Eddington's two tables are a paradigm motivation for this. See Eddington (1929). Of course, nowadays it's not so much atoms in the void, but subatomic particles, fields, and other things, but still not ordinary objects.

[26] Precise boundaries, or determinate precise location, and so on, are some of the candidates for such features. This situation is not that different from classical arguments against ordinary objects. For example, simply because objects like bars of metal are believed to contain no space free of metal, it doesn't mean that there are no bars of metal if this is false. It just means that we were wrong about what they are like.

concerned with molecules, and molecules have parts. Chemistry is not just concerned with molecules in case there happen to be any, but it implies that there are molecules. It has established beyond anyone's doubt that pure oxygen and pure hydrogen mixed together and exposed to a certain amount of energy results in H_2O. And H_2O has two hydrogen atoms and an oxygen atom as parts. Now, a water molecule is not an ordinary object, but it is composite, in the sense that it has parts. A water molecule might not be an object in an ordinary sense. But many other sciences directly talk about things that are not just objects and composite, but midsize objects that are in metaphysical standing on a par with paradigm ordinary objects. Although there isn't a science of chairs or tables, scientific evidence supports the view that there are midsize composite objects. It is compatible with science that there are no chairs or tables. Something might only be a table if it was constructed with the appropriate intentions, and it is compatible with science that no one had the appropriate intentions. But that there are no midsize composite objects is not compatible with many well-established sciences. Here is one example:

Materials science is the discipline that investigates the properties of materials, that is various pieces of matter divided into kinds like metals, polymers, ceramics, and so on, with an eye on industrial applications. It is a highly successful and established part of the sciences that draws on physics and chemistry to explain and predict the properties of pieces of matter at a macro level. One of the concerns of materials science is when materials fracture. This is obviously of great practical application and so much work goes into what factors of different materials make them more or less fragile. Similarly for many other topics in materials science. The study of polymers is all about the study of very large molecules that give rise to macro objects, and how various features of such molecules relate to various macro properties of these objects.

Materials science is unquestionably highly successful and it has made many important discoveries. Many nice things we take for granted are a consequence of the results of materials science. Materials science not only theorizes about macro objects in case there happen to be any, it implies that there are macro objects, ordinary objects that can break and have parts. If we are to believe in materials science, we have to believe in ordinary objects, in the sense of composite macro objects.

Materials science is just one example, and in no way the only discipline that implies the existence of ordinary objects. Arguably astronomy does, generously counting celestial objects as ordinary, and biology does, with living things, and so on and so forth. The question, of course, remains how all these different theories go together to make one account of the world. But simply because this question remains we should not think that these sciences are not correct, in the sense that they give a true description of the world. Even though the question of how all the sciences fit together remains open, the evidence overall is in favor of ordinary objects. The results of many highly successful sciences imply that there are ordinary objects, and even though microphysics doesn't imply that there are ordinary objects, it also doesn't imply that there aren't any. Overall, the evidence of the sciences is in favor of ordinary objects.

This is in contrast to the view often presented in discussion of the manifest and the scientific images of the world. Here there is supposed to be a tension between the picture of reality drawn in physics where allegedly all there is is particles in the void and the ordinary conception of the world containing macroscopic objects. But the physical picture should not be understood as stating that all there is is just particles in the void, even if we for the moment imagine a classical atomistic physics. At best it should be seen as stating that all the smallest things are particles, or all the physically most fundamental things are particles.[27] This is perfectly compatible with there being other things that are made up from these particles, either simply the sum or fusion of some particles or something that is constituted by these particles or their sum/fusion. Overall science not only allows for macro objects, it implies their existence.[28]

There is an important objection to this position which maintains that science can't distinguish between there being a metal bar and things being arranged metal bar-wise. It is in a sense the science-level version of the claim that perception can't distinguish between these options discussed above. The objection can come in two forms, analogous to two ways it has been argued scientific theories are underdetermined by evidence.[29] The first form affects the content of the scientific theories, the second affects the epistemic support these theories have. It is reasonable to think that scientific theories are underdetermined in both aspects to some extent or other. The content of a theory is not fully determined, and neither is the theory fully confirmed by the evidence. To ask for that would be asking for too much. After all, for ordinary assertions outside of science we get underdetermination of both kinds as well without much bad effect. Although underdetermination in general is undeniable, the question remains whether scientific theories are underdetermined with respect to there being composite objects as opposed to simples arranged object-wise. And here the answer is that they are not. The contents of our thoughts and utterances are not indeterminate this way. The propositional content of our thoughts distinguishes between these options. Our theories understood as primarily mathematical statements by themselves do not distinguish between them, and they have an interpretation of being about objects as well as of being about simples arranged object-wise. But our theories are not just mathematical statements. They have a content beyond that, related to how we describe the situation that we hope to capture in our theories, and that content is one about objects, not simples arranged object-wise. What precisely the content of our theories is is not clear, but it is clear that it is not indeterminate between simples arranged object-wise and objects. This might not be so for parts of physics, but it is so for almost every other part of science. Most of it is determinedly about composite things

---

[27] We will discuss in detail the possible role of a notion of fundamentality in metaphysics, and whether ontology should be concerned with just the fundamental things, in chapter 13. Whether being physically fundamental gives rise to a kind of metaphysically special status will be one of our questions there.

[28] A more detailed critical discussion of some arguments for science being in conflict with ordinary objects can be found in Thomasson (2007).

[29] For a discussion of scientific underdetermination and many further references, see Norton (2008).

of various kinds. Whether or not we have evidence in favor of our theories which are about composite objects as opposed to an alternative with a different, simples arranged object-wise, content is quite a different question, one we will turn to now.

The thought behind an argument for the underdetermination of our theories by the evidence, but not in their content, goes like this: even though our present theories have a content that implies that there are ordinary objects, these parts of the theories are not properly confirmed by the evidence. We can give alternative theories, ones that do not imply that there are ordinary objects, and which are equally well confirmed by all the available evidence. These alternative theories would reformulate all talk of objects into talk of things being arranged object-wise, using the common van Inwagen translation.[30] And then, the objection goes, we would have to recognize that there is no evidence at all that would distinguish the object version of a theory or whole scientific discipline from the things arranged object-wise version. Thus the evidence does not support one theory over the other, or so the objection goes. But this is mistaken. There is lots of evidence that supports the object theory over the things arranged object-wise theory. The object theory predicts that there is a bar of metal in the lab, the object-wise theory doesn't predict it. That there is such a bar can be confirmed with the observation that there is such a bar of metal in the lab. That is, it can be confirmed empirically by looking and forming the defeasible warranted belief that there is such a bar on the basis of perception. This speaks in favor of the object theory, and similarly in a thousand other cases. Of course, the person who thinks that the two theories are empirically indistinguishable will object that I described the evidence in terms of objects, namely that the content of my observation is about objects. They will insist that the evidence has to be spelled out in neutral terms. For example, Dorr and Rosen, in Rosen and Dorr (2002: 164), object that describing the evidence in terms of objects is simply an illegitimate appeal to common sense. Without a neutral description of the evidence one is just begging the question, or so they object. But this objection is unjustified. When I present the evidence in terms of objects I present it how perception entitles me to believe what the world is like. I am defeasibly entitled to this description, and I am thus defeasibly entitled to take this perceptual observation as a point in favor of the object theory, which predicted it, as opposed to the arranged object-wise theory, which didn't predict it. Describing the evidence in terms of objects is not simply a reliance on common sense, in the sense of merely shared beliefs or beliefs we simply happen to take to be true. It is a reliance on how perception presents the evidence to me, something which I am defeasibly entitled to believe to be the case. There is nothing illegitimate in drawing on our perceptual beliefs, beliefs we are defeasibly entitled to have. Although our theories might be underdetermined by the evidence in general, our evidence nonetheless distinguishes between the object versions and the things arranged object-wise versions. It is here that the position about perceptual evidence for objects discussed above infects the issue of scientific evidence for objects.

---

[30] See van Inwagen (1990) and Rosen and Dorr (2002).

Since perception distinguishes between the two options of either objects or simples arranged object-wise, in the sense that we get evidence for one but not the other, this difference gets furthered in the scientific case in the way outlined here. Thus overall there is lots of empirical evidence for the existence of ordinary objects. Such evidence comes from both perception and science, and in both cases the evidence is in favor of the existence of ordinary objects, and thus objects in general.

7.3.3 Metaphysical Arguments as Further Evidence

Although there is empirical evidence in favor of ordinary objects, this does not settle the question of whether or not we should believe in ordinary objects. Other evidence might override the empirical evidence we have in favor of ordinary objects, and a number of arguments from the philosophical literature can be seen to try to do this. Maybe the empirical evidence is weak, and the arguments given in metaphysics against them are to be favored in the end. There are a number of arguments in the metaphysical literature against ordinary objects. They include arguments related to vagueness like the problem of the many, the causal overdetermination arguments, and arguments derived from the relationship between objects and their parts, as well as others.[31] The question remains whether in light of these arguments as well as in light of the empirical evidence in favor of objects we should, all things considered, accept or reject ordinary objects.

In general the metaphysical arguments are not brought up to override empirical evidence. It is widely held in the metaphysical debate about ordinary objects that there is no empirical evidence that favors them over simples arranged object-wise, since in the end the question of the existence of ordinary objects is not an empirical, but a distinctly metaphysical one. The metaphysical arguments in that presumed scenario don't have to trump empirical evidence, they are considered the only source of evidence for and against objects. In fact, the general setup is that common sense, somehow, is attached to objects, but rational arguments make them problematic, and thus overall we have dogma vs. reason, and reason wins that one. But this is not so. Belief in ordinary objects is not simply dogmatic common sense, but a belief we are entitled to on the basis of perception as well as further empirical investigation. The philosophical arguments against objects thus are not the only reasons in town to be considered, but they need to compete with the empirical evidence. How then do these arguments stack up overall?

Let us take one paradigmatic case: the causal overdetermination argument.[32] It goes something like this: if there are objects like baseballs then they must have causal powers, thus if there is a baseball at all then it must have been what caused the window to break. But the window's breaking (or the simples arranged window-wise dispersing)

---

[31] See van Inwagen (1990) and Merricks (2001) for discussions.
[32] See Merricks (2001) for someone who uses it to argue against objects.

is already caused by the atoms, or simples, that make up the baseball. The baseball in addition does not do any further causal work. So there is no such baseball.

To be sure, this is an intriguing argument that raises a number of important issues about causation. But to accept the conclusion of the argument on the basis of it would be to give it more weight than all our empirical evidence in favor of ordinary objects. And here the balance strikes me to go clearly in favor of the empirical evidence. The overdetermination argument raises some worries about ordinary objects based on intuitively plausible claims about causation. Whether these claims in the end turn out to be true remains to be seen, and how much weight their intuitive plausibility should be given remains unclear. But these concerns are outweighed by the fact that perception entitles us to believe in objects and that science confirms this belief. Since most sciences entail the existence of composite objects, to accept the causal overdetermination argument is to reject most sciences. This would be a complete overreaction to a puzzle about causation. Similar remarks apply to arguments from vagueness and the like. Overall, the weight of reason sides with an acceptance of composite and ordinary objects. The purely metaphysical arguments, like the overdetermination argument, have force if one holds that there is no empirical evidence for ordinary objects. But once the empirical evidence is acknowledged they pale by comparison. Not because metaphysics can't correct what seems to be the case empirically, but because of the comparative strength of these particular arguments on the one hand and the empirical evidence on the other.

7.4 Metaphysical Questions About Objects

The question of the existence of ordinary objects, and thus objects in general, is answered in the affirmative, given the available evidence. What then is left for a metaphysics of ordinary objects to do? Some think that all of the work is left open by this, since the issue relevant for metaphysics never was whether such objects exist or whether there are such objects, but rather whether such objects are fundamental, or a part of ultimate reality, or the like. We will consider this more general approach to metaphysics, which affects all questions in ontology, in detail in chapter 13. But even for those, like myself, who take the primary ontological question about objects to be the question of whether or not there are any, many more questions remain for metaphysics. First, there are questions about the relationship between objects and properties, as outlined in section 7.1. But, as mentioned there, we are not in a position to answer that question until we discuss properties in chapter 11. Second, there is the question of how objects relate to other objects, in particular their parts, see section 7.1. Here, in particular, the special composition question of van Inwagen (1990) remains open, i.e. the question of when some things compose something. We can by now rule out nihilism: the view that composition never occurs and thus everything is simple. Ordinary objects have proper parts, and so at least some things have proper parts, and thus at least some things compose something, or so we have good empirical reasons to believe. But other than that, most further questions are still left open. Do any things compose something, no matter how scattered? This universalist option and its denial are compatible with everything we have seen so far. Furthermore, the question of how the parts of an object relate to the object is also left open. Is the object simply the sum or fusion of the parts? Is it merely constituted by the sum of the parts, but different from that sum? These and other related questions are so far simply not addressed.

Although there is empirical evidence for the existence of ordinary objects, it is not so clear if there is empirical evidence against the existence of arbitrary sums of objects. Is there empirical evidence that there is no sum of my laptop and my coffee mug? This relates to the more controversial question of what the contents of our perceptual beliefs can be. Not every concept can occur in a perceptual belief. The crucial tiebreaker above in favor of objects was that our perceptual beliefs are about objects and we are defeasibly entitled to those beliefs. If our perceptual beliefs can also have the content that there are only two things on the table, then we would also be defeasibly entitled to such beliefs, and thus this would defeasibly speak against universal composition. But whether our perceptual beliefs can have this content is not so clear. Such beliefs might be inferred from perceptual beliefs in a way that might lose the defeasible entitlement to them. What we have seen here does not answer these questions.

If perceptual beliefs have contents that there are exactly two things on the table, then this raises a question about the parts of these things: how can there be only two things if these things have many parts? But here there might be the possibility of dealing with such cases as cases of a contextual restriction of what an object is. Maybe there are exactly two maximal objects on the table, objects that are not themselves proper parts of other objects. On a more permissive conception of an object it might well turn out that there are more than two. Which notion of an object figures in our perceptual beliefs is itself an issue for debate, and it might well be that the notion of an object is polysemous along various dimensions.[33] I take these to be substantial issues in the philosophy of perception that I can't hope to properly address here. The question of whether there is empirical evidence for or against universal composition is thus one we have to leave open. And similarly for a number of other options, other than nihilism, about how one might want to answer the special composition question.

Similar considerations apply to questions about constitution and the relationship between an object and its matter. It is unclear what empirical evidence we have for any view in this debate, and how strong it is. It is thus unclear what questions remain open and how strong the metaphysical arguments need to be for one view or another. I do feel confident, however, that there will be several questions left open that we won't be able to settle empirically, although it is unclear to me at present which ones these are. Such questions are then properly subject to further theorizing, and this rightly proceeds just the way metaphysical questions about objects have been traditionally

---

[33] See the discussion in Sidelle (2002).

dealt with: try to come up with a general theory, compare it to others in terms of overall attractiveness, support it with thought experiments, and so on. In an area where empirical considerations remain silent this is the right way to go. In an area where we have good evidence from perception and the sciences, as in the case of the existence of ordinary objects, such considerations will have little force. What questions are properly addressed in metaphysics is thus not unrelated to what questions are addressed in other parts of inquiry, ones that we have in general good reason to listen to. We will return to this issue in more generality as well as more detail in chapter 12. For the moment we can simply conclude that we have good reasons to think that there are ordinary objects, and that these reasons are largely empirical. The primary ontological question about objects has an affirmative answer, the secondary ontological question about them is largely left open. The answers to the ontological questions about ordinary objects are thus in stark contrast with the answers to those questions about natural numbers that we defended in chapters 5 and 6. And a crucial part of why these two are so different is that internalism is true for talk about natural numbers, while externalism is true for talk about ordinary objects. But to have a fuller picture of metaphysics, the role of ontology in it, and how to think of a world that contains at least objects, we will need to look at two further, but central, cases: properties and propositions. We should turn to them now.

8

Talk About Properties
and Propositions

We have seen that internalism is true for talk about natural numbers, that the primary ontological question about them has a negative answer, and that this is crucial for the solution to a number of central questions in the philosophy of arithmetic. In contrast, externalism is true for talk about objects, the primary ontological question about objects has a positive answer, on largely empirical grounds, but this leaves many of the metaphysical questions about objects open. It is now time to look at our other two main ontological questions: the ones about properties and propositions. As we will see, they are similar to each other, but quite different in various ways from the ontological questions we have looked at so far. However, both of them are very important for metaphysics, and the consequences we will draw from our discussion of them will be the most significant and wide-reaching ones in this book. In the present chapter we will look at what we do when we talk about properties and propositions. In the following chapter we will deal with an important refinement, and in the two chapters after that, chapters 10 and 11, we will be in a position to draw a number of conclusions for large-scale metaphysical questions related to properties and propositions.

8.1 That-Clauses and Property Nominalizations

The term "proposition" in the philosophical literature is a technical term. It is tied to the use of that-clauses, in particular as they occur in the ascription of content. When someone has a belief, say, then that belief has a certain content, which is what the person believes. The content of a belief is ascribed to the person with a belief ascription, a sentence like

(290)   Sue believes that the restaurant is closed.

The content of Sue's belief, i.e. that the restaurant is closed, is what it represents the world as being like. A proposition is simply whatever that content is, whatever the that-clause "that the restaurant is closed" stands for or refers to. Propositions in this technical sense are thus whatever that-clauses stand for or refer to, if anything, of course. One could also think of propositions differently, it's a technical term after all. For example, we could take propositions to be whatever things play a certain theoretical role or have a certain property. But this is not how the term is generally understood, and not how we will understand it for now. If there are such things as propositions, then they are whatever that-clauses refer to or denote. We will see for what this matters shortly.

It might seem prima facie counterintuitive to think that that-clauses refer to anything. Reference is also a technical term in philosophy, but it expresses a familiar relation that has some clear and paradigmatic instances. It is paradigmatically the relationship that holds between a proper name and the object it names. We saw above, in section 7.2, that although the standard view of names takes them to be referential, there is a standard alternative to it that takes them to be descriptions, and thus only referential in the broader sense that includes denoting expressions. And there is a more radical alternative to even that which takes names not to be referring even in the broader sense, but only takes speakers to refer with uses of names. As we saw in section 7.2, we are entitled to take names to be at least broadly referential, certainly in many of their uses where they are in subject position. Reference is thus paradigmatically the relationship that holds between (certain uses of) proper names and objects.

But that-clauses prima facie are nothing like names. On the face of it they do not stand for an object, but specify the content of a belief. They do not refer to the content, but say or specify what that content is. After all, that-clauses are clauses, and clauses aren't anything like names. But there are a number of good arguments that have been very influential in philosophy that suggest that that-clauses refer to objects or entities, and the entities that that-clauses refer to are thus propositions. Most prominent among these arguments are quantifier inferences:

(291)   Sue believes that the restaurant is closed. So Sue believes something.

(292)   Sue believes everything Bill said. Bill said that the restaurant is closed. So Sue believes that the restaurant is closed.

It would seem that the validity of these inferences requires the referentiality of that-clauses. If there is something Sue believes, then that thing must be the content of her belief, and thus there must be such a thing as the content of her belief. We can however dismiss this reasoning as invalid in light of our discussion of quantification in chapter 3. Such quantifier inferences are perfectly valid on the inferential reading of the quantifiers. And the validity of this inference is independent of the referentiality of that-clauses.[1] Whether such quantifier inferences are also valid in the domain conditions reading of the quantifier is a further question, one that is closely tied to the question of whether that-clauses are referential. And this is the question we thus have to make progress on in this chapter.

---

[1] For an account in a similar spirit of these inferences, see Schiffer (1987b).

That-clauses do not occur exclusively in the ascription of content, as in belief ascriptions. They also occur in subject position of what appears to be a regular subject–predicate sentence:

(293)   That the restaurant is closed is surprising.

It would be premature to connect that-clauses too closely to ascriptions of content. We will look at them in more detail shortly.

Although it is prima facie implausible to think of that-clauses as referential, it is prima facie plausible to think of some property nominalizations as referential. Property nominalizations, as we use the term here, are expressions like "wisdom," "being wise," and "the property of being wise." At least "wisdom" can naturally be taken to be a referential term. It looks a lot like a name: it is a simple word, not a complex phrase, and it naturally occurs in subject position of a simple subject–predicate sentence, as in

(294)   Wisdom is desirable.

Given this prima facie difference it might seem strange to discuss that-clauses and property nominalizations in the same chapter, but number words in a separate one. It will turn out, however, that the issues that arise with that-clauses and property nominalizations are quite similar, and completely different from those that arise with trying to understand number words or names for objects. For that reason, which hopefully will become clear soon, talk about properties and propositions is treated together in this chapter, while talk about natural numbers got its own chapter, and talk about objects merely got part of a chapter.

8.2 The Referential Picture of Language

The question of whether or not that-clauses or property nominalizations are referential is different than the question of whether or not certain names refer. In the latter case there is a real question about whether or not names like "Sherlock Holmes" or "Santa" refer. Are these names truly empty names, names with no referent at all, or do they succeed in referring to something, maybe some abstract object, or maybe some special kind of object? This question is a question about whether or not a referring expression succeeds in referring. That is to say, it is the question of whether or not an expression which has the function to refer succeeds in carrying out that function. Names, in their most straightforward uses, paradigmatically aim to refer to an object. Some names succeed in carrying out that function, some fail. Those that fail are empty names, names with no referent. When we wonder whether or not fictional names refer, the question is generally not whether or not these names are referential, that is have the function to refer, but whether they succeed in carrying out that function, and thus succeed in referring.

The debate about that-clauses is different. Here the main issue is whether or not that-clauses are referential in the sense of whether or not they even aim to refer to something. The debate is thus about whether or not that-clauses have a similar semantic function as proper names, not whether or not they succeed in carrying out whatever function they have. What then is the function of that-clauses and property nominalizations? Are they referential expressions? Do they aim to refer?

This leads to the more general questions of which phrases are referential, and how to find that out. Although we already discussed whether number words are referential in chapter 5, this question arises more broadly and in a different way in this chapter. In the case of number words there were some uses of number words that appear, on the face of it, to be referential, and some that appear not to be referential. The question there was how they go together, and the non-referential uses won out. But in the case of that-clauses and nominalizations it is simply not clear from the start whether the relevant phrases should be seen as referential. We saw in chapter 2 above that in a certain limited range of cases they are not referential, but the considerations given there only applied to the limited range of cases. We should thus look at this issue more generally now.

We need to discuss which words or phrases are referential. Being referential, again, is understood not as succeeding in referring to something, but as aiming to refer to something, having reference as one's semantic function. Reference, in turn, is the relationship that paradigmatically holds between a name and the object named. Although "reference" is a technical term, the relation it expresses is perfectly familiar, at least in its paradigm cases. The question we need to make progress on thus is

(295)   Which phrases are referential?

Our goal is not to answer this question in general in this chapter, but to find out how this question could be settled in principle, and then try to settle it for that-clauses and property nominalizations. To make progress on this we will consider an improving series of answers to (295).

Some first attempts to answer (295) reflect a commitment in different degrees to what we could call *the referential picture* of language. This picture takes being a referential expression to be a paradigm case of either the semantic function of all expressions, or at least of all expressions of a certain kind. In this spirit is a first, and most radical, attempt of an answer to (295):

(A1)   Every expression in natural languages refers. They simply differ in what they refer to.

This answer reflects a picture of language which is often called the "Fido"–Fido theory of meaning.[2] This theory has some formidable objections against it. It takes every expression to be equal in its semantic function. Every expression picks out an entity.

---

[2] See, for example, Schiffer (1987a).

Different expressions pick out different entities, and maybe different categories of expressions pick out different categories of entities, but the general semantic function of all expressions is the same: to pick out and refer to an entity. But if every expression has the semantic function paradigmatically had by a proper name, how does stringing these expressions together give rise to a truth evaluable content? In general one does not get truth conditions when one strings a sequence of names together. And this is so even if the names name a variety of different things, i.e. if we, for example, combine some names for objects and some names for properties into a sequence of names. And besides this difficulty, the view that every expression refers is very implausible. There seems to be a crucial difference in the semantic function of names and of modifiers. The word "very" doesn't pick out an object, but modifies an adjective, and those are different things. We can thus put the radical answer aside and maintain that not all expressions are referring expressions.

A next attempt to answer (295) is less radical and seems to avoid the problems that the most radical answer had. It first acknowledges a fundamental difference between singular terms and predicates. Singular terms pick out an object, predicates don't pick out an object but complement singular terms semantically in that they make a claim about the object picked out by the singular term. Together they form an indicative sentence with truth conditions, since the "unsaturated" part of the predicate gets saturated by the object the singular term picks out. Predicates and modifiers are not referential, but singular terms are. This division between predicates and singular terms marks a fundamental difference in semantic function between different expressions in our natural language, but they also contain two claims of uniformity: predicates and singular terms form two uniform classes of expressions when it comes to semantic function. In particular we get a second answer to our question (295):

(A2) Every singular term in natural languages refers. They simply differ in what they refer to.

This answer is also incorrect. Although it is unclear what precisely a singular term is, among even the paradigm examples of singular terms are non-referential expressions. An uncontroversial example are quantifier phrases, which can be singular and in subject position of a sentence: "some man" and "every woman" are examples of clear cases. A further, slightly more controversial case of a quantifier that is not referential are definite descriptions. Singular terms are paradigmatically proper names, and other expressions that are syntactically like proper names, in particular in that they can occur in subject position of simple subject–predicate sentences. Such sentences nicely accommodate what was missing in the most extreme referential picture. The predicate can be taken to be non-referential, and instead having the function of saying something about what the subject term stands for. The subject term on the present referential picture has the function of picking out an entity, the thing the predicate says something about. But quantifiers refute (A2). They are paradigmatic singular terms, appear in simple subject–predicate sentences, but do not refer to an object.

This naturally leads to another hypothesis about singular terms which is relevant to our question (295). Since the best and at present least controversial counterexamples to (A2) are quantifiers this gives rise to the conjecture that all singular terms are either referential or quantificational, and thus leads to a method for settling whether a singular term is referential: if it is not quantificational, then it is referential.[3] We can label this *the bifurcation hypothesis*:

(BH)   All singular terms are either referential or quantificational.

This hypothesis is widely accepted, at least implicitly, although hardly ever made explicit, and it is very influential in judging which terms should be taken as referential. It explicitly rules out a further category of non-referential, non-quantificational singular terms.[4] If (BH) is false, then it might well be that there is a range of further singular terms with a variety of different semantic functions, but believers in (BH) have none of this. So far the common referentialist picture.

The referential picture, captured in the bifurcation hypothesis (BH), favors seeing that-clauses and nominalizations as referring expressions. After all, they seem to appear in subject position of simple subject–predicate sentences, and they don't seem to be quantificational. Although both of these appearances can be deceiving, on a first pass the referential picture is in favor of seeing these expressions as referential. But on the other hand, the issue whether these expressions are referential is one about the referential picture itself. The referential picture takes reference to be a paradigm case of the semantic function of singular terms. Leaving quantifiers aside, that is what singular terms do. If we grant for the moment that that-clauses are not quantifiers, but they nonetheless appear in true subject–predicate sentences in subject position, then the assessment of the referentiality of that-clauses is closely tied to the assessment of the referential picture itself. It competes with *the non-referential picture* of language. This non-referential picture takes reference to be a more unusual and more atypical semantic function even of syntactically singular terms. Singular terms come in possibly a quite diverse variety of kinds: some are referential, some are quantificational, and some are other things, possibly many other things. Of course, the believer in the non-referential picture will hold that these other kinds of singular terms have not been properly classified and studied as such to this day. When we thus ask: "but what are these other kinds, and what do they do if not refer or quantify?" the proper answer at this stage will have to be: "we don't know yet." Sure, we could give it some other name: they "specify," but don't refer, or the like, but that would just be a placeholder for what hasn't properly been spelled out yet. The question for us here will be how we can decide whether an expression is referential or not, and how this question turns out for our two cases: that-clauses and property nominalizations. We

---

[3] This hypothesis is endorsed by Stephen Neale in Neale (1993). It is critically discussed in Hofweber and Pelletier (2005).

[4] In Hofweber and Pelletier (2005) such terms were called "encuneral."

will need to address the general question first, how to decide, and then see how it turns out for these cases.

This question of reference is not trivial. It is not sufficient, for example, to note that the relevant terms are about something. "Wisdom" is about wisdom, uncontroversially, but this does not settle the question of whether "wisdom" is referential. We should remind ourselves of the two senses of aboutness: a referential and a topical sense, discussed in section 4.1. It is in the referential sense that a name is about the object named. But aboutness also has a topical sense. In this latter sense we can talk about aliens, whether or not there are any aliens. Aliens here are merely the topic of our conversation, what we are talking about. In this sense it is uncontroversial that sentences containing "wisdom" are about wisdom, but this doesn't answer the question of whether or not "wisdom" is referential. We will have to look somewhere else to settle the question.

8.3 Semantic Facts and Semantic Values

Whether or not a word or phrase is referential is a semantic fact about the phrase. It is on a par with the semantic fact of what the truth conditions of a certain sentence are, except it is a fact at a sub-sentential level. And since being a referential expression is a semantic fact about expressions that are referential, it would seem natural to look at semantics to find out which expressions are referential. What, then, does semantics say about it?

Semantics can be understood broadly, to include the study of all semantic facts. And understood this way it is, of course, in the domain of semantics to find out whether or not an expression is referential. However, semantics as it is practiced these days in linguistics is generally not that broad. Although semantics might lay claim to that larger domain of facts to investigate, in fact it focuses on a narrower set of questions. Mainstream linguistic semantics generally is divided into two parts: compositional truth-conditional semantics and lexical semantics. As we will see, neither one of them is giving the question of reference much thought, and it seems to be quite irrelevant to their concerns. To see this, let us look at these two parts of semantics in turn.

Compositional semantics is generally carried out within the larger project of giving a truth-conditional semantics for natural languages. Such a semantic theory assigns semantic values to various expressions in order to generate the truth conditions of the sentences in which they can occur correctly. Compositional semantics thus focuses on how the composition of expressions of various kinds leads to the truth conditions of the resulting sentences. The central task of a compositional semantics is to do just that.

(296)   **Central Task**: assign semantic values, and ways to compose or combine them, such that the truth conditions of every sentence in the language are generated correctly.

What, though, are the constraints on assigning semantic values in such compositional semantics? And what can we read off from the semantic values that are assigned in our best compositional semantic theories about whether or not the relevant expressions are referential?

To answer this question we should distinguish between *correctness conditions* and *excellence conditions* of a compositional semantics. Correctness conditions are the conditions that a semantic theory must meet in order to be correct. They are a minimal requirement for the theory to be acceptable. This is not a matter of degree. The theory is either correct or it isn't. Among the obvious candidates for correctness conditions is to generate the right truth conditions for the sentences in the language. Excellence conditions are conditions that can distinguish different theories that meet the correctness conditions. Correct theories can be better or worse overall. Excellence is a matter of degree: a theory can be more or less excellent. Among the excellence conditions are such things as simplicity, the use of mathematical tools that are familiar and accessible, and so on. As for the correctness conditions it is clear that the compositional semantic theory has to generate all the right truth conditions of the sentences of the language. If it predicts the wrong truth conditions, then the theory is clearly not correct. The question is if there are also other constraints on a theory for it to be correct. An answer to this question that is widely reflected in the practice of semantic theorizing is "no." This can be seen as the Minimalist Approach to the correctness conditions for compositional semantics:

(297)   **Minimalism:** The correctness conditions are exhausted by the Central Task.

Minimalism, in particular, allows that any semantic value can be used, as long as it leads to the right truth conditions. And it turns out that the right truth conditions can be gotten with a variety of different semantic values. I will illustrate this in just a second.

Semantic practice favors Minimalism, or at least a version of it. We will focus on the simple version of Minimalism stated above since it is the one central for our main concern here. A slightly different version of Minimalism is to add an explanatory requirement to the correctness conditions. A correct semantic theory must not only generate the right truth conditions, but also explain why various sentences can have different readings and how they get them. On such a conception of the correctness conditions of a compositional semantics, a correct semantic theory must explain why and how different readings of a sentence arise. But this further requirement won't affect our main question here about the relationship between semantic values and being referential, and we will thus just take Minimalism in its simple formulation. In particular, there are no constraints on the choice of semantic values as long as they fit the minimalist requirement, and there are no constraints on semantic values that are tied to the relevant expression being referential or non-referential. To illustrate this, consider the following:

First, every expression will get assigned a semantic value, even ones that are uncontroversially not referential. In type-theoretic based approaches expressions like "very" get assigned a certain function as a semantic value just like predicates, and even proper names. These functions are of different types for different categories of expressions, but all expressions get assigned some semantic value or other. Thus simply having a semantic value does not mean the expression is referential.

Second, expressions that are uncontroversially referential can get semantic values that are not their referents. This is made vivid by Montague's treatment of proper names in his Montague (1974). Proper names there are "type-raised" to have the same type as quantifiers. This allows for a simpler treatment of complex noun phrases like "Sue and some man." Here both "Sue" as well as "some man" get the same kind of semantic value, basically a set of properties. And the complex noun phrase gets a semantic value of that kind as well, one determined by the semantic values of the two conjuncts. This has great semantic advantages, but it has the consequence that a proper name gets something other than its referent as its semantic value. "Sue" does not refer to a set of properties, but to a person. For this to work and to carry out the central task of a semantic theory the set of properties has to be associated with the referent in a close enough way so that the truth conditions come out correctly at the end. But this is simply done by taking the set of all of Sue's properties, and then using it appropriately to generate the truth conditions. The correct truth conditions can be generated by assigning "Sue" its referent, or a different semantic value closely related to its referent. Semantic practice does not reject a semantic theory for doing the latter. To the contrary, Montague's assignment of sets of properties as semantic values of proper names is seen as a great idea. That it doesn't assign the referent as the semantic value of a referring expression is no legitimate criticism in semantic practice.

Third, in assigning semantic values to expressions as that matters for the correctness of such assignments is how these assignments further the central task. In particular, it does not matter whether we assign things that exist as semantic values or things that don't exist, as long as the truth conditions are generated correctly. If we could solve a tricky problem in semantics by assigning unicorns as semantic values to certain expressions, then this would be real progress. It would be progress even though we all agree that there are no unicorns. As with intentional transitive verbs in general, one can admire Santa, without Santa existing, and one can assign Santa to an expression without Santa existing. For something to be a suitable semantic value it only matters that it can do the job it needs to do: generate the truth conditions correctly. Here things that exist are no better than things that are merely pretended to be there, or assumed for the moment to be there. This is a striking contrast with the use of theoretical entities in other parts of inquiry. If there are no neutrinos, then this is ground for criticism in actual physical practice of any theory that uses neutrinos. But in actual semantic practice this is not so. If it would work to assign unicorns as semantic values to generate the truth conditions of sentences that previously resisted such semantic treatment then this would be progress. Of course, if certain things don't exist since their characterizing properties are incoherent, then this would make these things unsuitable to be assigned as semantic values. They won't be able to help with the central task, but not because of their non-existence but because of their incoherence. It is arguable whether the excellence conditions of a semantic theory put more restrictions on what we can assign as semantic values, but the correctness conditions do not.

Compositional semantics thus has surprisingly little to do with the question of whether or not certain expressions are referential, at least not when it comes to the use of semantic values. Referential and non-referential expressions get semantic values, even semantic values of the same kind. Referential expressions can get semantic values other than their referent, and which semantic values can legitimately be assigned is independent of the existence of what is assigned. If Minimalism is correct, as semantic practice suggests it is, then there are no constraints on semantic values other than what is required to carry out the central task. In particular, no conclusions about referentiality can be drawn from the semantic values assigned. Minimalism thus naturally supports the independence thesis:

(298)   **The Independence Thesis:** the appropriateness of assigning semantic values is independent of the existence of what is assigned.

In particular, no conclusions can be drawn about the existence of certain things from their use in correct semantic theories. Such things would be just as good if they didn't exist, as long as their supposed features are coherent. For their role in semantics it is just as good to assume or pretend that they are there as their being there.

The above considerations are more general than we need for our main concern, but they have significant implications for our main question: whether or not that-clauses and property nominalizations are referential expressions. There is a common argument when it comes to the semantics of attitude ascriptions, like belief ascriptions, which concludes that that-clauses must be referential.[5] The argument is this: the only way to accommodate belief ascriptions in a compositional semantics is to treat them as relational. "A believes that p" has to be understood as being of the form Bel(A, that p). And thus we need to assign semantic values to the that-clause, which is one of the arguments of the belief relation. These semantic values are propositions, and that-clauses refer to them. Thus to see what propositions are like we need to see how we can treat belief ascriptions in a compositional semantics. Or so goes the argument.[6] We can now see that this argument is mistaken. It might well be true that the only

---

[5] Stephen Schiffer argued in Schiffer (1987b) that if natural languages have a compositional semantics then that-clauses refer. However, there are no suitable referents for them, so natural languages don't have a compositional semantics. The first part of this argument is widely accepted, while the second part is not. I hold that the first part of this argument is mistaken.

[6] For versions of this, see in particular Schiffer (1987b) and Schiffer (1992).

way to treat belief ascriptions in a compositional semantic theory is to treat them relationally, and to assign semantic values to both arguments of the relation, including the that-clause. But this does not mean that the that-clause is referential, nor that the semantic value assigned to the that-clause is a proposition. The that-clause might well be an argument, syntactically, of the belief relation, but this does not mean that it is a referential expression. And the semantic value assigned to the that-clause doesn't have to be its referent even if it is referential. Similarly, even if belief ascriptions need to be treated in a compositional semantics by assigning separate semantic values to the that-clause, to "believes," and to the phrase standing for the believer, it does not follow that belief is a relation between a person and a proposition, in the sense that it is a relation between the believer and the entity which is what is believed. "believes," the verb, might have two argument places, but this is to be understood syntactically, as what other phrases or complements it needs to form a complete sentence. It should not be inferred from this alone that believing is relational in the controversial sense: that it relates two entities: a person to a content. Of course, it might be that the that-clause is referential and the semantic value is its referent. But the fact that the that-clause is, or even has to be, treated that way in a semantic theory doesn't settle this question. The question of reference is left open by this. How then can we settle it?

If compositional semantics doesn't settle this question, maybe lexical semantics does? This is not very plausible for that-clauses, since they are complex, but it might be more plausible for property nominalizations. Being a referential expression is a semantic fact of such phrases, but the question remains if it is one that is captured in contemporary semantics. And here the answer seems to be "no." Lexical semantics is concerned with a number of features of words, how words lead to new words, and so on. But it doesn't seem to address the question of whether the resulting words are referential. And whatever aspects of the lexicon captured in lexical semantics feed into compositional semantics, being a referential expression doesn't seem to be one of them. And even if lexical semantics were to try to find out whether an expression is referential, the question would remain how it settles this issue. This is the question we hope to address. We can't just look at semantic theory to read off the answer to our question of reference. But that, of course, doesn't mean that semantic considerations aren't relevant for settling this question. Although we can't just read off the answer from semantic theory, in the following we will see some semantic facts that point to a particular answer.[7]

---

[7] The considerations in this section are to be distinguished from the argument that semantics can't inform metaphysics, since it only deals with how we represent the world, not how the world really is. Semantics, in other words, by itself doesn't settle what representations are true, and only true representations tell us what the world is like. But on the above argument we can add that various sentences are true, and all of our considerations would still go through. Even for true representations we have freedom in what semantic values we assign in order to carry out the central task. The above is thus different than, although compatible with, the views of Emmon Bach in Bach (1986) and Jeff Pelletier in Pelletier (2011).

8.4 Evidence Against Referentiality

Despite the fact that much of linguistic semantics does not speak to the question of whether or not a phrase is referential, there are some important considerations that directly address this question. In this section we will look at them for our present cases: that-clauses and property nominalizations.

One way to test whether an expression is referential is to see if it has the features that one would expect of a referential expression. First and foremost among them is its substitution behavior. A referential expression has as its semantic function to pick out an entity, be it a person, property, proposition, or whatever it might aim to pick out. On the most popular view of referential expressions, which we will be working with for the moment, a referential expression contributes the object it refers to to the content and truth conditions of the sentence in which it occurs. It contributes the object referred to, as opposed to a "mode of presentation" of the object, some descriptive content, how the object is thought of, or the like. This is reasonably uncontroversial for "direct" contexts, but controversial in special contexts, in particular inside a that-clause. For all occurrences of referring expressions in direct contexts we should thus predict that if two referring expressions refer to the same object, then replacing one for the other in a direct context should not matter for the truth conditions. Thus, if "Fred" and "Mr. Jones" refer to the same person, then "Fred is tall" and "Mr. Jones is tall" should have the same truth conditions. In general, if "a" and "b" are both referring expressions and refer to the same object, then "F(a)" should be true just in case "F(b)" is.[8] This substitution behavior is not restricted to referential expressions. If we take descriptions to be quantifiers and thus not to be referential, we can still expect them to interact with referential expressions in substitutions as long as what the description describes is what the referential expression refers to. So, if Pat is the wisest principal, then it should follow that, if Pat is tall the wisest principal is tall. If we say that a description denotes the object it describes, then substitution should go through as long as the denoted object is what is referred to with the substituted expression.

So, if an expression is referential it should have the above substitution feature. The question thus is whether that-clauses and property nominalizations have that feature.[9]

---

[8] The controversial cases of this are names in belief contexts and as (parts of) arguments of intentional transitive verbs. Thus one might think that even if Fred is Mr. Jones one shouldn't expect that if Sue believes that Fred is tall that this implies that she believes that Mr. Jones is tall. After all, she might not realize they are the same. This issue, however, is orthogonal to our concerns here. See Salmon (1986) for more on this debate. We will in the following restrict ourselves to cases that are direct context for this debate, and drop the qualification. An important challenge to the idea that failure of substitution is to be explained by the direct vs. indirect context distinction is presented by Jennifer Saul in Saul (1997) and Saul (2007). Saul does not reject that co-referring expressions can be substituted for each other. Instead she gives an explanation of why we sometimes reject such substitutions. This explanation, however, won't apply to the cases we will consider below.

[9] The substitution argument was put forward by Kent Bach in Bach (1997) and Friederike Moltmann in Moltmann (2003b), among several others.

There is a well-known class of cases where substitution seems to fail. The first observation to make is that if that-clauses are referential, then we should expect the clause *that p* to pick out the same proposition as the phrase *the proposition that p*. Whatever proposition "that p" refers to, it must be the proposition that p, and thus the phrases "that p" and "the proposition that p" must pick out the same proposition. It is left open by this whether "the proposition that p" is itself a referring expression or a denoting expression, for example a description. This is a question we will return to below, but for now it is not relevant. Even if "that p" is a referring expression and "the proposition that p" is a description, the proposition that one refers to will have to be just the proposition that the other denotes. Substitution of one for the other should thus be expected to hold. But just this doesn't seem to be the case throughout. A case where substitution seems to hold is belief ascriptions:

(299) a. Fred believes that the restaurant is closed.
        b. Fred believes the proposition that the restaurant is closed.

It is arguable that these two are equivalent, or at least that there is a reading of (299b) which is equivalent to (299a). But on the most natural reading they seem not to be equivalent. The most natural reading of (299b) has Fred believing a proposition like he believes Sue, that is he believes what Sue is telling him. In any case, no similar attempt to save substitution is at all plausible for cases like the following:

(300) a. Sue fears that the restaurant is closed.
        b. Sue fears the proposition that the restaurant is closed.

(301) a. Sue expects that the restaurant is closed.
        b. Sue expects the proposition that the restaurant is closed.

(302) a. Sue mentioned that the restaurant is closed.
        b. Sue mentioned the proposition that the restaurant is closed.

These pairs of sentences all differ in truth conditions, and in these cases the second member of the pair does not have a reading that is equivalent to the first. The question is how a believer in the referentiality of that-clauses can account for these substitution failures.

Although the problem of substitution failure is most widely discussed in the case of attitude ascriptions like the above ones, the phenomenon also arises when that-clauses are in subject position of the sentence. Consider

(303) a. That the restaurant had to close is disgusting.
        b. The proposition that the restaurant had to close is disgusting.

These examples are prima facie problems for the referential picture. If "that p" is referential, and "the proposition that p" is at least a denoting expression, then substitution should be expected to be valid in all the cases above. But on the other hand, if at least "that p" is not referential, then we do not have to expect substitution to be valid. If a that-clause specifies the content of an attitude, rather than refers to it, we don't have to expect substitution to go through. Substitution is only to be expected for terms that pick out the same object. If the terms don't pick out the same object since they don't pick out any objects, then we don't have to expect substitution to hold. What then can a defender of the referential picture of that-clauses say in response?

One possibility for those who believe that that-clauses are referential is to reject that substitution has to hold for referential expressions. This line was suggested by Stephen Schiffer in his Schiffer (2003). There he argues that substitution arguments are "based on a confusion about substitutivity *salva veritate*" Schiffer (2003: 93). In fact, he thinks there are clear counterexamples to this substitution principle, and thus it can give no trouble for the referential theory of that-clauses. Schiffer says:

For example, if Pavarotti is the greatest tenor, we still can't substitute "the greatest tenor" *salva veritate* for "Pavarotti" in

The Italian singer Pavarotti never sings Wagner.

since

The Italian singer the greatest tenor never sings Wagner.

isn't even well formed. Schiffer (2003: 93)

However, we should distinguish two kinds of substitution failure: one syntactic, one semantic. *Syntactic substitution failure* occurs when the replacement of a term in a syntactically well-formed sentence with a co-referential one makes that sentence syntactically non-wellformed. The above example with Pavarotti is one of syntactic substitution failure. *Semantic substitution failure* occurs when replacing a term in a syntactically well-formed sentence with a co-referential one leaves that sentence well-formed, but changes its truth value. Apparent cases of semantic substitution failure are Frege's examples of substituting co-referential names in belief ascriptions.

Syntactic substitution failure is philosophically unproblematic. Simply because two terms refer to the same object does not mean, and shouldn't be expected to mean, that they have the same syntactic features. For Schiffer's Pavarotti example we have a fairly straightforward explanation, at least in outline, why the syntactic substitution failure occurs. The appositive phrase "the Italian singer" is headed by a determiner, "the." If it is appositive to a phrase that is itself headed by a determiner we get ungrammaticality. The two determiners clash. That's why it isn't grammatical to say "the Italian singer the greatest tenor," but it is grammatical to say "the Italian singer Pavarotti," or "my hero the greatest tenor," and so on. To be sure, why you can't have two determiners in this way might be a substantial question in syntax, and different syntactic theories will have different answers in the details, but there is no philosophical puzzle here.

Not so with semantic substitution failure. If that-clauses refer to propositions, then any other reference to the same proposition should be just as good. If it is not, then maybe we have to conclude that that-clauses don't refer to propositions, and that we thus shouldn't expect that we can substitute for a clause which does not refer a term which does refer. The examples above are ones of semantic substitution failure, if that-clauses refer. Schiffer is correct to point to syntactic substitution failure, but this does not help him to explain semantic substitution failure, or to reject a substitution principle where grammaticality is preserved. In the examples above grammaticality is preserved, but the truth value changes. It is a case of semantic substitution failure, and thus an example that threatens to refute the theory that holds that that-clauses refer.[10]

A different attempt to answer these problems has been pursued by Jeff King in King (2002). According to King, attitude verbs come in two kinds: those that have only one relevant reading, and those that have two readings. Among the former is "believes," among the latter are "fears," "expects," and "mentions." King claims that with verbs of the latter kind a particular reading is triggered by the syntactic form of the complement it takes. Thus when "fear" is followed by a clausal complement it is forced to have one reading, and when it is followed by a noun phrase complement it is forced to have another reading. Substitution failure is therefore not a problem and that-clauses can be referential. Our apparent counterexamples above, from (300a) to (302b) are not counterexamples, since the verbs "fear," "expect," and "mention" have different readings in these pairs of examples. So, that they have different truth conditions is not a problem for the view that that-clauses are referential, or so King.

King's view has several problems. First, as he notes himself, it gives the wrong result with quantifier inferences. For him the inference from

(304)   Sue fears that the restaurant is closed.

to

(305)   Sue fears something.

should not be valid. "something" is a noun phase, on the present way of classifying things, and thus should force one reading of "fear." But "that the restaurant is closed" is a clause and thus forces a different reading of "fear." The readings of "fear" in the premise and conclusion are thus different, and so the argument is not valid. King holds that, for some reason to be worked out, some quantifiers are different and don't force the reading that noun phrases in general force. He acknowledged this to be true for "something," "everything," and "nothing."[11] But this list isn't complete since similar inferences work with "two things," "most things," and "many things."[12] It is one thing to acknowledge these counterexamples, another to maintain the account in light of them. Even if "something" is special in various ways, it still is a noun phrase and not a clause, and so it still should force one reading, but it doesn't. That Sue fears something

---

[10] This objection to Schiffer's attempt to deal with the substitution arguments was made in Hofweber (2006b). See also Rosefeldt (2008) for a similar diagnosis.

[11] These are also the "special quantifiers" in Moltmann's technical sense. See Moltmann (2003a).

[12] Suppose all Fred fears is that p, that q, and that r. Sue fears that p and that q. Then it is valid to infer that Sue fears most things that Fred fears. Similarly, Sue fears two things that Fred fears.

is implied by both (300a) and (300b). But on King's view it is unclear how this is possible.[13]

A second concern about King's account of substitution failure is that it does not seem to be required to have one or the other kind of complement with "fear." One can have mixed cases instead. Consider

(308)   John at the same time feared his mother and that she might leave him.

(309)   Putnam mentioned logicism and that social justice is elusive in the same sentence.

Here the additions of "at the same time" and "in the same sentence" make clear that these cases can't be understood as some form of conjunction reduction. (309) can just be seen as short for something that contains two occurrences of "mention," one with a clausal complement, and one with a noun phrase complement. These are examples where the complement of "fear" is a mixed conjunction. For King this again should not be possible, but it is.[14]

Finally, King's account fails to deal with substitution failure outside of attitude ascriptions. Our examples using "disgusting" above, (303), exhibit substitution failure, but don't involve attitude ascriptions. To explain these cases of substitution failure in a similar way King would have to claim that a whole variety of verbs are polysemous, and their reading is forced by the syntactic form of the subject argument. This is very implausible, though, and faces all the same problems listed above. The substitution arguments thus speak against the view that that-clauses are referential.

---

[13] In Rosefeldt (2008), Tobias Rosefeldt has tried to defend King's view from this objection by arguing that quantifiers like "something" are ambiguous between a nominal and a non-nominal reading. The quantifier inferences are thus valid on one reading of the intentional verb with the corresponding reading of the quantifier: a clausal complement premise interacts with non-nominal "something," while a NP complement interacts with nominal "something." But I find this explanation hard to swallow. On King's view it is the syntactic category of the complement of the verb that triggers one reading or another. NP complements trigger one reading, clausal complements another. It is not clear to me what would make the quantifier in (305) non-nominal. Simply because it takes the place of a clausal complement does not mean it is non-nominal, unless it is supposed to be of a different syntactic category in this case. But I take it that the suggestion can't be that "fear" in one of its readings syntactically requires a clausal complement, and that "something" in one of its readings is a clause that meets this requirement. "something" is never a clause. But without claiming this, I am not sure how the suggestion helps King. When "something" is in a "non-nominal position," as for example in

(306)   He is something I hope to be one day.

this is not to be understood via an ambiguity in the quantifier "something," but the result of a type shift into predicative position. This would be no different than

(307)   He is the man I hope to be one day.

which does not support an ambiguity in "the man," which can, of course, also appear in nominal, or subject, position. Quantifiers in general can clearly be non-nominal, e.g. adverbs of quantification like "sometimes," but this does not support that "something" is ambiguous, nor does it help King's view.

[14] See also Pryor (2007: 230f.).

Substitution failure is not restricted to that-clauses. It in particular also occurs with property nominalizations. Property nominalizations contain phrases like "wisdom," "being wise," and "the property of being wise." And these are quite different when it comes to their contribution to the truth conditions. Here, too, we have examples using attitude verbs and examples where substitution failure occurs in subject position:

(310) a. Pat fears wisdom.

      b. Pat fears the property of being wise.

These differ in truth conditions since (310b) can be true in a case of general property phobia, while (310a) might be false. Similarly for property nominalizations in subject position:

(311) a. Wisdom is disgusting.

      b. Being wise is disgusting.

      c. The property of being wise is disgusting.

These have different truth conditions, as (311c) can be true in a case of disgust of properties, while (311a) is false, and wisdom is in fact searched out while properties are seen as being among the disgusting things. Thus the substitution failure arguments carry over quite directly from the case of that-clauses to the case of property nominalizations. In particular, it can't be that all three phrases in subject position in (311) refer to the property of being wise. It does remain a possibility so far to hold that only "the property of being wise" refers to a property. The other two phrases might refer to something else. Maybe "wisdom" refers to a type of state and "being wise" refers to something else, a universal. But this won't help with the problem since now we get substitution failure with those phrases:

(312) a. The state type wisdom is disgusting.

      b. The universal of being wise is disgusting.

The problem is not what kind of thing these phrases are supposed to refer to, but that they are supposed to be referential in the first place. Substitution failure thus speaks against the referentiality of property nominalizations and that-clauses.[15] It doesn't help in dealing with substitution failure to claim that "that p" or "being F" refers to something other than a proposition or a property, as we saw above. Whatever it might be, the substitution failure will reemerge with the phrase "the universal of being F," "the trope Fness," or whatever "being F" is supposed to refer to. The substitution arguments thus are good arguments that property nominalizations and that-clauses do not refer to properties or propositions, nor anything else.

---

[15] A more detailed discussion of substitution failure, with many more examples, is in the work of Friederike Moltmann. For her latest and most detailed position, see Moltmann (2013a). Moltmann's own view is a version of a "mixed view," which will be critically discussed shortly.

But this is not enough to establish internalism in general about talk about properties and propositions. It leaves open the possibility of a mixed view: that-clauses don't refer, but "the proposition that p" does refer, or at least denote. Similarly, in the case of property nominalizations one might hold on to a mixed view: "being F" doesn't refer, but "the property of being F" does refer or denote. This view doesn't seem to be subject to further substitution arguments. Since "being F" isn't referential, on this mixed view, we shouldn't expect that substituting a referential term for it preserves truth conditions, and the other way round.[16] The mixed view is contrasted with the pure view. The pure view holds that none of the relevant phrases are referential. Thus for the case of talk about properties, neither "being F" nor "the property of being F" nor "Fness" are referential. For the case of talk about propositions, the pure view maintains that neither "that p" nor "the proposition that p" are referential.

Should we prefer a mixed view, or the purely non-referential view? The substitution arguments so far don't decide between them. As long as one of them is not referential or denotational, substitution can not be expected to go through. Whether both are non-referential is thus a further question. Either way, we have to understand how, for the case of talk about propositions, that-clauses and "the proposition that p" relate to each other. And here the mixed view has a problem. Suppose we hold that "that p" is not referential or denotational. It is simply a clause that specifies what the content of a belief is, but doesn't refer to it or denote it. How does "the proposition that p" relate to the that-clause? The natural idea is this: "the proposition that p" is a definite description. Just as "the composer of *Tannhäuser*" states there is a unique x who composed *Tannhäuser*, "the proposition that p" makes a similar contribution to the truth conditions. But this is problematic in several ways. First, but not foremost, there is an analogous problem to that discussed in section 2.3.1 for the case of "the number of Fs" being a description. There is a crucial difference between "the proposition that p" and "a proposition that p." The latter is generally quite awkward and almost ill-formed, but if "the proposition that p" is merely an ordinary description this should not be expected, since generally the determiner "the" can be replaced with the determiner "a." "a composer of *Tannhäuser*" is perfectly fine, and differs from "the composer" in that it leaves open that *Tannhäuser* has more than one composer. As we saw in section 2.3.1, this is not a knock-down objection, since sometimes this replacement is not possible with descriptions, as in "the richest man." Here the uniqueness built into "richest" clashes with "a" in "a richest man." But this explanation for a failure to be able to replace "the" with "a" doesn't seem to work with "the proposition." We will see shortly what explains this better. Overall this does cast doubt on the description view.

---

[16] Mixed views along these lines are endorsed by Moltmann (2013a), Rosefeldt (2008), and Pryor (2007), among others. Congenial is also Kemmerling (2016), in particular with respect to that-clauses in belief ascriptions vs. other terms for propositions. These issues arise for number terms as well, and we discussed mixed and pure views about number terms in section 5.4.6.

Second, and more importantly, it is not clear how the description is to be spelled out. "the proposition that p" can't just be "the unique thing x such that x is a proposition and x is identical to that p." On the present mixed account "that p" is not a referring expression and so the latter identity claim would be ill-formed. The quantifier binding "x" here is to be understood as an external quantifier ranging over a domain. The "x" on the left-hand side of the identity statement is thus a variable that is referential (or bound externally) whereas the right-hand side contains a non-denoting expression. Although identity statements are well-formed when both sides are referential, and when both sides are non-referential, they are ill-formed when one side is and the other isn't. If "the proposition that p" is supposed to be a description, then this can't be how it is to be understood. It also doesn't help to spell out the alleged description "the proposition that p" as "the unique x which is identical to the proposition that p." That would just be circular. How about then "the unique x which is a proposition and has the content that p." Now we don't make an identity statement with a non-referring that-clause, but we say that propositions *have* contents. But if there are propositions then they *are* contents, they do not have them. And furthermore, the issue just gets pushed to unpacking "the content that p," which presumably is also a description and thus faces the same problem. There might, of course, be other options, but no satisfactory ones have been given to the best of my knowledge. It is not clear if any more elaborate attempts to flesh out the description can be seen as a semantic analysis as opposed to an insistence on the claim that these phrases are descriptions.

The pure view has a better account to offer here, and thus is overall to be preferred. Although "the proposition that p" starts out with "the" it is highly questionable that this phrase is a definite description. A more plausible account of their relationship fortunately is possible. It takes "the proposition" to be *an apposition* to "that p." An apposition is a comment on a phrase, one that doesn't affect the semantic function of the phrase it comments on. "the famous composer Richard Wagner" still leaves "Richard Wagner" as a proper name, just as "Richard Wagner, the famous composer," does. Similarly, "the proposition that p" contains a non-referring clause "that p" with an appositive phrase out front: "the proposition." But that apposition doesn't turn a non-referential clause into a referring phrase, it rather makes a comment on the non-referential clause. The apposition view can also explain, in outline, why "a proposition that p" is problematic by pointing to parallel problems with uncontroversial appositions. Although "the composer Richard Wagner" is fine, "a composer Richard Wagner" is not. However, "Richard Wagner, a composer, . . . " as well as "Richard Wagner, the composer, . . . " are both fine, and similarly for appositions to that-clauses. Here the apposition to a that-clause can be different than just "the proposition," it can include "the fact," "the conjecture," "the widely debated hypothesis," and so on. And just as in the case of uncontroversial appositions, we get that "a conjecture that p . . . " is bad, "the conjecture that p . . ." is fine, and "that p, a conjecture, . . . " is also fine, at least with some contrasting or elaborations like "that p, a hotly debated conjecture, ..."

Taking "the proposition" to be an apposition, not the beginning of a definite description, makes sense of how "that p" and "the proposition that p" go together under the assumption that "that p" is non-referential. The mixed view has a hard time connecting the two, but the purely non-referential view does not. I thus side with a pure view that takes singular terms for properties and propositions to be non-referential throughout.[17]

Overall, we can assess the situation as follows: there are two apparently good arguments for that-clauses and property nominalizations being referential, and two apparently good arguments that they are not referential. The former are the argument using quantifier inferences and the fact that that-clauses have semantic values. We have seen that both of these arguments should be rejected. The quantifier inferences by themselves do not support the view that these expressions are referential, since quantifiers in general, and in this particular case as well, have an inferential reading that validates the respective inferences whether or not that-clauses refer or denote anything. The need for semantic values for that-clauses in a compositional semantics does not show that these clauses are referential, nor does it support the claim that these semantic values are propositions. There are thus no good arguments for the claim that that-clauses refer. On the other hand, there are two good arguments against them being referential. First, there is the argument that our pre-theoretic judgments about the semantic function of clauses support a difference between clauses and more paradigmatically denoting or referring terms. This has some force since we are competent speakers of our language and, although there is room for theory as a tool for discovery of semantic function in a large range of cases, there must also be room for what seems clear to us we do when we use certain phrases. And here that-clauses stand out: they are used to specify what the content is, not to refer to something which is a content. Similar considerations also hold for at least certain property nominalizations, although this second argument is probably stronger in the case of clauses. Second, and more importantly, there are the substitution arguments. We have seen that those are just the kinds of arguments that can show that a phrase is not referential, and that the attempts to block these arguments fail. These arguments are exactly parallel for both that-clauses and property nominalizations, and they are the most forceful arguments in this debate for or against referentiality. Overall, we should conclude that that-clauses and property nominalizations are not referring or denoting expressions. Since the pure form of the non-referential approach was to be favored over the impure, mixed form, this supports internalism about talk about propositions and internalism about talk about properties, at least when it comes to the use of the relevant singular terms. What is left open by all this so far is whether such an internalist account can be properly extended to quantified talk as well.

---

[17] See also the above discussion in section 5.4.6 of the relationship between "the number two" and "two." For an opposing view, which takes appositions to be descriptions, see Schnieder (2006a).

8.5 Quantification over Properties and Propositions

Looking at quantifier-free talk about properties and propositions suggests that internalism is true for such talk. But before we can conclude this we need to see whether quantification over properties and propositions can be understood along internalist lines. Here we face two problems, an easy one and a hard one. The first is to say how to coherently formulate such quantification. Quantifiers in particular over properties are different in various ways from the quantifiers we focused on above in chapter 3. We will be able to adjust our account of quantification to deal with this slight modification in this section. The second, much more serious, problem is the problem that it seems that quantization over properties and propositions can't be understood along internalist lines, since this gets the truth conditions wrong. The particular account of the truth conditions of quantified statements that was given in chapter 3 seems to give the wrong truth conditions for quantifiers over properties and propositions. We will discuss this problem in detail in chapter 9, where we will also be able to solve it. The solution given there will have significant consequences, which we will work out in detail in chapter 10. But first, let's think about the proper formulation of quantification over properties and propositions. And for us here the natural thing to consider is how to apply the view of quantification developed in chapter 3 to the case of quantification over properties and propositions. In chapter 3 we focused on quantification over objects, which could be put as: quantifiers that interact with variables in broadly noun phrase position. But quantification over properties and propositions is not quite like that, and we thus need to extend our account from chapter 3 to include it.[18]

---

[18] There are also other options in the literature for combining non-referential that-clauses or property nominalizations with quantifiers. One could hold that quantifiers in these cases are substitutional, see Schiffer (1987b), although Schiffer holds for reasons to be discussed in chapter 9, that this can't always be the case, see footnote 6 in Schiffer (1987b: 288f.). This is largely congenial with the present proposal, at least in spirit, although the modification to be made in the next chapter might be more in conflict with this spirit.

One could also hold that such quantifiers are simply primitive, not to be understood in terms of quantification over objects, and not to be spelled out semantically other than in terms of themselves. See Prior (1971). But simply to say that it is primitive doesn't really help in saying why it is primitively different from quantification over objects. If there are two primitive kinds of quantification in natural language, then why should we think they are different? It should be accounted for somehow that there is such a difference, and simply to declare it primitive isn't going to do that.

One could furthermore try to understand quantification into positions occupied by that-clauses and property nominalizations to be a different kind of quantification, non-nominal quantification, and hold that it does not range over a domain of objects that is referred to by the relevant singular term (since those don't refer), but rather over a domain of objects which are the semantic values of these non-referring terms. See Rosefeldt (2008) for a proposal of this kind. But this, too, is unsatisfactory. First, it doesn't explain why quantifiers when they interact with one syntactic category, nominals, do not range over semantic values, but over objects that are referred to, while otherwise they range over semantic values. If they always ranged over semantic values, then we would get the wrong inferences from sentences that contain non-denoting terms which nonetheless have semantic values. Second, distinguishing two kinds of quantification based on what category the quantifiers interact with doesn't help with that-clauses in subject position, as in examples like (303). Here non-nominal quantification does not seem to help. Not that there is anything wrong with non-nominal quantification as such, of course. See Rayo and Yablo (2001).

It is thus better to rely on a general story about quantification that is motivated independently and to apply it here. The account of quantification defended in chapter 3 allows us to do this.

Let's look at the simpler case of propositions first. We will again take recourse to the notion of a (grammatical) instantiation of a quantified statement, as we did in chapter 3, to spell out the truth conditions of quantified statements in their internal reading. So, for a statement like

(313)   He fears something I believe.

an instantiation is

(314)   He fears that p and I believe that p.

To make this notion precise one will have to address a number of issues, for example about the scope of quantifiers, a task not unique to the internalist. Furthermore, we will take recourse to the notion of a (grammatical) sentence. Finally, we assign to a quantified statement infinitary sentences that are equivalent in truth conditions as follows:[19] Suppose that "S [something]" is a statement with a particular quantifier over propositions, and "S [that p]" is an instantiation of the former. Let "$\bigvee_p$ S [that p]" stand for the disjunction over all instances of "S [that p]," whereby for every sentence of our language there is one instance of "S [that p]," replacing "p" with that sentence. For illustrative purposes only, think of the quantifier as a (particular) substitutional quantifier with the substitution class being all the sentences in our language.[20]

The case of quantification over properties is a little more complicated since quantification over properties can apparently occur in subject as well as in predicate position, and there is an issue about higher-order predication. Whether or not there ever is quantification into a predicate position is controversial.[21] Possible examples are sentences like

(315)   He is something I am not, namely rich.

It is not clear whether this is quantification into predicate position since the "is" of the predicate is still present. In any case, an internalist can specify the truth conditions of such utterances quite directly, again by taking recourse to the notion of an instantiation of such a quantifier, and that of a predicate of our language. In this case an instance will be

(316)   He is F and I am not F.

where "is F" is replaced by a predicate in our language. The truth conditions of the quantified statement are then simply the disjunction over all the instances formed with predicates in our own language. In this case it would be $\bigvee_F$ (He is F and I am not F).

---

[19] I will only describe the case of the particular quantifier. The universal quantifier is analogous. Extending it to generalized quantifiers is analogous to our treatment of them in the appendix to chapter 3, a topic we will also return to in appendix 2 of chapter 9.

[20] The difference between internal readings of the quantifier and substitutional quantification was discussed in 3.5.1.

[21] See Rayo and Yablo (2001) and Künne (2003: 360ff.).

Quantification over properties, however, often is not of this kind. Often the instances of quantifiers over properties do not directly involve predicates, but expressions like "the property of being F" or "being F." One example would be

(317)   There is a mental property which is not a physical property, namely the property of feeling pain.

Here the instances would be something like

(318)   Being F is a mental property and being F is not a physical property.

We will have to expand our account to include such cases of quantification over properties as well. This can be done quite directly by exploiting the connection between a predicate, like "is F," and its nominalization "being F" or "the property of being F." In cases where a quantifier over properties is a quantifier into a subject position we still form a conjunction (or disjunction) where there is a conjunct corresponding to each predicate, but now the predicate appears in its nominalized form. Let "$[_{nom}F]$" stand for the nominalization of the predicate "F." Then the infinitary disjunction assigned to a quantified statement that quantifies into subject position is the disjunction of all the instances such that there is an instance for every predicate in its nominalized form. In the case of "Some property is G" it is the disjunction "$\bigvee_P([_{nom}P]$ is G)."

One final issue is higher-order predication. Since properties themselves can have properties there is a well-known division in the theory of properties between those who take a typed and those who take a type-free approach. For the former, all properties implicitly come with a type, and every quantifier over properties only ranges over some type of properties or other. In cases of higher-order predication properties of higher type are predicated of properties of lower type. There are many options one has in spelling this out in some more detail, and we will not get into them here. For a type-free approach one denies that properties come in types and claims that quantifiers over properties range over all of them. In predication properties can be applied to all others, even to themselves, in principle. To do this the predicate occurs nominalized in the subject position, and regularly in the predicate position, as in

(319)   Being a property is a property.

In our notation we can write self-application as

(320) $[_{nom}P]$ is P

or

(321) $P([_{nom}P])$

We will discuss this topic more in the next chapter, in connection with the paradoxes. An internalist has both of these options available as well, and the hard work that needs to be done in a typed approach, like assigning types to particular occurrences of quantifiers, can be carried over to a typed internalist approach as well. Instead of types being assigned to properties directly they will be assigned to predicates, and disjunctions and conjunctions will be formed involving only predicates of the appropriate type. In a type-free version this will not be necessary.

With the internalist truth conditions outlined above we can now see in outline how internalism about talk about properties and propositions assigns infinite disjunctions and conjunctions to a large variety of quantified statements as their truth conditions. For example

(322)   There is something we have in common.

will by the above account be equivalent to the disjunction over the instances formed within our language, or more realistically, a contextual restriction thereof.

In this case it is quantification into subject position, and thus we get

(323) $\bigvee_P$ (we have $[_{nom} P]$ in common)

which is equivalent to

(324) $\bigvee_P$ (you have $[_{nom} P]$ and I have $[_{nom} P]$)

which in turn, granting the equivalence between "x has the property of being P" and "x is P," is equivalent to

(325) $\bigvee_P$ (you are P and I am P)

Whether or not a predicate occurs nominalized or regular in a disjunction is determined by the grammar of the sentences in which it occurs. To avoid confusion, but at the price of extra notation, we will make this explicit in the following, using the above notation.

We thus have a coherent formulation of internalism about talk about properties and propositions. This is coherent for both parts: the quantificational part, and the singular terms part. Reflection on whether that-clauses and property nominalizations are referential supports internalism. Internalism can easily account for the trivial quantifier inferences that lead many to believe in externalism. However, the biggest and most serious problem for internalism about properties and propositions is still open. Although the trivial quantifier inferences are easily dealt with, this does not mean that all quantification over properties and propositions can be dealt with. Do the truth conditions for the internal reading of the quantifier outlined above correctly capture the truth conditions of quantifiers over properties and propositions in general? Here is very good reason to think the answer is "no." This issue is significant enough to warrant its own chapter: chapter 9. We will discuss this problem for internalism in detail not only because it is a powerful argument for rejecting internalism, but especially also because this problem and its solution are closely tied to an important large-scale philosophical issue which we will focus on in chapter 10.

9

Inexpressible Properties
and Propositions

Internalism about talk about properties and propositions was defended above by considerations about natural language.[1] But there is one objection to internalism that seems to refute it right away, and it isn't directly tied to the considerations that seem to speak in its favor. This objection applies to internalism about properties and propositions in particular, with no similar objection carrying over to internalism about talk about natural numbers. The objection is that internalism about properties and propositions is incompatible with something which is clearly true: that there are inexpressible properties and propositions. Apparently internalism implies that every property is expressible in our present language, but this seems to be quite clearly false. The present chapter is devoted to this objection. I will argue that the objection is indeed a very good objection and that it shows that I simplified in an illegitimate way so far in my formulation of internalism. However, with a proper formulation of internalism, one that does not simplify this way, we can see that the objection can be answered.

The main part of this chapter deals with a straightforward version of the objection from inexpressibility, how it requires giving up a simplification made so far, and how to state internalism properly without that simplification. More technical versions of related objections are discussed in an extended appendix. This chapter does not simply aim to sidestep the objection to internalism and then move on. The lesson we will learn from this objection will have an important positive upshot and it will be a crucial part of what is maybe the most radical proposal in this book, which we will get to in chapter 10. We should thus carefully look at this objection and what follows from it.

9.1 The Problem for Internalism

Externalism takes talk about properties and propositions to be talk about a domain of entities, likely independent of our language and thought. Internalism takes talk

---

[1] This chapter is based on parts of Hofweber (2006a). The overlapping text has, however, been updated, revised, and reorganized.

about properties and propositions to involve non-referring terms, and quantifiers over them to generalize over the instances, instances tied to predicates and sentences, in our own language. Internalism can thus in part be characterized with the metaphor that properties are shadows of predicates, and propositions are shadows of sentences, our predicates and sentences. They are not things independent of language and thought. Instead talk about properties and propositions is tied to our own predicates and sentences. Of course, it is only a metaphor, but it gets something right. Since the issues to come are in essence the same for properties and for propositions, let's focus on properties for now as our example, and return to propositions later.

Whether internalism or externalism is true about talk about properties seems to be a substantial and difficult question, one that is closely tied to issues about the semantic function of phrases like that-clauses and property nominalizations, and to what our aim and goal is in talk about properties and propositions in general. Internalism about (talk about) properties and propositions was formulated in the previous chapter as a coherent alternative to externalism, and we have seen evidence that speaks in favor of internalism. However, there is one very powerful argument against internalism that is especially applicable to internalism about properties and propositions. This argument has nothing to do with the semantics of that-clauses or property nominalizations, but instead addresses the truth conditions of the quantified statements. The inferential reading of quantifiers gives rise to truth conditions that are tied to their instances. Thus according to internalism, when I say that

(326)   There is something we have in common.

then this is equivalent to the disjunction

(327) $\bigvee_P$ (you are P and I am P)

with one disjunct for every predicate in our language. But that means that whatever we have in common must be something that is expressed by a predicate in our language. In the relevant sense here, the predicate "is F" expresses the property of being F. When we have a property in common, according to internalism as stated, there must be a predicate in our language that expresses that property. This might be fine for the case of (327), but it won't work in general. For example:

(328)   Some properties are not expressible in contemporary English.

seems quite clearly true, since after all, why should we believe that English just happens to express all properties? But (328) comes out false on the internalist truth conditions outlined above. It would be equivalent to a disjunction over the instances in contemporary English. But every instance

(329)   [$_{nom}$ P] is not expressible in contemporary English.

is false. For every instance in contemporary English of "P," [$_{nom}$ P] is expressible in contemporary English, simply with the predicate "is P." So (328) is false on the truth conditions given to it by internalism, but it seems clearly true. Thus internalism can't be right. Properties can't just be shadows of our predicates.

Externalism has no problem with (328). For externalism quantification over properties is quantification over a domain of entities. One does not have to assume that these entities are tied to predicates, except that some of them are presumably expressed with predicates. There is no assumption that all of them are expressed with predicates and thus there is no obstacle to holding (328). Internalism seems to have a problem with (328), but externalism doesn't. Thus externalism seems to be the only view compatible with the truth of (328).

The internalist has a number of options in light of this argument. First, they could deny (328), of course. But (328) seems to be quite compellingly true. We will discuss shortly why we generally accept (328), and that there are a number of good arguments in its favor. I will thus put this option aside for the moment. Second, the internalist could hold that we gave the wrong truth conditions for the internal reading of the quantifier. The motivation for the internal reading of the quantifier was the need we have for it in general in communication, to communicate information in various informational states we might be in. The function of this reading of the quantifier is to inferentially relate to instances, to have a certain inferential role. And the simplest truth conditions that give this reading of the quantifier that inferential role are the disjunction or conjunction over the instances.[2] But maybe there are other options for the truth conditions to the inferential reading of the quantifier. Why not others that can accommodate inexpressible properties? It is tempting to try to solve the apparent problem with inexpressible properties that way. But, as discussed in chapter 3, it is not clear how this could go while keeping the spirit of the inferential reading of the quantifier. We want to have that reading of the quantifiers to deal with a certain need in communication, and that need is served with the internal truth conditions. This motivates that the quantified statements on their internal reading have the simplest truth conditions that give them that inferential role, and those simplest truth conditions are the generalizations over our own instances. It is hard to see how one can motivate more complex truth conditions that would also support the inferential role and generalize not merely over our instances, while keeping the spirit of internalism alive. What else would they generalize over if there is no domain to draw on? It would be ad hoc to demand a change in the truth conditions of internal quantifiers simply to accommodate our judgments about the inexpressible. To make progress here we need to understand this objection and how it affects the spirit of internalism. Maybe it shows that internalism is wrongheaded, or maybe it shows something else, but it certainly shows something.

---

[2] The more general case was discussed in the appendix to chapter 3 on generalized quantifiers. It is sufficient to consider the simpler case right now; we will return to generalized quantifiers in an appendix to this chapter.

9.1.1 Inexpressible Properties and the Inductive Argument

I take it that we all believe that there are properties inexpressible in English. However, it is not so clear why we accept this. After all:

1. For a property to be inexpressible in a language means that no predicate (however complex) expresses it. Simply because there is no single word in a certain language for a certain property doesn't mean it isn't expressible in that language.
2. We can't be persuaded that there are properties inexpressible in English by example. One can't say in English without contradiction that the property of being Φ isn't expressible in English.

So, why again do we believe that there are inexpressible properties?

There are a number of different arguments for there being inexpressible properties. We will look at several of them in this chapter. The simplest and, as we'll see, most important argument is the following:

Even though we can't give an example of a property inexpressible in English, we can give examples of properties not expressible in older, apparently weaker languages. For example, the property of tasting better than Diet Pepsi is not expressible in Ancient Greek. So, there are properties expressible in English, but not in Ancient Greek. In addition, we have no reason to believe that English is the final word when it comes to expressing properties. We can expect that future languages will have the same relation to English that English has to Ancient Greek. Thus, we can expect that there are properties inexpressible in English, but expressible in future languages. In short, there are properties not expressible in present-day English.

Let's call this argument for inexpressible properties *the inductive argument*. It is a powerful argument. The main task for the next few pages will be to see whether or not the inductive argument refutes internalism about talk about properties.

9.1.2 Some Distinctions

What does it mean for a property to be expressible in English? Well, that there is a predicate of English that expresses it. But that could mean at least two things. On the one hand, it could mean that there is a predicate of English that expresses this property in the language English. On the other hand, it could mean that there is a predicate of English such that a speaker of English expresses this property with an utterance of that predicate. Which one of these we take will make a difference for the issue under discussion. To illustrate the difference, consider:

(330)   being that guy's brother

This predicate does not express a property simpliciter, it only expresses one on a particular occasion of an utterance of it, that is, in a particular context. In different contexts of utterance it will express different properties. However,

(331)   being Fred's brother

expresses a property independent of particular utterances, or better, expresses the same one in each utterance.[3] If "that guy" in an utterance of (330) refers to Fred, then this utterance of (330) will express the same property as any utterance of (331) will. However, there might be contexts in which an utterance of (330) will express a property that can't be expressed by an "eternal" predicate like (331).

So, when we ask whether a property P is expressible in a language $\mathcal{L}$ we could either ask

1. whether there is a predicate Φ (of $\mathcal{L}$) such that in every context C, an utterance of Φ (by a speaker of $\mathcal{L}$) in C expresses P, or
2. whether there is a predicate Φ (of $\mathcal{L}$) and a context C such that an utterance of Φ (by a speaker of $\mathcal{L}$) in C expresses P.

Let's call expressible in the first sense *language expressible* and expressible in the second sense *loosely speaker expressible*. The latter is called *loosely* speaker expressible because it only requires for there to be a context such that an utterance of Φ in that context by a speaker of $\mathcal{L}$ would express P. Any context is allowed here, whether or not speakers of that language actually ever are in such contexts. We can further distinguish this from what is *factually speaker expressible*. Here we allow only contexts that speakers of that language actually are in.[4]

We can grant that Ancient Greek does not allow for the expression of the property

(332)   tasting better than Diet Pepsi

in the sense of being language expressible. However, it seems that it is expressible in Ancient Greek in the sense of being loosely speaker expressible. In the context where there is Diet Pepsi right in front of a speaker of Ancient Greek they could simply utter the Ancient Greek equivalent of

(333)   tasting better than this

while demonstratively referring to Diet Pepsi. But since there was no Diet Pepsi around during the time when Ancient Greek was a living language, this context is not allowed when considering the question of whether or not this property is factually speaker expressible. In that case it seems that the property is not factually speaker expressible in Ancient Greek, just as it is not language expressible in Ancient Greek. In general, being language expressible implies being factually speaker expressible, which implies

---

[3] I assume that "Fred" and "brother" are disambiguated, i.e. with respect to whether we talk about a monk or a sibling, and whether it's Fred Dretske, Fred Astaire, Fred Flintstone, or any other Fred.

[4] To simplify, we consider someone only as a speaker of their native language. This is also implicitly assumed in the inductive argument. We can also ignore complicated issues about identities of languages over time. For present purposes it does not matter what the details are about how long and under what conditions a language continues to be the same. Intermediate notions between factually speaker expressible and loosely speaker expressible can also be formulated depending on how strictly one takes "actually." This is of no consequence for our discussion, though.

being loosely speaker expressible, and none of these implications can be reversed (or so we can concede for now).

What all this shows is that both the inductive argument and the above account of the internalist view of talk about properties were too simplistic. In the latter it was simply assumed that contextual contributions to content do not occur and that the truth conditions of talk about properties can simply be given by infinite disjunctions and conjunctions of eternal sentences of the language in question. But that's not always so. Sometimes predicates express properties in some contexts that can't be expressed with eternal predicates. To say this is not to deny that properties are shadows of predicates, just that they are shadows of eternal predicates. Let's call a version of internalism about talk about properties *extreme internalism* if it claims that quantification over properties is equivalent to infinite disjunctions and conjunctions formed with eternal predicates. And let's call a form of internalism *moderate internalism* if it accommodates contextual contributions to content. What we have seen so far is that extreme internalism can't be right. An internalist will have to endorse moderate internalism. But how is this form of internalism supposed to be understood? How can internalists accommodate contextual contributions to what is expressed by a predicate while at the same time holding that quantification over properties is merely generalizing over the instances in our own language?

9.1.3 The New Problem for Internalism

Here is the problem: even if an utterance of a sentence with demonstratives in Ancient Greek in the right context would express the property of tasting better than Diet Pepsi, it is quite a different story to extend this to an internalist account of the truth conditions of quantification over properties. In fact, it seems that this can't be done.

Let's suppose, for the sake of the argument, that the only property of beer that interests Fred is that it tastes better than Diet Pepsi. So

(334)   There is a property of beer that interests Fred.

An internalist account of quantification over properties has to get this to come out true. But it seems that this requires that there is a disjunct in the infinite disjunction that corresponds to

(335)   tasting better than Diet Pepsi

This is no problem for our language, English. But if the disjunctions have to be formed in Ancient Greek, then it doesn't seem to work. To be sure, and as we have seen above, the property of tasting better than Diet Pepsi is loosely speaker expressible in Ancient Greek. But how can this be used in the infinite disjunction? After all, merely having

(336)   tasting better than this

as part of one of the disjunctions won't do unless the demonstrative refers to Diet Pepsi. But how could it? The referent of a demonstrative is fixed at least partly by the intentions of the speaker using it. And in an utterance of (334) there are no such intentions that could back this up. For one, one can utter (334) while having no idea what property it is that interests Fred. And second, speakers of Ancient Greek will have no idea what Diet Pepsi is, nor will they have any around to demonstratively refer to. So, there is no way such speakers can fix the referent of such a demonstrative to be Diet Pepsi. Thus the truth conditions of quantified statements can't be equivalent to infinite disjunctions and conjunctions over the instances, even if the instances may contain demonstratives. Thus it seems that internalism is refuted, after all, even given the above distinctions.

9.2 The Solution to the Problem

An internalist claims that quantification over properties is merely a generalization over the instances of the quantifier. And it contrasts internalism with externalism, which claims that such quantification ranges over some mind and language independent domain of entities. The above considerations suggest that extreme internalism should be rejected. Extreme internalism is in trouble since not every object is referred to with an eternal term, and thus what properties can be expressed with eternal predicates is strictly less than what properties are expressed with context-sensitive predicates, namely in cases where a demonstrative refers to an object that isn't the referent of an eternal term. These considerations show that what objects there are, not merely what objects are referred to with eternal terms, matters for what properties and propositions there are. The case of demonstrative reference to an object that isn't the referent of any eternal term illustrates this. We should however not give demonstrative reference too central a role in this. What objects there are matters, not what objects can be referred to even with a demonstrative. So, if there are any objects that can't be referred to with a demonstrative, for whatever reason, these objects would nonetheless be relevant for what properties there are. We thus have to take "speaker expressible" liberally here. Any object has to be able to be contributed in a context. Context-sensitive expressions can have terms in them that in a context can stand for an object. Demonstrative reference is one way in which this can happen, but we will more liberally consider the notion of a context contributing any object as the value of a "demonstrative" or context dependent singular term. Thus "loosely speaker expressible" has to be understood as expressible with a predicate where context may contribute any object whatsoever as the value of a demonstrative or otherwise context-sensitive singular term. This will properly accommodate the above insight that what objects there are available for reference matters for what properties and propositions are expressible. So, an internalist will have to claim that quantification over properties is a generalization over all the instances of context-sensitive predicates, with context being allowed to contribute any object whatsoever, but without requiring referential intentions on the part of the speaker. We need to look at this more closely and draw on the help of artificial languages to clarify the situation.

Let's assume that the truth conditions of a fragment of a natural language without quantification over properties are correctly modeled with a certain interpreted formal language L. Adding quantification over properties to that language should give us an infinitary expansion of L, according to the internalist. Now, to accommodate demonstratives, we can do the following. Add infinitely many new variables $v_1, v_2 \ldots$ to L, which model the demonstratives. Build up formulas as usual, but don't allow ordinary quantifiers to bind these new variables. To accommodate talk about properties, we represent the truth conditions of quantification over properties with an infinite disjunction or conjunction as before, with one difference. Whenever we form an infinite disjunction or conjunction we also existentially or universally (respectively) bind all these new variables. Thus, now we do not simply represent "there is a property such that Φ" as the infinite disjunction over all the instances "$\Phi([_{nom}P])$," i.e. as "$\bigvee_P \Phi([_{nom}P])$." Now we take this disjunction and add existential quantification on the outside binding all the new variables. So, we now represent "there is a property such that Φ" as "$\exists v_1, v_2, \ldots \bigvee_P \Phi([_{nom}P])$." These extra quantifiers range over the domain of all objects, whatever they may be, i.e. they are external quantifiers.

The new free variables play the role of the demonstratives in this account, and the quantifier binding them plays the role of the arbitrary contexts that we allow in loose speaker expressibility. For example, the disjunction that spells out the truth conditions of (334) will contain a disjunct corresponding to

(337)   tasting better than v_i

Now there will be an existential quantifier that binds v_i from the outside. Since it will range over Diet Pepsis this disjunct will be true, and thus the disjunction will be true. And this will be so independently of there being a referring expression that refers to Diet Pepsi in the language in question.

However, there is no finite upper bound on how many of these new variables will occur in these disjunctions. Since we allow, and have to allow, every predicate to occur in the disjunction, we can't give a finite upper bound on how many variables can occur in these predicates. So, in the infinite disjunction there will be infinitely many variables that have to be bound, all at once, from the outside. But this can be done. We just have to go to a stronger infinitary logic. Not only do we need infinite disjunctions and conjunctions, we need quantification over infinitely many variables. Before we only used a small fragment of what is called $L_{\omega_1,\omega}$, now we use a small fragment of L_{ω_1,ω_1}. This latter logic also allows for quantification over countably many variables.[5]

---

[5] $L_{\omega_1,\omega}$ is an infinitary logic that allows conjunctions and disjunctions over countable sets of formulas, but only quantification over finite sets of variables (as in regular first or higher order logic). L_{ω_1,ω_1} allows for both conjunctions and disjunctions over countable sets of formulas, plus quantification over countable sets of variables, i.e. a single quantifier to bind all these variables at once. The basic language is usually the one of first-order logic, but one can define infinitary expansions of other languages just as well. See Keisler

Given this improved internalist picture of talk about properties we have the following:

- Properties are shadows of predicates, but not just shadows of eternal predicates.
- Talk about properties gives rise to an infinitary expansion of the original language, but not just to a small fragment of $L_{\omega_1,\omega}$, rather to a small fragment of L_{ω_1,ω_1}.

So, the property of tasting better than Diet Pepsi is not expressible in Ancient Greek either in the sense of language expressible nor in the sense of factually speaker expressible. It is however, expressible in Ancient Greek in the sense of loosely speaker expressible. And by the inductive argument we get that we have reason to believe that there are properties that are not expressible in English, but we get that only when "expressible" is understood in the sense of either language expressible or factually speaker expressible. However, according to the present version of internalism, quantification over properties has to be understood as being based on what is loosely speaker expressible with predicates. Therefore it will be true that

(338)   There are properties that are not expressible in English.

if expressible is understood as being language expressible or factually speaker expressible, but false, according to the internalist, if it is understood as being loosely speaker expressible.

Internalism is thus not trivially refuted by the inductive argument. Once we distinguish between extreme and moderate internalism, and between different notions of expressibility, we can see that moderate internalism is not refuted by the inductive argument given above. A moderate internalist should endorse the inductive argument as showing something interesting and important about a difference about what is language expressible or factually speaker expressible in different languages. The argument doesn't refute internalism, but instead is a guide to its proper formulation. On the proper formulation of the internalist truth conditions, quantifiers generalize over the instances, properly understood. This proper understanding and the new, modified internalist truth conditions will from now on be our official understanding of how the internalist truth conditions of quantification over properties and propositions are to be understood. Internalism from now on is thus moderate internalism.

The inductive argument given above is not the only argument against internalism from considerations about inexpressibility. There are also other versions of the inductive argument, and there are a number of more technical arguments, first and foremost cardinality arguments. These arguments aim to show that not every property can be expressible in our language, since there are more properties than there are predicates

---

(1971) and Barwise (1975) for much more on this. In our case here we only use very small fragments of these logics. All these fragments will be finitely representable, for example, and smaller than the smallest fragments studied in Barwise (1975) or Keisler (1971). A closer look at the technicalities of this is in the appendix to Hofweber (1999).

in our language. This argument would be compelling if every predicate could only be used to express one property, but with the distinction of what is speaker expressible and language expressible at hand the argument becomes less compelling. After all, the very same predicate can be used to express many different properties: "being taller than this" expresses a different property depending on what "this" refers to. On our proposed account of the truth conditions of internal quantifiers over properties it turns out that for every object o, there will be an instance which is the property of being taller than o. And this is central for the answer to the cardinality arguments. In an appendix to this chapter I discuss several such arguments in more detail. They include cardinality arguments of various kinds, more inductive arguments, and arguments related to the paradoxes. The reader who would like to see more on this topic is encouraged to consult this appendix. We will see there that the distinctions made above and the formulation of moderate internalism given are the key to answering all these arguments. Since some of the arguments are more technical, and since the answers to them are in some sense versions of the answer given above to the inductive argument presented there, this further discussion is moved to an appendix. I argue there that these modified arguments are also no problem for internalism, and thus overall that internalism, when properly formulated, is not refuted by considerations about expressibility.

In light of all this, and in light of the considerations we have seen in chapter 8 in favor of internalism, I conclude that internalism about properties and propositions is indeed correct. Since properties and propositions are central for many metaphysical debates this should have significant consequences for metaphysics, and I will argue in the following that it indeed does. In the next chapter, chapter 10, we will discuss the significance of internalism for metaphysical issues tied to propositions, and in the chapter after that, chapter 11, we will do this for properties. But before we get to that, we are now in a position to return to our discussion of ontology more generally, in particular our three puzzles about ontology. After having seen whether internalism or externalism is true for our four cases of natural numbers, ordinary objects, properties, and propositions we are now in a position to solve the second puzzle about ontology. This is the place to explicitly present the solution to the second puzzle, since we are now done considering what we do when we talk about numbers, objects, properties, or propositions, and these considerations are central for the solution to the second puzzle. After spelling this out we can move to the metaphysical consequences of what we have seen above.

9.3 The Solution to the Second Puzzle

To understand ontology as a philosophical discipline we have to answer three puzzles about it. First there was the puzzle of how hard ontological questions are. The ontological question about numbers seems to be the question of whether or not there are any numbers. And it seems that this question has a trivial affirmative answer, even though it was intended as a substantial question about reality. How that could be so was the first puzzle. We have seen that the trivial arguments are valid, but they do not answer the ontological question. The question "Are there numbers?" has more than one reading, and only on one of them, the internal reading, is it answered trivially. The other, external, reading expressed the ontological question. And it is not trivially answered. That was the solution to the first puzzle.

The second puzzle arose from the question of how important ontological questions are. On the one hand, they seem to be merely academic and of concern only to philosophers in their metaphysical theorizing. In the larger scheme of things they thus don't seem all that important. On the other hand, it seems that the truth of mathematics depends on the ontological question about numbers having one answer, and the truth of ordinary belief–desire psychology depends on the ontological question about propositions having a certain answer, and the truth of almost everything depends on the ontological questions about objects and properties having a certain answer. These ontological questions thus seem to be of the greatest importance.

The second puzzle is closely tied to the issue of whether or not our talk about numbers, objects, properties, and propositions implies or presupposes a certain answer to the ontological questions about them. In our solution to the first puzzle we saw that the trivial arguments don't answer the ontological questions. But the issue remained whether or not other considerations answer that ontological, external question. Do the results of mathematics answer the ontological question about numbers? Does (the truth of) our talk about objects, properties, and propositions, talk that occurs everywhere including in the sciences, answer the ontological question about them? If our ordinary, and scientific, talk about numbers, objects, properties, and propositions implies an answer to the ontological question, then this question is of crucial importance: that the answer turns out the way as is implied by our ordinary and scientific talk is crucial for the truth of our ordinary and scientific talk. If such talk implies that numbers, objects, properties, and propositions exist, then it is of the utmost importance that this is correct. After all, if it isn't, then all talk that implies that they do exist is mistaken. Of course, if this implication indeed holds, then there is good reason to believe that they exist. But this is not what the second puzzle is about. The second puzzle is about just how important it is how the answer turns out, not what reason we have for the answer to go one way. If our ordinary talk implies the answer to go one way and thus is only true if it is the right answer, then it is of the greatest importance that the answer goes that way.

The key to a solution to this puzzle is thus to figure out whether internalism or externalism is true for a particular domain. If externalism is true about our talk about numbers, then such talk implies that numbers exist, and thus implies an answer to the external question about numbers. If externalism is true, then regular utterances of "there are infinitely many prime numbers" involve external quantification over numbers, and thus imply "there are numbers" in its external reading. Thus externalism guarantees that what we say implies an answer to the external question. And thus that this answer is correct is required for the truth of our talk about numbers. If externalism is true then the ontological question is of the greatest importance. Without the answer going the way it is implied by our talk, our talk will be based on a big mistake.

But if internalism is true about a domain, then things are different. This holds in general, for any domain, although we have most explicitly discussed it in the case of talk about numbers, see also section 6.4 above. If internalism is true, then not only do the trivial arguments not imply an answer to the external question, our talk about numbers, say, in general does not imply an answer to the external question. The truth of the results of arithmetic, as well as the truth of ordinary talk about numbers, does not imply an answer to the external question. Although it is true that there is a number between 6 and 8, namely 7, if internalism is true, this does not imply that there are numbers in the external reading. On the internal reading we are merely generalizing over the instances, and in the instances number words are non-referring expressions. The question of whether there are numbers, understood in the external reading, is left open by this so far. And similarly for internalism about other domains of discourse, talk about properties and propositions for example. Thus if internalism is true, then an answer to the ontological question is not immediately implied. And so it is of less significance how it turns out. The truth of talk about numbers does not assume that the ontological question goes one way. And thus if internalism is true, then the ontological question is left open by our talk and thus is of less significance.

We have seen so far that internalism is true about talk about numbers, properties, and propositions. For these domains no answer to the ontological question about numbers, properties, and propositions is presupposed in our ordinary and scientific talk about them. However, externalism is true for talk about objects. In our talk about objects we must indeed count on the answer to the ontological question being affirmative for our ordinary talk to be true. This solves our second puzzle. Although the ontological questions about numbers, properties, and propositions would be of the greatest significance if externalism were true, these questions in fact don't have this significance since internalism is true. But since externalism is true about talk about ordinary objects, the ontological question about them is of the greatest significance. If the answer turns out to be that there aren't any then almost everything will have to go.

9.4 Appendix 1: Further Arguments Against Internalism Using Expressibility Considerations

This appendix deals with further arguments against internalism from considerations about expressibility. It can be skipped, since the details are not required for what is to come. However, these arguments are important arguments against internalism, and to see why they are not successful when internalism is properly formulated is important as well. Nonetheless, the detailed discussion of them is moved to an appendix, since it can be skipped by anyone who wants to move on to the next topic.

9.4.1 Modified Inductive Arguments

The regular inductive argument against internalism was defeated since the relevant properties were loosely speaker expressible in the old and new languages, but only factually speaker expressible and language expressible in the newer language. But could there not be stronger inductive arguments that point to a change in what properties are even loosely speaker expressible? To attempt such an argument one could try to find a property that is loosely speaker expressible in English, but not loosely speaker expressible in, say, Ancient Greek. Good candidates for such properties are ones that relate to an area where there is a substantial difference between Ancient Greece and us, like scientific understanding of the world. A tricky case is

(339)   being a quark

It might seem that it isn't even loosely speaker expressible in Ancient Greek. Whatever one's prima facie intuitions about this are, we should note that since this property is language expressible in present-day English, but presumably not in the English of 1600, something must have happened in the recent history of English that allowed for the language expressibility of this property. So, how did we come to be able to express it? That certainly is a hard question, related to some difficult issues. Two possibilities come to mind, however, namely:

- "being a quark" is a theoretical predicate of physics. It is at least in part implicitly defined by the physical theory that uses it. Thus we can express it because we have the theory.
- We can express the property of being a quark because we have been in contact with observable phenomena that are caused by quarks, like effects they have on some measuring instrument.

If either one of these is the correct account, then there is no problem for the internalist. The reason is simply the following. If the first is correct, then the problem of expressing the property of being a quark reduces to expressing the theory that implicitly defines "being a quark," plus making the implicit definition explicit. Simply put, the property of being a quark is the property of being such that the theory truly describes you. Thus the problem is pushed back to the properties used in the implicit definition of "being a quark," that is in the formulation of the theory that implicitly defines it. In general, though, if the apparent increased expressive power of new theoretical concepts comes from their implicit definition in scientific theories (or from mixing those with the above second point), then internalism is not in trouble.

If the second possibility is the right one, then the increased expressive power does come from being in contact with more objects. If we introduced "being a quark" as

(340)   being the kind of thing that causes these effects on the measuring instrument

or something along this line, then being a quark is loosely speaker expressible in older languages, though not language expressible. This case thus essentially reduces to the case of the standard inductive argument.

To be sure, these are only rough outlines of how this can work. How such predicates work in general is very difficult to say. We should, however, keep the fact in mind that something must have happened in the last few hundred years that made the change from speakers of English not being able to language express this property (at least not with a simple predicate) to their being able to language express it with just a few words. One easy explanation of how this might have happened is that speakers were able to express the property before, after all, either with a complex eternal predicate, or with some non-eternal predicate in the right context. If this is so, then there is no puzzle how we can now express it with a few simple words: we just introduced a word to stand for a property that we could express already, though only with a complex predicate, or only in special circumstances. But if this isn't so, what might have happened that made the difference? One answer is holism, and it is hard to see what another answer might be. The issue how we can understand expressive change over time in connection to these issues is discussed in more detail in Hofweber (2006a). There I put forward the expressibility hypothesis: the hypothesis that expressive change in human languages over time reduces to decontextualization and lexical addition. Decontextualization is making certain contents available to be expressed in any context that earlier were only available in particular contexts, paradigmatically introducing a name for an object that before could only be referred to demonstratively. Lexical addition is allowing for the expression of something in a simple way that could only be done in a complicated way before, paradigmatically introducing a simple word for something that before could only be said with many words. The expressibility hypothesis is independent of internalism, but congenial with it. Similarly, it is independent of internalism, but congenial with it, that all human languages have the same in principle expressive power, a topic discussed, for example, in von Fintel and Matthewson (2008). We won't discuss expressive change or expressive difference among human languages here. Many of these issues are empirical and we would have to remain in the realm of hypotheses, some of which are explored in Hofweber (2006a). However, we will consider in detail how we should conceive of differences of what can be represented by human beings and other creatures, real or imagined, in the next chapter, chapter 10.

9.4.2 Cardinality Arguments

A second strategy to argue against internalism using considerations about inexpressible properties takes recourse to cardinality considerations. Such arguments try to establish that the set of all properties is strictly larger than the set of all expressible properties. Here is a paradigmatic cardinality argument against internalism:

Our language has only a finite base vocabulary, and only finite combinations of it are allowed to form predicates that express properties. Thus overall we can form countably many predicates.

But there are uncountably many properties. There are, for example, uncountably many real numbers. And for every real number there is the property of being larger than that real number, and for any two different numbers those properties are different in turn. Thus there are uncountably many properties. So, internalism has to be false.

This is a prima facie very plausible argument. But once we take into account the distinctions that were drawn above we can quite easily see that it is flawed. The argument would work against extreme internalism, which holds that properties are shadows of eternal predicates. But, of course, this is not the form of internalism we are discussing now. Internalism has to be understood as moderate internalism, which holds that properties are shadows of predicates, though not just of eternal predicates.[6] The truth conditions of sentences that contain quantification over properties are captured by infinitary disjunctions as well as infinitary first-order external quantification. In particular, what is in the domain of the first-order external quantifiers will matter for the truth conditions of quantification over properties.

And once we consider this formulation of internalism we see that the above argument provides no problem for it. If we grant, as is presupposed in the above argument, that real numbers are objects in the domain of first-order quantification, then it will be true according to moderate internalism that

(341)   For every real number there is a property which is the property of being larger than that real number.

According to internalism the truth conditions of this sentence can be spelled out as (in semi-formal notation):

(342) $\forall r \exists \vec{v} \bigvee_p([_{nom}P(v_i)] = \text{being larger than } r)$

And (342) is true, as can be seen as follows. Fix an arbitrary number r. One of the disjuncts in the disjunction will be

(343)   (being larger than v_i = being larger than r)

with variable v_i bound from the outside by an (infinitary) existential quantifier. Since this variable ranges over real numbers, in particular number r, there is a value to the variable that makes this disjunct true, namely r. So, (342) is true.

Real numbers, as objects of the domain of first-order quantification, can be the values of the variables that occur in the infinitary disjunctions, which are bound from the outside by the (infinitary) existential quantifiers. So, the more objects there are in the domain of first-order quantification, the more properties are loosely expressible, and the stronger is quantification over properties. Arguments against internalism that rely on a certain set of things of large cardinality and then proceed to argue against internalism that there is a connection between that set of things and how many properties there are, and finally that thus there are more properties than the internalist can

---

[6] See section 9.1.2 for the distinction between extreme and moderate internalism.

allow, won't work. They rely on internalism being understood as extreme internalism, and thus based on using eternal predicates as the basis of expressibility. They do not affect the moderate version of internalism, which connects what properties there are to which objects exist.

Different versions of cardinality arguments are possible as well, but the reply to them is in essence the same. I won't discuss any further ones here, but more on this is in Hofweber (2006a). That paper also discusses whether the present proposal leads to paradoxes, since it allows for too many properties. Every predicate expresses a property, since "is F" expresses the property of being F. But then we seem to get paradoxes, like the Grelling–Nelson paradox, since we would have the property of being heterological, i.e. the property of not applying to oneself. Thus internalism leads to paradoxes, while externalism does not. However, as I argue in Hofweber (2006a), this diagnosis is mistaken. First, the paradox can be restated in terms that don't talk about properties at all, only about predicates, and no one should think that there is no such predicate as "is heterological." Second, several approaches to the paradoxes are compatible with every predicate expressing a property, for example Hartry Field's, in his Field (2004) and Field (2008). Another one is in Hofweber (2008). What the paradoxes show in the end is, of course, a completely different question, but to reject a close connection between properties and predicates because of them is an overreaction, just as rejecting a close connection between sentences and propositions for that reason. For more on paradoxes and internalism, see Hofweber (2006a).

9.5 Appendix 2: Generalized Quantifiers, Once More

Our treatment of generalized quantifiers in section 3.7 relied on the simple form of internalism. Here context-sensitive elements were not considered, in part because it wasn't central to deal with the examples of quantifiers that were our focus there. Now we need to at least outline how the account of generalized quantifiers given in section 3.7 extends to our present, and proper, treatment of the truth conditions of quantified statements on their internal reading, in particular when they range over properties and propositions. This appendix can also be skipped. It presupposes the appendix on generalized quantifiers of chapter 3.

The key for dealing with generalized quantifiers in the simple case was to consider the instances of an internally quantified statement, up to true identity claims. This leads to equivalence classes which can then be used to specify the truth conditions of generalized quantifiers as spelled out in Table 3.1. But now the instances contain new free variables, bound from the outside by external particular or universal quantifiers over objects, in the case of the particular or universal internal quantifiers ranging over properties or propositions. But a modification of our old approach is possible in a quite simple way.

First, take the domain D which is the domain over which the external quantifiers range. The truth conditions of a sentence in English are given by a sentence in a language $\mathcal{L}$ which we assume captures every aspect of English relevant for us now, minus the internal quantifiers. To give the truth conditions of those, expand $\mathcal{L}$ to $\mathcal{L}^+$ by adding new constants for every object in the domain D. Now the instances of the internal quantifiers in $\mathcal{L}^+$ are treated as context-insensitive instances, skipping the external quantifiers that bind the new variables discussed above. Instead there are many more instances, since $\mathcal{L}^+$ is most likely not a countable language any more, since it contains a new constant for every member of D, which could be very many. We can think of internal quantifiers in $\mathcal{L}$ as generalizing over instances that involve demonstratives that can refer to any object whatsoever, and internal quantification in $\mathcal{L}^+$ as generalizing over context-insensitive instances, but with a name for any object whatsoever available to the instances. The truth conditions in each case remain the same. Still, in $\mathcal{L}^+$ we can now carry over our truth conditions given in Table 3.1. That we are potentially dealing with higher cardinality cases won't affect the truth conditions given there. Let's take the case of quantification over properties to illustrate. We can say that two instances P and Q belong to the same equivalence class just in case the identity statement of the corresponding nominalizations is true:

(344) $[_{nom}P]$ is identical to $[_{nom}Q]$

Let P_{IS} be the equivalence class around P formed in this way. We can then again, as in section 3.7, talk about equivalence classes being realized in a sentence and use considerations about how many are realized in the truth conditions of the generalized quantifier statements, just as spelled out there, allowing for the adjustments that we are potentially dealing with different categories of instances.

The idea, overall, simply is to get rid of the new external quantifiers binding new variables in favor of new names and a higher cardinality language, and then use our treatment of generalized quantifiers as given in section 3.7 for that higher cardinality language. Without repeating the details, I hope it can be seen that such a treatment of generalized quantifiers is possible even with the newly specified truth conditions of the internal readings of quantifiers.

Besides extending the account of quantification over properties and propositions to generalized quantifiers, there are other issues that we have not addressed here. We did not properly discuss plural quantification, we did not discuss other quantificational devices like adverbs of quantification, we did not discuss cross-linguistic consideration about how quantification is different across languages, and how the account presented can be carried over to them, and so on and so forth. All these are substantial further issues that we can't hope to resolve or properly address here.[7] For each one of them we

---

[7]   For plurals alone the list of issues to consider would be vast, and include such different issues as those discussed in, for example, Boolos (1984), Link (1998), Linnebo and Nicholas (2008), and many more.

would need a good bit of further work, as they are complex topics in natural language semantics. Just to take plurals as an example, we would need to carry over some of the insights from the semantics of plural quantification and plurals in general to formulate the truth conditions of plural quantifiers. This would have to be done not only for plural quantification over objects, so to speak, as in

(345)   Some numbers have their sums among them, for example, the even numbers.

but also for plural quantization over properties and propositions:

(346)   Some properties exclude each other, while others are compatible.

To deal properly with plurals when quantifying over properties and propositions we would have to allow not just our new singular variables discussed in this chapter, but also new plural variables, corresponding to "plural demonstratives" as opposed to merely the singular ones we focused on here. And we would need to accommodate collective and distributive predication, and so on and so forth.

Another issue we are not able to look into in more detail here is the apparent non-wellfoundedness of the way in which the truth conditions of the internal quantifiers are specified, in particular when quantifying over properties and propositions. For a simple illustration, take a universal quantifier over propositions, which is equivalent to the conjunction over the instances. But the instances themselves will contain universal quantifiers over propositions. And its instances will again contain such quantifiers, and so on. This might sound like a big problem, but it is closer to a fact of life when talking about propositions and their truth on anyone's view. On an externalist account of quantification over propositions, these quantifiers range over all propositions, many of which make universal claims about all propositions, including themselves. Such issues are familiar in the literature on the semantic paradoxes, in particular in the discussion of truth and the non-vicious circularity and non-wellfoundedness that the evaluation of sentences with a truth predicate in them can give rise to. The options to deal with this phenomenon for us here are similar to those discussed in the literature on paradoxes. I have discussed some of this in a more technical context in the appendix to Hofweber (1999), where comparisons to hierarchical as well as non-hierarchical approaches in the theory of truth, as well as to weak fragments of infinitary logic, are made.

All these issues are good and important things to work out, but I won't be able to do this here. I hope that what we have seen in chapter 3 and in this chapter motivated that we have an internal reading of quantifiers, that it is operative in quantification over properties and propositions, that expressibility considerations do not refute this, and that we can see for a range of cases what the truth conditions of sentences with quantifiers in this reading are. Further issues will have to be postponed for another time.

9.6   Appendix 3: Revisiting the Internal–External Distinction

The discussion of internal quantification over properties and propositions made clear that we need to distinguish a context-dependent instance of a quantifier from a context-independent one. In particular, we need to allow for demonstratives or free variables to be part of the instances, which are then bound with external quantifiers from the outside. As we saw, these considerations are crucial for understanding internal quantification over properties and propositions. We should now consider how this issue affects our original distinction between internal and external readings of quantifiers, in particular also quantifiers over objects. Here, too, we can distinguish instances that are context-insensitive from those that are not, in particular those that contain demonstratives like "the mother of that boy" or simply "this." Such instances of internal quantifiers over objects would then need to be bound by external quantifiers, also over objects. But this might seem worrisome. Does it collapse the distinction between internal and external readings of quantifiers? And would it affect our discussion of talk about natural numbers and the internalist position about arithmetic defended above? In this appendix I would like to show briefly that this is not so. Modifying the account of the internal reading of quantifiers over objects along the lines suggested in this chapter does not collapse the distinction between internal and external readings of quantifiers, nor does it affect our conclusions about internalism. But it does affect some smaller points made above.

Suppose then we do modify internal quantification over objects along the lines developed in this chapter. We thus allow instances with demonstratives or variables, and bind them with the proper external quantifier over objects from the outside. Thus

(347)   Something is F.

has the truth conditions, on the internal reading, that

(348) $\exists \vec{x} \bigvee_t F(t[\vec{x}])$

where $\vec{x}$ are variables some of which may occur freely in term t. Here $\exists \vec{x}$ ranges externally over the domain of objects, whatever they may be. But even on this understanding of the truth conditions of the internal reading of quantifiers, they differ from the truth conditions of the external reading, although not quite as much as we had before. Now the external reading implies the internal reading (for the particular quantifier), but not the other way round: if some object o in the domain satisfies F and thus the external reading of "something is F" is true, then the internal reading is guaranteed to be true as well: the disjunct "this is F" is true when "this" has o as its value. On the other hand, the internal reading does not imply the external one: if "t" is a term that does not stand for an object in the domain, but "F(t)" is true while nothing in the domain is F, then the internal reading is true while the external one is false. The example of Fred admiring only Sherlock still shows that there is something Fred admires is true on the internal reading, but false on the external one. The situation with the universal quantifier is exactly the other way round, and *mutatis mutandis* for other quantifiers. These truth conditions of the internal quantifier don't collapse the distinction between the internal and the external readings, but do require a revision to something I said above in chapter 3. There the claim was that neither reading implies the other. This is true on the context-insensitive version of the internalist truth conditions, but not on the context-sensitive one. Now we do have some implications, but not in both directions.

Should we take the truth conditions of the internal reading of quantifiers in general to be the simple ones given in chapter 3, or the context-sensitive ones outlined here based on the considerations in this chapter? The truth conditions of the internal reading were motivated by the need for a certain inferential role. These different truth conditions correspond to different notions of an inferential role. The eternal, context-insensitive truth conditions given in chapter 3 give the quantifier a language-internal inferential role, one invariant for different utterances of the sentence in different contexts. The truth conditions outlined in this appendix give an utterance of a quantifier an inferential role, placing it inferentially among utterances in any context of the quantifier-free instances. Utterance inferential role can then be distinguished between factual and loose utterances, just as we above distinguished between factual and loose speaker expressibility. This distinction will correspond to whether the external quantifier in the internal truth conditions is restricted to objects actually available to speakers. In the most permissive and unrestricted sense we should thus see the truth conditions of the internal reading of the quantifier over objects to be the ones tied to the loose utterance inferential role, just as we did in this chapter with the truth conditions of quantification over properties and propositions. This extra complexity is central for a proper understanding of quantification over properties and propositions, but it isn't central for our concerns about quantification over objects, and, in particular, for quantification over natural numbers. Here the simple truth conditions were enough to formulate internalism properly, and the more complex truth conditions won't take away from that. All the arguments for internalism, in the case of natural numbers, as well as against internalism, in the case of ordinary objects, were based on considerations about quantifier-free expressions, which were then completed with an account of quantifiers. Here both the simple version we used in the relevant chapters as well as the more complex version developed now would do. In the case of properties and propositions the simple version will not do, however, and only the proper, more complex one gives us a coherent internalist picture.[8]

---

[8] Thanks to Jody Azzouni and Otávio Bueno for discussions of the issue in this appendix.

10

Ineffable Facts and the Ambitions of Metaphysics

Reality, broadly understood, can be thought of in two different ways, maybe corresponding to two ways of making an underspecified notion more precise. First, reality can be seen as everything that is the case—the totality of all facts that obtain—and, second, reality can be seen as everything there is—the totality of all things that exist. These two ways of thinking of reality are clearly related. For example, for every thing that exists the fact that this thing exists obtains. And for every fact that some particular thing exists, that thing needs to exist. Still, there might nonetheless be important philosophical differences in these ways of conceiving of reality. Reality as the totality of things is closely tied to ontology, since ontology tries to find out what exists. Reality as the totality of facts is closely tied to inquiry in general, since inquiry in general tries to determine which facts obtain. In this chapter we will consider whether there are any limits to inquiry, as carried out by human beings, which are tied to limits of what we can represent conceptually. We will in particular consider how such a limitation might affect the ambitions of metaphysics. This will have implications for how the two conceptions of reality are related in philosophically significant ways.

When we try to find out, in inquiry, what reality is like, we try to find out which facts obtain. To do this we need to do two things: first, represent a fact in thought or language and, second, determine that this representation indeed represents reality accurately. Both tasks, when completed, are an achievement. To be able to represent a fact is not always trivial, it requires certain conceptual resources, the ability to form a certain conceptual representation of what is supposed to be the case. And to determine that a representation is accurate is, of course, also not trivial. And both aspects of this task lead to a possible limitation in inquiry for human beings: first, there might be certain facts that we human beings are unable to represent since we don't have the relevant conceptual resources, and second there might be certain facts where, even though we can represent them, we are unable to determine whether these representations are indeed accurate. The former is a representational limitation, the latter is an epistemic limitation. Whether and how much we are limited in what we can know due to a limitation in what we can find out to be true, i.e. whether we are limited epistemically, is widely discussed, and mostly has its own field in epistemology. But whether we are limited in what we can represent in the first place is not as widely discussed, but also very significant, in particular for metaphysics.

Any fact that we human beings are incapable of representing in thought or language we can call an *ineffable fact*. If there are any ineffable facts then all of them together would form an aspect of reality that is ineffable for us, beyond what we can even consider or entertain to be the case. Whether or not parts of reality are ineffable for human beings in this way is crucial for what we can hope to achieve in inquiry. In particular, it is crucial for whether ambitious metaphysics can hope to achieve what it aims to do. If some aspects of reality are ineffable for us, then how can we hope to arrive at a large-scale picture of what reality is like? Some parts of reality we can't even entertain in thought or language, so how can we hope to come to a large-scale picture of all of it, including the ineffable parts? Prima facie, the ineffable seems to undermine ambitious metaphysics and point to a limitation we face in inquiry. The question of whether there are any ineffable facts is thus closely tied to questions about how our minds and their representational capacities fit reality and the totality of facts.

In this chapter I will argue that the answer to these questions depends on whether or not internalism or externalism is true for talk about facts and propositions. Either way, there will be substantial consequences from the answer, and I will in particular focus on the consequences we get from the internalist answer, since I defended internalism in the previous chapters. The internalist answer will unsurprisingly be that there are no ineffable facts. After all, ineffable facts are naturally tied to inexpressible propositions: for any fact that p there is a true proposition that p, and for any true proposition that p there is a fact that p. Whether or not true propositions and facts are the same, or just closely correlated, is inessential for such a connection. Whatever their more precise relationship, it is clear that internalism and externalism come as packages for facts as well as propositions. All the reasons we had for (and against) internalism about talk about propositions carry over to reasons for (and against) internalism about talk about facts. We could simply speak about internalism with regard to *the propositional*: propositions, facts, truths, and similar things that are specified with a that-clause. Thus our topic from chapter 9, whether there are any inexpressible propositions, is closely connected to our present question, whether there are any ineffable facts. And it will thus be no surprise that, properly understood, the internalist will say that there are no such ineffable facts. But the questions we are trying to answer here are not just the same as the ones we considered in chapter 9. We are now approaching a similar issue from a different angle, and I hope to make clear what the significance of the position defended so far is when seen from that angle. The general issue largely left out in chapter 9 can be put as follows: questions about the ineffable are about the relationship between thought and reality. But how could it be that considerations about language, about that-clauses and quantifiers, could support that our representational capacities are fully up to the task of representing all of reality? This will need to be explained and explored in more detail to really make sense, and we will do this here. The more technical considerations that appear to support inexpressible propositions, and by analogy, ineffable facts, were considered in chapter 9. Now we must consider the larger philosophical questions tied to this issue.

Although I defended internalism above, both the externalist as well as the internalist answers to the question about the ineffable have significant consequences, in particular for metaphysics. I will thus not only pursue the internalist take on this topic, but also the externalist alternative. I will argue that the answer to the question of whether any aspects of reality are ineffable depends on whether internalism or externalism is true about the propositional. To do this we will first look a bit more at the connection of ineffable facts to the ambitions of metaphysics. Then I will make a proposal about what an externalist should say about the question of whether any aspects of reality are ineffable, and what follows from it for metaphysics. After that we will see what an internalist should say, and what follows from that. Since I defended internalism above I will defend its consequences below as well, and the consequences of internalism about talk about propositions are quite substantial and significant.

10.1 The Dark Vision

Metaphysics hopes to find out what reality is like in general ways. So understood it is an ambitious project, and the question remains whether we human beings are suitable to carry it out. One familiar worry about our possible inability to carry it out is epistemic: we might be able to consider the right answer, but we will be unable to find good enough evidence to support it over some of its alternatives. We might have to remain ignorant since we can't tell which of various options is the right one. A more substantial worry supports a stronger kind of limitation: we might not even be able to conceive of the right answer. These possible limitations apply not just to metaphysics, but to inquiry in general. Nonetheless, the second kind is especially pressing for metaphysics. We could thus distinguish two kinds of ignorance we might face. *Regular ignorance* obtains when we can ask the questions whether something is the case, but we can't figure out what the answer is. This presumably is the case for us presently for example with the question of whether an asteroid caused the extinction of the dinosaurs, and we can thus give examples of regular cases of ignorance. On the other hand, *deep ignorance* obtains when we don't know something and we can not even ask the question whether it is the case. Contrary to regular ignorance, we cannot give any examples of deep ignorance, since otherwise we could ask the question whether this example is indeed a fact. Deep ignorance has as its source a representational limitation, a limitation in what we can say or think, and if we are deeply ignorant of some parts of reality then this might affect metaphysics. Our minds might be too limited to carry out metaphysics in its ambitious form. The worry here is that some aspects of reality are so alien that a mind like ours simply can't represent them in thought or language, that is, represent them with a conceptual representation. The concepts available to a mind like ours might be limited in a way that not all of reality can even in principle be represented with them. A limitation in our representational capacities might have the result that some facts are bound to be ineffable for us, and some truths are thus beyond what we can even conceive, not to speak of come to know. We would be deeply ignorant of them. If there are ineffable facts, then this presses the question of why we should think that human beings can hope to be able to find out what reality is like in various general ways. After all, some facts that obtain and thus are part of reality are beyond what we can even represent. As such they not only can't be known by us to obtain, we can not even consider whether they obtain. To consider whether a fact obtains requires representing that fact in thought or language and then to wonder whether it indeed is a fact. Ineffable facts, if there are any, are completely beyond us, unknowable and beyond what we can consider or entertain. Ineffable facts thus can be more hidden from us than merely unknowable ones or merely incomprehensible ones. All ineffable facts are unknowable and incomprehensible, but not the other way round. We will never know whether the number of grains of sand on earth exactly 500 million years ago was odd or even, but we can represent both options. And we might never comprehend or understand why anything exists at all, even though we can easily represent this fact. If we had good reason to believe that there are ineffable facts and thus that reality outruns what we can represent, then we might well also have reason to hold that a good part of reality outruns our capacities to represent it in this way. It would seem dubious that we could still hope to succeed in ambitious metaphysics if this were our situation.

We should clarify the notion of an ineffable fact to see more clearly whether ineffable facts undermine metaphysics.[1] First of all, the issue is about conceptual representation, not other kinds of representation. Conceptual representations are involved in thought or language. Such representations are true or false. Other kinds of representations, for example paintings, photographs, or other kinds of pictorial representations, are accurate or inaccurate, but not true or false. Conceptual representations are employed in thoughts or utterances, and thus they are central when it comes to knowledge and inquiry, which aims for knowledge. If there are any limits to our conceptual representations, then they are tied to limits of inquiry as we can hope to carry it out. Our focus is thus whether we are limited in our conceptual representations.

Second, a fact is ineffable if we *cannot* represent it conceptually, that is, represent it in thought or language.[2] But this characterization is rather imprecise, since "can" can be understood in many different ways. On one straightforward understanding it is clear that there are many facts that we human beings cannot represent conceptually. To give one example, consider "the sand-metric of planet earth": the precise distance of every grain of sand on earth to every other one (right now). This is an enormously complex fact, since there are supposed to be something like 10^{24} grains of sand on earth. No

---

[1] The characterization of an ineffable fact is discussed in more detail in Hofweber (2016a), but the extra details aren't central for us here.

[2] Of course, it is not enough to represent a fact to have a representation about the fact, as in: John's favorite fact is amusing. Rather, to represent a fact one must have a representation that captures the fact, as in: "Dogs bark" represents the fact that dogs bark.

human being will live long enough or has a brain big enough to represent the sand-metric fact. But even though this is a limitation in what we can represent, it is not one that undermines the ambitions of metaphysics. That fact is simply the conjunction of many simple facts, each one of which we can represent without problem: grain x is 7.3 cm from grain y, and so on. To make "can" more precise in the characterization of the ineffable we need to see what we should allow to vary and what we should keep fixed. Any way of doing this will give us a perfectly good notion of the ineffable, but we need to focus on the most interesting one for our purposes. We should thus focus on a notion of the ineffable as the *completely ineffable*: something that we can't represent in principle, even allowing ourselves more time and more memory. The completely ineffable would be a fact that a mind like ours is simply unsuitable to represent, not simply because of limitations of time or memory, but even if we relax those constraints. The notion of the ineffable as the completely ineffable is still not very precise, but at least it is an improvement and gets closer to the most relevant precise notion, which we will hope to get even closer to shortly.

Similarly, the notion of a fact should not be understood too finely when we wonder whether metaphysics is overly ambitious in light of there being ineffable facts. If we individuate facts so finely that the fact that I am hungry is different from the fact that TH is hungry, and the fact that it is sunny here is different from the fact that it is sunny in Chapel Hill, then there will be lots of ineffable facts, and different ones for different people, places, and times. If facts are that finely individuated, then it matters who, where, and when one is to be able to represent a fact. Such facts can be ineffable for a person, time, or place, but they are not in principle ineffable for human beings. There will be lots of ineffable facts, finely understood, but it won't motivate that ambitious metaphysics that tries to come to an understanding of the general features of reality is overly ambitious. For that it matters whether there are ineffable facts even when understood more coarsely.

We hope to see whether our minds are up to representing all of reality, that is, whether there are any facts that human beings are in principle incapable of represent-ing conceptually. And one way to consider this is to see if such a fact would pass *the incommunicability test*: some other creature, be it an advanced alien or a god, would be incapable of helping us to form such a representation even if they could represent the fact and they tried to be as helpful as possible. They couldn't tell us what this fact is no matter how hard they tried. Our minds just wouldn't be the kinds of minds that could represent that fact if this were the case. All the aliens could tell us is that unfortunately we are not the kinds of creatures that can represent this fact, and their best effort to help us would be in vain.[3] An ineffable fact, in the most interesting sense, should pass the incommunicability test. This has as a consequence that ineffable facts aren't just those that involve objects that we might have a hard time referring to, say unusual abstract objects, or material objects outside of our light cone, or what have

---

[3] For a science fictional description of a case like this, see Varley (1977).

you. Any alien who can represent a fact involving an unusual object *o*, say the fact that *o* is unusual, can help us represent that fact as well, by allowing us to piggyback on the aliens' representation of *o*, i.e. their name for *o*. In general we are able to exploit the referential abilities of others to expand our representational repertoire. This is how we refer to Socrates and other past or faraway objects, after all. We can say that the interesting notion of the ineffable is *object permitting*: conceptual representations of particular objects are not a relevant limitation. That is to say, we can for present purposes grant that we can have a representation of any object, via a name for that object. The real worry is that there are some facts that require a representation of a completely different kind than we can have. These would be facts that aren't the fact of an object having a property, or something equally familiar, but a completely alien fact, a kind of a fact that a mind like ours just can't in principle represent, since it requires a way to represent that is beyond us. This is the issue relevant for us here and for metaphysics.

Let us call the *Dark Vision* the worrisome possibility that we are bound to be deeply ignorant of large-scale features of reality, since we are limited in what we can in principle represent in thought or language. If the Dark Vision obtains, then we are limited in what ambitions metaphysics can have for human beings due to a limitation of which facts human beings can in principle represent conceptually. Our minds would be too simple to carry out metaphysics in its ambitious form. It wouldn't conflict with metaphysics in general. Locally metaphysics might well be in a position to answer its questions. But it would be in tension with the ambitious form of metaphysics that tries to find out what reality in its large-scale features is like, that is: what all of reality is in general like, including the parts we can not even conceive. If the Dark Vision is correct then human beings seem to be incapable of completing ambitious metaphysics successfully. We might be able to make some progress on it, but the project is beyond what we can carry out successfully. An important question tied to our assessment of ambitious metaphysics is thus whether there are any truths or facts ineffable for human beings. Are there some facts which are in principle beyond what human beings can represent in thought or language?

There are two possible answers to this question: yes or no. They correspond to two theses:

The effability thesis: Everything is effable.
The ineffability thesis: Something is ineffable.

Which one is it? And what must our minds and reality be like for that to be the right one? I will argue that the answer to this question depends on whether internalism or externalism is correct about talk about propositions. If internalism is true, then the effability thesis is true, and consequently there are no restrictions on the ambitions of metaphysics from these considerations. However, the proper way to understand why our representational capacities perfectly match the facts leads to significant consequences, which we will discuss in detail below. On the other hand, if externalism is true, then we should think that some facts are ineffable for us and thus that the ineffability thesis is true. We should conclude from this that ambitious metaphysics has to be approached with a good dose of modesty and humility. I will argue that modesty doesn't simply follow from there being ineffable facts, but rather from how we should think the ineffable and effable facts are related. I would like to start by considering the issue from an externalist point of view, despite the fact that I argued in earlier chapters that externalism is false. Nonetheless, it will be crucial for understanding the consequences of the internalist position on this issue to contrast it with the externalist alternative. Furthermore, I hope to make the case that there are important consequences for the externalist picture that I do not believe are fully appreciated by all those who believe externalism to be correct.

10.2 Externalism and the Ineffable

Suppose now, for the sake of this section, that externalism is correct and some of our talk about propositions is true. Talk about propositions is thus about a domain of entities. It is natural to take this domain of entities to be simply there, independent of our talking about it: it is just one aspect of reality that is there for us to discover.[4] Thus externalism will be true not just for talk about propositions in particular, but about the propositional more generally, including talk about facts and truths, which closely correspond to true propositions, as discussed above. The externalist picture of the propositional thus takes reality, understood as what is the case, to consist of a language-independent totality of facts. Such a domain of facts is simply there, for us to represent. Thus for there to be ineffable facts two things have to come apart: what facts there are in this domain, and what facts we can represent in principle in thought or language. I will argue that an externalist should believe that there are ineffable facts. Of course, no examples of such alleged facts can be given, at least not explicitly, but nonetheless, there are a number of good arguments that establish that what is the case and what we can in principle represent to be the case come apart. The most important ones are as follows.[5]

10.2.1 Arguments for the Ineffable

BUILT-IN COGNITIVE LIMITATIONS

The thoughts we can think must fit into our minds. And our minds think a certain way; they have a certain cognitive setup. So, any thought we can think must have a

---

[4] An exception to this externalist picture is Schiffer (2003). For Schiffer, propositions exist, but they are not language independent in a sense he aims to make clear. I take the more standard externalist picture on board here where propositions exist and are independent of us.

[5] I won't reintroduce other important arguments now which we already discussed in appendix 1 to chapter 9, for example the cardinality arguments. The arguments in this section are mostly focused on the large-scale relationship between what we can represent and what reality is like.

content that a mind like ours can represent. But our minds didn't develop with reality as a whole as their representational goal. They developed to deal with situations that creatures like us have to deal with to make it: midsize objects that are reasonably stable and have stable properties, some of which need to be eaten, some that need to be avoided, and so on. The question is why we should think that a mind that developed to deal with problems in this limited situation and under those selection pressures should be good enough to represent everything there is to represent about all of reality. We know that not all of reality is like the world of stable midsize objects. The very small is very different than that, for example. We should thus expect that a mind that has developed like ours won't be suitable for all of reality. Our biological setup imposes a constraint on how we must think. This constraint arose in response to selection pressures that came from a special kind of an environment. We can expect minds like ours to be good enough to deal with the situation they evolved to deal with, but why should they be good enough to deal with any situation whatsoever? Thus we should expect that our mind has a hardwired constraint on what it can represent and that this likely won't be good enough for all facts.[6]

THE ARGUMENT FROM ANALOGY

Although we can't give an example of a fact ineffable for us, we can give examples of facts that are ineffable for other, simpler creatures.[7] Take a honeybee, which can represent various things about its environment like where the nectar is, but is in principle incapable of representing that there is an economic crisis in Greece. Its mind is just not suitable to represent such facts, even though it can represent other facts. That there is an economic crisis in Greece is a fact ineffable for the honeybee, but not for us. But we can imagine that there are other creatures that relate to us like we relate to the honeybee. We can imagine that there are vastly superior aliens or gods, say, who look down at us just as we look down at the honeybee. And analogously, they would say that we humans can't possibly represent that *p*, while they clearly can. They would be able to give examples of facts that are ineffable for us, but not for them. Whether or not there really are such aliens or gods doesn't matter for this argument. The point simply is that this analogous reasoning makes clear that we should expect there to be such facts that could be mentioned by the aliens or gods as ineffable for us, but not for them. The facts are there, whether the aliens and gods are there or not.[8] What we can

---

[6] See Fodor (1983: 119ff.), Nagel (1986: 90ff.), and Noam Chomsky at various places, for example Chomsky (1975). Chomsky's views on this matter are more carefully discussed in Collins (2002), which contains many references to particular passages of Chomsky's work.

[7] The argument from analogy is a more radical version of the inductive argument discussed in chapter 9. The inductive argument in essence was motivated by expressive change over time among humans. The argument from analogy considers differences across species, and thus more radical differences. The inductive argument was used to point to the importance of context dependence of what can be expressed. The argument from analogy is now used to point to more fundamental differences in what different creatures can represent.

[8] This argument can be found in Nagel (1986: 95f.).

represent is somewhere on a scale, with the honeybee towards the bottom, cats and dogs above that, us better still, but there is plenty of room at the top. Even if we are the best so far when it comes to representing reality, there is room to grow, and other creatures might well be much above us.

Maybe the honeybee's representations are not conceptual representations and thus might not have propositional contents at all, but merely indicate something about the world. They might carry information, but not have propositional content.[9] But this doesn't really change the situation. Just as we can point to information that the honeybee can't carry with its representational capacities, aliens might point to propositional contents that we can't represent conceptually, besides higher forms of representation that they have in addition. In the end we should think of what can be represented by us as being somewhere on a scale, with the honeybee on one side of us, and other creatures, real or imagined, on the other. And what those further over on the scale can represent is ineffable for us, and thus we should think that some aspects of reality are ineffable for us.

EXPLAINING EFFABILITY

If nothing is ineffable, then what facts obtain and which facts we can represent are exactly the same. But that would mean that two very different things exactly coincide. What we can represent is one thing, what reality is like is a quite different thing. If they coincide, then we should ask for an explanation of why these two coincide. It is certainly conceivable that they do coincide. Maybe our representational capacities have reached the limit of what can be represented. Maybe we just made it to the top, while other creatures were still on the way up, unlikely as all this may be. But even if everything is effable, we should ask for an explanation of why it is so. And if no such explanation is forthcoming, we should accept that it isn't so, barring good reasons to the contrary. But what would explain that what reality is like and what we can represent about it coincide?

The most natural way one might try to explain why what reality is like and what we can represent coincide is via a connection of what reality is like and our representational capacities. One route for such a connection is of limited use: what reality is like affects what we can represent. This route can explain why our representations are sometimes accurate, but not why they exhaust all of reality. That reality affects and forms our representational capacities makes plausible that we sometimes represent correctly, but it doesn't explain why we can represent all of reality. The other route does better here: what reality is like is affected by our representational capacities. In particular, what there is to represent about reality is somehow due to us and our minds. This is a version of idealism, and it is in a sense the natural companion of the effability thesis. Reality is effable by us in its entirety since we, in particular our representational

---

[9] See Dretske (1981).

capacities, are responsible for what there is and what it is like. No wonder our minds are good enough for all of reality, since reality is not independent of our minds.

Idealism could in principle explain why the effability thesis holds, but we have good reason to think that this form of idealism is false. That is, we have good reason to think that reality do not depend on us in the alluded to sense. What there is and what it is like do not depend on us in a natural sense of dependence. There would have been electrons and they would be like what they are in fact like even if we didn't exist, and so electrons and what they are like don't depend on us. Furthermore, there was a time before there were human beings when reality was otherwise pretty much as it is now. So, in a natural sense of dependence, reality doesn't depend on us globally. These are simple and possibly naive arguments, but if idealism should explain why the effability thesis holds it will need to be spelled out in a way that makes sense of a dependence of reality on us that can support and explain the effability thesis. There certainly are some options here that could be pursued. One is to try to analyze the content of statements of dependence in a way that makes them acceptable to idealism. Or one could develop the idealism in a way that places us in some sense outside of time, and connect time and the temporal aspect of reality to us as well. One version of this is well-known, Kant (1781), but it is not clear whether it is coherent, what a coherent formulation would look like, and whether it is compatible with other things we take ourselves to know to be true. An idealist explanation is in principle possible, but ones along the lines outlined above don't seem to have much going for them.

The idealist strategy to explain the effability thesis outlined above in effect connects two versions of idealism. Of those two, one is reasonably taken to be false and the other is closely tied to our present topic. Let us call *ontological idealism* the view that what there is, and what it is like, are mind dependent, in a sense to be made more precise. Let us call *conceptual idealism* the view that what can be said or thought about reality depends on us, in a sense to be spelled out. Thus what the range of the conceptual or propositional is that can be employed in principle to apply to reality is connected to us. Conceptual idealism is in essence the view that the effability thesis holds not by mere accident, but for a reason tied to us. Conceptual idealism combines the effability thesis with a certain explanation of why it holds. Ontological idealism might support conceptual idealism, but ontological idealism is false, or so we have good reason to think. The question remains whether conceptual idealism is nonetheless true, for other reasons. Could conceptual idealism be true even though ontological idealism is false? For that to be so the effability thesis has to be true, and it has to be true for a certain reason, not just by accident. So far we have seen no reason why that should be so; to the contrary, we have seen that there is little hope to explain why the effability thesis might hold. Assuming externalism, we should thus side with the ineffability thesis and accept reality as outrunning what we can represent about it.

Putting all these together we get good reasons to hold, on an externalist conception of the propositional, that some aspects of reality are ineffable for us. These arguments support that there are ineffable facts even on an object-permitting notion of the ineffable, and that these are ineffable in principle, not just de facto. Some aspects of reality should be seen as being beyond us in principle, beyond what human beings can in principle represent in thought or language. The question now is what we should think follows from this for metaphysics, again, assuming externalism about the propositional. We will contrast this with the internalist account which we will attend to shortly.

10.2.2 The Hiddenness of the Ineffable

Suppose then that we should accept that some aspects of reality are ineffable for us, in the sense that the human mind is not in principle capable of representing them in thought or language. What should we conclude from this? Even though many, I suspect, would accept that we are limited in what we can represent, few draw much of a conclusion from this. Although the consequences of this are not as broadly skeptical as one might fear, they are nonetheless significant. To see what we should properly conclude from there being ineffable facts we need to get clear on one further aspect of the ineffable that is puzzling. This is the problem of why the ineffable is so well hidden from us. Any fact that is ineffable would make it hidden from us in the sense that we can't represent it. But the ineffable, if it is indeed real, is more hidden from us than this, and this extra-hiddenness is puzzling in a way relevant for what we should conclude from its existence.

A first thing to note is that we never seem to perceive something ineffable. This is not a triviality. We might well wonder why it is never the case that we perceive something that we simply cannot describe or represent in thought. In such a situation we would be perceptually connected to a fact, event, or something, which we cannot represent in thought or language, and we might even realize that our representational capacities give out here. Why does this never happen? It is not inconceivable that one day we open a door, look inside, and what we see is simply beyond what we can describe in words. We see something that we can't represent conceptually, and realize that this is so. It is hard to imagine what that would be like in more detail, in part because it never happens, in part because we might have to rely on our concepts in imagination. The question is why this never happens. One answer, of course, would be that the effability thesis is true and nothing is ineffable, but there are also other possible explanations. One of them could come from the philosophy of perception. It might be that everything we perceive has to be conceptualized, and what we perceive has to be tied to the conceptual content of the perception. Since conceptual content, involving our concepts, can be represented by us, it is no wonder that what we perceive can be represented by us.[10]

But even if we could explain why we never perceive the ineffable without relying on the effability thesis, the question remains why we nonetheless never encounter the ineffable in other ways. It might be that what we perceive has to be conceptualized, and thus is effable, for reasons having to do with how perception works. But this

---

[10] One way this might be is in Kant (1781).

doesn't answer the question of why we nonetheless apparently never encounter the ineffable in other ways. Even if we never perceive it directly, we might realize that there is something ineffable right here, behind this door. In such a case we could realize that what we can perceive of the situation is effable, but there is more to it than what we can perceive, in the sense of represent in perception. This would be a scenario where we encounter the ineffable, and we recognize that what we encounter is ineffable. This might not just happen in unusual situations, but it could conceivably be part of regular everyday experience or scientific theorizing. We might recognize regularly that now we are approaching the limits of what we can represent, and that the answer to our problems lies beyond it. But why does this never seem to happen, given all the reasons we have seen for there being ineffable facts? To be sure, some people think they do encounter ineffable facts in various ways. Maybe visual art or music can present facts ineffable in language. Whether we encounter ineffable facts in artistic experiences is a worthwhile issue to discuss, but we don't seem to encounter them in ordinary inquiry and everyday perception. Why is the ineffable so well hidden from us and apparently so irrelevant for our attempts to understand various parts of the world in inquiry? The ineffable seems to fall out of the picture and for all intents and purposes every fact we encounter is effable. This naturally gives rise to the impression that the effability thesis should be true, even though, on the widely shared externalist assumptions, we have good reason to think it is false. To properly appreciate the significance of the ineffable we need to get clear on why the effable facts seem to be all the ones we ever encounter.

10.2.3 The Sub-Algebra Hypothesis

If we accept the arguments for there being ineffable facts then we need to understand how they might be related to the effable ones, in particular why the ineffable facts seem to be as irrelevant to ordinary inquiry as they seem to be, and why they are so well hidden from us. What could explain that the possibly vast range of ineffable facts is systematically hidden from us in a stronger way than is suggested by their being merely ineffable? There is a way to understand this, which seems to me to be the best way to make sense of it, and it is best spelled out with a mathematical analogy.

Consider a simple mathematical structure, say the integers with addition, multiplication, and subtraction:

$$\ldots -3, -2, -1, 0, 1, 2, 3, \ldots \qquad (+, \times, -)$$

The integers are closed under these three arithmetical operations. The sum, product, or subtraction of any two integers is always another integer. The integers with these operations thus form an algebraic structure. This structure corresponds to a language suitable to describe it: it has a name for each integer, and function symbols for each operation, and additional basic logical vocabulary.[11] This we could call the language of (the particular instance of) the structure.

---

[11] The additional vocabulary would in our case just include "=" and Boolean connectives. We don't allow variables or quantifiers here for the sake of the example.

Now consider the world, so to speak, from the point of view of an integer, where integers represent the world with the language just outlined. They will be able to capture the world in terms of the language of their structure. Not only can they represent their structure completely, from their point of view it will seem perfectly natural, even obvious, that the integers is all there is. After all, the sum of any two of them is always another integer, and so is the product or subtraction. And all those integers can already be represented by them. In particular, any question the integers can ask about the world in their language will have an answer that can be stated in their language. What's 7×8? It's another integer: 56. We can say that in their situation they enjoy *question–answer completeness*: if you can state the question, then you can state the answer. From the point of view of the integers, they naturally take it that they can capture all of reality there is to capture. If we were integers with those representational resources it would seem compelling to us that the integers is all there is.

All this is so even though the integers are embedded in different, larger structures, for example the rational numbers with the above operations as well as division:

$$\ldots - 3 \ldots - 2 \ldots - 1 \ldots - \tfrac{1}{2} \ldots 0 \ldots \tfrac{1}{2} \ldots 1 \ldots 2 \ldots 3 \ldots \qquad (+, \times, -, \div)$$

Here between every two integers there are infinitely many other numbers, getting arbitrarily close to them. The rational numbers, just like the integers, are closed under addition, subtraction, and multiplication. In addition the rational numbers are closed under division (leaving out division by 0, as usual), whereas the integers are not. Although the integers are embedded in the rational numbers, the other rational numbers are completely hidden from them. Since the integers form a sub-structure, or sub-algebra, of the rational numbers with addition, multiplication, and subtraction, these other rational numbers can never be reached from the integers with those operations: they always lead back to the integers. But if the integers just had access to division, say, then they could get out of their sub-algebra and reach the rest of the rational numbers. And if we would add just one more rational number, say $\tfrac{1}{2}$, to them, then they could use that to reach lots of other rational numbers: $1\tfrac{1}{2}$, $2\tfrac{1}{2}$, and so on. But from the point of view of the integers they are all there is. And this will seem clear and compelling to them, since after all, given their resources, what else could there be? Sums and products of integers are always integers. This is so even though the integers are surrounded by, and thinly spread among, things that are not integers. These other aspects of numerical reality are completely hidden from them.

This could be our situation. The parts of reality that we can represent might form a sub-algebra of all of reality. That is to say, it might be that we can represent certain objects, events, facts, and propositions, and certain relations or operations on them such that whenever we apply the operations to things we can represent we get something that we can represent as well. Whenever we can represent an event, say, then we can represent the cause of that event. Whenever we can represent a fact we can represent the explanation of that fact, and so on. Our representational system might be a closed structure analogous to an algebraic structure. And that structure might be a sub-structure or sub-algebra of all of reality. This hypothesis we can call *the sub-algebra hypothesis*: the hypothesis that what we can represent forms a proper sub-structure of all there is to represent. It is based on the analogy to an algebraic structure and a sub-algebra of it. If it were correct, then we would be much like the integers. We can never represent aspects of the much richer reality we are part of, but it would seem to us that the parts of reality we can represent are all there is. After all, all operations and relations that we can represent and that hold among things we can represent lead to things we can represent. So, from our point of view, what else could there be? Everything else would be systematically hidden from us. Similarly, we should expect that we have question–answer completeness, and so for any question we can ask we can state the answer (whether or not we can know if this is the correct answer). If this is our situation, then it would seem to us that everything is effable, even though what we can eff is surrounded by, and possibly thinly spread among, the ineffable. We can never get there from our point of view, and it will be completely hidden from us.

In considering this hypothesis we can note right away that it can't be quite our situation. First of all, we don't enjoy complete question–answer completeness in the way the integers would. If the effability thesis is true, then, of course, we can state any answer to any question, since we can state everything. But if the ineffability thesis is true, then we can ask questions where we know we can't state the answer, for example "What are all the ineffable truths?" Still, we might enjoy *large-scale question–answer completeness*, which is question–answer completeness leaving aside questions that deal with the ineffable and related questions. In general, and for almost all cases, it might be that if we can state the question, we can state the answer. And just this seems to be the case. When we ask what caused something, or what explains something, then we can in general at least state the answer, even if we don't know that it is the right answer. This fact we can understand on the sub-algebra hypothesis even if the ineffability thesis is true: causal and explanatory relations are ones under which our structure is closed. We can represent the explanations of what we can represent, and we can represent the causes of what we can represent. And again, all this could be true even though we are surrounded by the ineffable. And just like the very same arithmetical operations of multiplication and addition apply to the integers as well as the rational numbers, so causal and explanatory relations might hold among the facts or events we can represent as well as those that are ineffable for us. The ineffable aspects of reality might be very much like the ones we can eff, just outside of our sub-structure, or they might be completely different.

The sub-algebra hypothesis would explain why the ineffable is systematically hidden from us even though it is possibly a vast part of reality. Since it is outside of our sub-algebra we won't encounter it via causal or explanatory relationships. We can expect our sub-algebra to be closed under causal and explanatory connections. And we can expect our sub-algebra to be properly integrated with those things that causally effect us, for example in perception. Such a sub-algebra would be a very stable resting point for a representational system. There is no immediate need to develop it further, even if it only captures a small part of what there is to represent. On the other hand, if a representational system does not form a closed structure, or at least something reasonably close to it, then we would expect it to develop further if it develops at all. But once it is reasonably closed it rests at a stable place. The sub-algebra hypothesis explains why the ineffable is systematically hidden from us even if the ineffability thesis is true. It accommodates what needs to be accommodated, and we have good reason to think it holds, assuming externalism. The question remains what follows from it.

10.2.4 Ineffability and Modesty

The sub-algebra hypothesis makes clear in what sense the ineffable matters, and in what sense it doesn't matter. It doesn't matter locally, i.e. for particular questions of fact that aren't concerned with reality as a whole. If I ask an ordinary question like why there is a sandwich on the table, who ate my apple, or why the sky is blue, then I should expect that I can state the answer. Causal and explanatory relationships are part of my representational system, and thus are what my algebra is closed under. For ordinary local questions the ineffable will fall out of the picture.[12]

But not so for global questions about all of reality. Here, too, our reasonable question–answer completeness will likely allow us to state the answer if we can state the question. But the ineffable and how it is hidden from us will often mislead us into accepting the wrong answer. This is worrisome for questions about materialism and naturalism in particular. From our point of view it can seem perfectly compelling that everything is material and all there is fits into the natural world. So the questions "Is everything material?" and "Is the natural world all there is?" are questions we can state, and whose answers (yes or no) we can state as well. But our representational limitations might lead us to accept the wrong answers. If the sub-algebra hypothesis is correct then we might be in the materialist/naturalist sub-algebra in a largely non-naturalist world. If this is our situation we would naturally, but incorrectly, hold that materialism or naturalism is true. Just as the integers would find it compelling to think that all there is are integers, even when there are infinitely many other numbers before we get to the next integer. We might similarly find naturalism compelling even though the non-natural is infinitely close to us, and all around us, but systematically hidden from us. The ineffable is locally irrelevant, but globally central. And it is, in particular, central for the large-scale metaphysical questions about all of reality.

The sub-algebra hypothesis is not a skeptical hypothesis, in the sense that it is not a scenario that we can't rule out to obtain, and which invalidates our entitlement to our ordinary beliefs. On the contrary, it is a scenario that we have reason to believe obtains, but on this scenario our ordinary beliefs are taken to be true. The parts of reality

---

[12] Quantified claims, when relevant to local issues, should be taken to be restricted to the locally relevant domain, and thus they won't range over all of reality, in contrast to those that are explicitly intended to be unrestricted, like the ones that are intended to make claims about all of reality.

that we can represent we do represent mostly correctly, or so we can grant here. But when we aim to make claims about all of reality then we reach a limit. The sub-algebra hypothesis does not warrant a rejection of trying to answer large-scale metaphysical questions about reality, but it does warrant a form of modesty. We must recognize that our situation is one where these questions are to be approached with a sense that whatever answer seems compelling to us might simply reflect our own limitations, and not how reality is. The ineffability thesis combined with the sub-algebra hypothesis in particular suggests modesty about global metaphysical questions. Modesty is not agnosticism or quietism, but it is a step in that general direction.[13] It does not justify the abandonment of grand metaphysical theorizing, but it does justify giving such theorizing a somewhat different epistemic status from other parts of inquiry. How different will depend on how strong the reasons are that we have for the ineffability thesis and the sub-algebra hypothesis. It will be a difference in degree, and to what degree is not clear so far. Modesty for grand metaphysics follows, to what degree is left open.

The argument for modesty given here is importantly different from the argument that we should be modest about judging how many people are in this room, since after all there might be lots of invisible and otherwise undetectable people all around us. We have no reason to think that there are invisible people around us; that is just a hypothesis we might not be able to rule out. But we do have reason to think that there are ineffable facts, and that these facts are systematically hidden from us. These reasons are not conclusive, of course, but they are good reasons nonetheless. Thus reflecting on our own situation should lead us to conclude that there is a reasonable expectation that we fall short of complete effability. We have reason to believe that what we can represent is less than what is the case. We also have reason to believe that the ineffable is systematically hidden from us and that the sub-algebra hypothesis is correct. Thus we can expect to be misled in our judgments about global features of reality, and so modesty in grand metaphysics is advisable, and the ambitions of metaphysics need to be toned down. At the same time we should expect the ineffable to be locally irrelevant. The consequence of all this is modesty for metaphysics, but it is insignificant for most of the rest.

Externalism about propositions is generally seen as a pro-metaphysical view. It gives one an ontology of propositions, and with them likely objectivity of the fact-like or propositional aspect of reality. Since there is a domain of facts existing independently of us, it is fully objective how the world is to be carved into facts and fully objective what facts obtain: just look at the domain of facts. But the externalist picture of the propositional supports modesty and humility in grand metaphysics, likely making ambitious metaphysics beyond our reach, or so we have reason to think. Although many, I suspect, accept externalism about propositions, few live up to the stance

---

[13] Modesty thus contrasts with the positions taken by Gideon Rosen in Rosen (1994) and Sven Rosenkranz in Rosenkranz (2007).

towards ambitious metaphysics that it suggests. The point here is not to scold any externalists, but to bring out the contrast between this view and the internalist alternative defended here. Internalism is not merely a view about the semantics of talk about propositions, it comes with a completely different picture of the propositional, the fact-like aspect of the world. We now need to work this out more carefully.

10.3 Internalism and the Ineffable

Externalism about the propositional supports there being ineffable facts, the sub-algebra hypothesis and suggests modesty and humility in grand, ambitious metaphysics. But externalism is false, or so I have argued in earlier chapters. Suppose now that these arguments were correct and internalism about the propositional is true instead. Internalism was defended by considerations about language, and these considerations are the appropriate ones, since it is a thesis about language and its use. I will give no further arguments for internalism here. Instead we will look at what consequences internalism has for our debate about the limits of conceptual representation. I will argue that internalism gives us a coherent and substantial alternative to the externalist picture, but it incorporates a completely different picture of the propositional aspect of reality.

10.3.1 Internalism and the Effability Thesis

Externalism about the propositional doesn't directly imply that there are ineffable facts, but assuming externalism we have good arguments for that conclusion nonetheless. Internalism about the propositional, on the other hand, directly implies that there are no ineffable facts. This simply follows from the internalist truth conditions of quantified statements. The effability thesis was simply this statement:

The effability thesis: Everything is effable.

The quantifier here ranges over propositions, facts, or truths, but in any case, over the propositional broadly understood. If internalism is true about quantification over the propositional, then, on the simple formulation of internalism, this statement is equivalent to the conjunction over all the instances, that is, all the instances in our language. Thus the effability thesis is equivalent to:

(349) $\bigwedge$ (that p is effable)

and this conjunction is true, since each conjunct is true. The simple formulation of internalism, however, is too simple, as we saw in chapter 9. On the proper formulation of internalism the instances can contain new variables that are bound from the outside with external quantifiers over objects. On this formulation the truth conditions of the effability thesis are not (349), but rather

(350) $\forall \vec{x} \bigwedge$ (that $p[\vec{x}]$ is effable)

Here the instances are sentences of the form "$p[\vec{x}]$ is effable," where $\vec{x}$ are some variables that are bound by an external quantifier outside of p. But then every instance is effable on the object-permitting notion of the ineffable. The object-permitting notion is the one discussed above, in section 10.1, which passes the incommunicability test and which allows reference to objects for free. Each instance is just a sentence in our language, possibly augmented by reference to certain objects. Thus on the proper formulation of internalism, and using the proper object-permitting notion of the ineffable, every truth and every fact is effable. The effability thesis is thus true. The notion of the ineffable that is relevant for our discussion is the object-permitting one, the one that passes the incommunicability test. The proper formulation of internalism understands the truth conditions in the way spelled out in chapter 9. These two are in perfect harmony in their support of the effability thesis.

The question remains how this could be. It is one thing to answer arguments against internalism using considerations about expressibility, as we did in chapter 9, but quite another to make sense of this situation, in particular when seen as one about the relationship between our representational capacities and what reality is like. Internalism is a view about language, but it brings with it a completely different picture of the propositional than externalism. I hope to fill in this picture in the remainder of this chapter and thereby to make clear how the effability thesis can possibly be true. To start getting there, let's see how an internalist would respond to the arguments for ineffable facts that should persuade an externalist that there are such facts.

10.3.2 The Arguments for the Ineffable Revisited

We considered three arguments for ineffable facts above: built-in limitations, the argument from analogy, and explaining effability.[14] Let's see how an internalist should respond to them.

What could explain that a mind like ours can represent everything there is to represent? The internalist has a simple answer: "everything" here generalizes over our instances. No wonder we can represent everything, since we can represent every one of our instances: every instance in our language of "that p." And relying on an object-permitting notion of effability we can eff every instance with parameters. It is no accident and no mystery that we can represent everything there is to represent. Internalism maintains that it is based on a mistake to wonder here why two separate things coincide: the propositional and our representational abilities. Instead the propositional and our representational abilities are not two unconnected things, if internalism is true. The truth of the effability thesis falls out of how talk about propositions is to be understood. It would be hard to explain why the effability thesis is true on an externalist picture, but it is quite straightforward on an internalist one.

---

[14] We also considered some further prima facie compelling arguments in the first appendix to chapter 9, in particular the cardinality arguments, but the answer proposed there is available to the externalist as well as the internalist, so we will not revisit them here.

What about our built-in limitations? Maybe our mind must think a certain way, maybe we are inflexible and fixed in how we have to think, due to how our mind evolved or how our brain is structured. This naturally supports that we are limited in what we can represent on an externalist conception of facts. If facts are simply there, as part of an independent domain, then we should expect a mismatch. But on the internalist conception there is no such domain. So even our lack of flexibility does not support ineffability, since it does not support that this lack of flexibility is a limitation. We can explain why the effability thesis holds even if our minds have a fixed setup.

Finally, let's consider the argument from analogy, which is maybe the most compelling and forceful argument for ineffability. Here the internalist answer is clear: although it appears to be coherent to imagine aliens or gods who relate to us like we relate to the honeybee, there in fact can be no such creatures. We can represent everything there is to represent, while the honeybee can not. Since we can already represent everything there is to be represented, no creatures can represent more. There can be more powerful creatures with better spaceships, but they cannot represent any more facts. But whether this answer is at all satisfactory, or merely an endorsement of an absurd consequence of a view, can't really be appreciated without looking more generally at the internalist picture of the propositional. And only with that picture clearer in view can we see that this answer is not in fact absurd. Although an externalist will take it to be extreme bullet biting, the internalist will take it to be a deep insight into the nature of the propositional and the fact-like aspect of reality. We can only assess who is closer to the truth once the full picture is on the table, to which we must turn now.

10.3.3 The Internalist Picture of the Propositional

The internalist does not simply establish the effability thesis with a semantic trick. Consider, as a contrast, a person who holds that they own everything. They say that the correct semantics of "everything" is that it ranges only over their things, and thus they own everything, since they own all their things. This is a bad view on several grounds, not the least of which is, of course, that this is not the correct semantics of "everything." But the crucial difference between this view and the internalist view of talk about propositions goes beyond that. On the internalist view it is not true that when we say that we can represent everything we say that we can represent everything that is in some sense ours. We do not restrict our quantifier to some subset of the propositions that are related to us, analogous to the universal owner who restricts their quantifiers to the subset of things that they own. Internalism does not restrict the quantifier, but instead embodies a different view of what such quantifiers do, which is tied to a different view of what we do with talk about propositions or facts. Such quantifiers are unrestricted inferential quantifiers. As such their truth conditions give them a certain inferential role in our language, and the simplest truth conditions that do this are the ones that are equivalent to generalizations over all the instances. That these instances are instances in our language, and thus ones we can represent, is not the result of some sort of a restriction, but simply a consequence of the simplest truth conditions that give us what we need. Talk about propositions or facts, on the internalist picture, is not talk about some independent domain, it is not talk about any entities at all. There are no such entities as propositions or facts at all. On the internalist picture, that-clauses do not refer, they do not aim to pick out any entities, and so talk about propositions is very different than talk about objects.

Still, the internalist defense of the effability thesis must seem unsatisfactory. It seems to involve too much language and not enough metaphysics and thus can't be a defense of the view that all of *reality* is effable by us. Reality played little role in this defense. But that doesn't mean that the defense was defective. The internalist's explanation of why the effability thesis is true relied on the connection between the quantifier "everything" that occurred in the statement of the effability thesis and the instances of such quantified sentences in our own language. That this was enough to see that the effability thesis is true is surprising, but that doesn't make it incorrect. That the explanation has to rely on reality in addition to what we do when we talk about all facts or truths is true on an externalist picture of the propositional, but not on an internalist one. For the latter, reflection on our language is enough. We can see that once we talk about the propositional at all, in the internalist way, the truth of the effability thesis follows. This is surprising, but it might just be true if internalism is indeed true.

It is tempting to say that on the internalist picture the propositional is not an independent part of reality, but somehow due to us. Although this is tempting, it is not clear how it can make sense. Which facts obtain is, of course, not in general due to us. But there is something right about this, although it is hard to put one's finger on what precisely it is. I will try to work this out in the next section, but first we should see more clearly whether all this is too shallow a victory for the effability thesis.

Let's grant for now that internalism is true, as I argued above, and thus the effability thesis, as stated, is true as well. What this might be taken to show is that we need to state differently the question we wanted to ask when we originally asked whether anything is ineffable. As formulated the question has a negative answer, but maybe we need to reformulate it so that it is more substantial and its answer is more about reality and less about language. After all, the question we intended to ask was not supposed to be settled by the semantics of quantifiers and the non-referentiality of that-clauses. Of course, we can't demand that the questions we ask are answered the way we expect or intend them to be answered, but still, maybe the lesson we should draw from all this is that we need to state the question we wanted to ask differently. Maybe internalism wins a shallow victory when it comes to the letter of the effability thesis, but it only pushes the real issue somewhere else.

This line of thought is indeed tempting, but, in the end, it is mistaken. The truth and recognition of internalism do not motivate that we should ask the question differently. In fact, there is no better way of asking it. Instead internalism shows us the answer to the question we wanted to ask in the first place. To see this, let's consider some attempts to ask the question differently and why they won't change the situation.

Assuming internalism about the propositional, it won't make a difference to ask about the effability of every fact, or every proposition, or every proposition-like thing. The truth conditions of internal quantifiers ranging over these guarantee that the answer is in line with the effability thesis. So, maybe we should state the question in a way where we quantify over something else, something where internalism is not true: sentences, or inscriptions, or thought tokens, or something along those lines. These are simply material objects (we can assume) and so internalism doesn't apply to talk about them. Should we thus ask instead whether there is some (actual or possible) concrete inscription that can't be translated into our language, or some thought token that has a different content than any such token we can in principle have? But this won't make a difference. If internalism is true, then we can conclude that there can be no such thought token or inscription. If such a token has a content at all, then it has a content which we can think as well. Anything else has no content. Thus if there are inscriptions, sentences, or tokens that we can't translate it is not because they have a content that is beyond us, but rather because they have no content at all. And there certainly is no failure of translation if you fail to be able to translate something devoid of content. This way of trying to restate the question won't change the issue.

Another attempt could go via truth. Maybe the aliens can say things truly which we can't say at all? Maybe true things accessible to them go beyond the truths accessible to us? But this, too, won't help. There is a bridge-principle that connects things that are true to contents:

(351)   x is true if and only if x has a true content.

It is hard to see, even inconceivable, how anything could be true, but not have any content. With this connection, moving the issue to truth doesn't change things, since contents belong to the realm of the propositional.

Finally, one might try to throw in the towel on truth, content, and propositions, and acknowledge that we are not limited when it comes to those, but that there is a limitation nonetheless, but we can't properly articulate our own limitations. To illustrate with the advanced aliens again, the idea is that although they are not doing better than us when it comes to truth and content, they are doing better when it comes to truth* or schmuth and content* or schmontent. When the aliens look down at us from their advanced spaceships, they will certainly take us to be limited, and maybe we can't quite say how, but they might think of us as missing out on some important truths* or contents*. Now, this is certainly right in many ways, but is wrong in the crucial way that matters here. We are clearly limited when it comes to the aliens, we don't have the spaceships, and their invasion of earth might be a walk in the park for them. But the crucial issue for us is whether we are limited in what we can represent conceptually. And here the thought experiment motivates no such thing. Truth*, whatever it is supposed to be, isn't truth, and content* isn't content. Whichever of our many limitations we consider, they are not limitations in what we can in principle represent conceptually. If internalism is true, then we can reason conclusively from it that every fact can in principle be represented conceptually by us. Although certain things can't be conceptually represented by us, these are not the kinds of things that conceptual representation is supposed to represent in the first place. Although we can conceptually represent facts about feelings, facts about landscapes, or facts about schmuth, we might not be able to capture feelings, landscapes, or schmuth with conceptual representations. But that is not what they are supposed to capture. They aim to represent facts or truths instead. And here they can do all that could in principle be done. Our capacity for conceptual representation is complete in its domain, and this indicates that there is no proper separation between what is to be represented conceptually and the representations that do the representing.[15]

Let us now return to the argument from analogy, which was somewhat postponed in its assessment above. When the aliens look down at us like we look down at the honeybee, then we are correct in thinking that they are superior to us in many ways, but incorrect in thinking that they can represent more facts than we can. When they think about us as these primitive creatures, then they do not think truly that these human beings can't represent all the facts. That simply isn't true, assuming internalism, of course. We might not be as good as the aliens in all kinds of things, but when it comes to representing the facts we are at least as good.

There is no question that there is the strong sense that we are limited analogously to the limitation of the honeybee. It might be hard to state what that limitation is supposed to be more precisely, but there clearly is the sense of a limitation motivated by analogy. However, assuming internalism, we can reason conclusively that there is no limitation when it comes to conceptual representation, i.e representation of facts, truths, and propositions. We will have to weigh the sense of a limitation against the reasons against a limitation. And since it is reasonable to go with reasons rather than a sense, it is reasonable to conclude that the sense is misguided and the argument from analogy, although powerful, is in the end mistaken.

We can thus conclude, assuming internalism about our own talk about propositions, that there is nothing the aliens can truly say that we can't say in principle as well. When

---

[15] The conclusion here is similar to one drawn by Donald Davidson in Davidson (1984), however, they are reached in quite different ways. Davidson held that due to his particular theory of meaning it was impossible for there to be a language that is in principle untranslatable into our language, and thus that there could not be variation across conceptual schemes, and so the notion of a conceptual scheme is based on a mistake. The present view holds that there could not be content that goes beyond what we can say in principle, not because of a theory of meaning, but because of what we do when we talk about propositions. A further similarity is to Hilary Putnam's argument in Putnam (1981) that we are not brains in vats, since the question of whether we are, as stated by us, is guaranteed to have a negative answer, something we can see by reflecting on language. Internalism is not tied to skepticism, but the fact that to focus on the language employed in the question can be a key to its answer is similar both in Putnam's argument as well as here. It should be noted that the internalist picture of the propositional is quite different than McDowell's view on the matter. McDowell explicitly contrasts his view with an "arrogant anthropomorphism" McDowell (1994: 39) which holds that we can represent all facts with our present conceptual resources. In contrast, he holds that what there is to represent about reality is not influenced by our conceptual resources, but an independent fact about it. See section 8 of Lecture 2 in McDowell (1994), which is devoted to this issue.

it comes to conceptual representation we face no limitation, and this is a consequence of what the propositional aspect of reality is like. Internalism does not give us reason to think we are limited, but it gives us a different picture of the propositional. The propositional is not part of reality in the sense that there is no domain of propositions or facts of which we might capture more or fewer. Instead, it reflects our employment of talk about propositions, and it doesn't and can't go beyond that. The propositional is a reflection of our talk about propositions. This might sound like a version of idealism, and it is.

10.3.4 Conceptual Idealism Vindicated

Idealism sounds bad, but it doesn't have to when properly stated. Idealism broadly understood holds that, in some form or other, minds, in particular minds like ours, are central to reality. The commonest and maybe most natural way to be an idealist is to hold that what reality is like depends on our minds. And this is naturally spelled out as ontological idealism: the view that what there is and what it is like depend on our minds in some way. Many of the historically significant idealists were idealists of this sort, but it is rather problematic and rightly widely rejected. It is not clear how the notion of dependence is supposed to be understood such that idealism so formulated is compatible with other things we know to be true. We know that the universe existed before there were any humans and we know roughly what it was like before we were around. It is not clear how this is compatible with what there is depending on us, on a natural understanding of dependence. In particular, it is not clear how to even state the idealist position without it leading to immediate conflict with many other things we know. Ontological idealism is false, but idealism might still be true.[16]

An alternative way of thinking about idealism is to tie it not to ontology but rather to the conceptual or propositional aspect of reality. Although what there is is independent of us, the fact-like or propositional aspect of reality is not. Although reality, understood as what there is, is independent of us, reality, understood as what is the case, is not. This way of understanding idealism faces the same problems at first as ontological idealism. How can we coherently state a notion of dependence for the propositional that doesn't immediately conflict with what we know to be the case? Here we have to distinguish two ways in which the propositional could be dependent on us. The propositional is *truth-dependent* on us just in case which propositions are true is dependent on us. In this sense it is, of course, false that the propositional depends on us. Which truths are true is not something that in general depends on us. We can make some propositions true or false by affecting the world, but we don't make the true propositions true

---

[16] A weaker form of idealism is to hold that even though what there is does not depend on us, nonetheless, there being minds like ours is no accident. Any world must contain some minds, and without minds there could be no material world at all. A version of this position was defended by Anton Friedrich Koch with an intriguing argument in Koch (1990) and Koch (2010). I don't believe that his argument works, and I have tried to say why not in Hofweber (2015). Another recent defense of a different kind of idealism was given by Robert Adams in Adams (2007).

in general. Alternatively, the propositional is *range-dependent* on us just in case the extent of the propositional is dependent on us, which is to say: what propositions there are as candidates for being true depends on us, somehow and in some sense. The natural way to understand this is that what can be represented conceptually in principle is, somehow, dependent on us and our conceptual resources. And a natural way to understand that is that it is guaranteed that all there is to represent about reality can be represented by us, in principle. And this in turn means that the effability thesis is true for a reason connected to us. The view that this is so was called *conceptual idealism* in section 10.2.1. In this way, internalism about the propositional supports conceptual idealism.

Conceptual idealism so defended is thus the view that the propositional is range-dependent on us. It is a form of idealism, but different from and independent of ontological idealism.[17] As discussed briefly above, the standard route to conceptual idealism, and with it the standard defense of the effability thesis, is via ontological idealism. But conceptual idealism might be true even if ontological idealism is false, and internalism about talk about propositions and facts supports just that option. If internalism is true, then the propositional depends on us, not for its truth, but for its range. Internalism thus supports idealism, not ontological idealism, but conceptual idealism.

Conceptual idealism avoids the incompatibility worries that ontological idealism faces rather directly. It is a promising candidate to provide a coherent formulation of idealism. Although an idealist might be tempted to claim more about how the world depends on us, it is not clear whether this would be coherent. It is not clear how we can on the one hand stick with our picture of a world that is there for us to discover and which has us as just a small part of it, while at the same time elaborating on how the world in the end depends on us. But we can formulate conceptual idealism coherently. And if there is at least this kind of dependence, range dependence of the propositional, then this supports a version of idealism that we can make sense of.[18]

---

[17] In Nagel (1986) Thomas Nagel takes something like conceptual idealism to be the defining mark of idealism. This is slightly unusual, but I believe, with Nagel, that the real issue about idealism is just that. Nagel, of course, rejects idealism so understood.

[18] Hilary Putnam, in Putnam (1981) and other places, has defended a view he calls "internal realism" that goes by the motto that there is no ready-made world. However, Putnam focused on ontology and hoped to argue that the world by itself is not carved into objects. What there is, for Putnam, is tied to our ways of talking about it, and thus his view is best understood as a version of ontological idealism. As Simon Blackburn has argued quite successfully in Blackburn (1994), this view leads to a conflict between the statement of the idealist position, on the one hand, and other things we take ourselves to know to be true, on the other hand. It thus turns the idealist position into an incoherent one, given what we know. Putnam's view focuses on the world not being ready-made when it comes to the objects that inhabit it. I find it more fruitful and promising to consider the conceptual aspect of the world to be, using the same metaphor, not ready-made. This can, I hoped to make clear, be stated coherently using the notion of the range dependence of the propositional. Even if the world of objects is ready-made, the world of facts is not, on the way to spell out the metaphor attempted here.

Reality can be seen as being either the totality of facts or the totality of objects. At the beginning of this chapter we briefly considered the connections between these two ways of thinking about it, and now, at the end of the chapter, we can see a crucial difference between them. Internalism about talk about the propositional supports the view that although what is the case is not truth-dependent on us, it nonetheless is range-dependent on us, and thus idealism about the propositional is correct in the form of conceptual idealism. Conceptual idealism concerns facts and thus reality, understood as the totality of facts, is not independent of us in one crucial sense. However, since externalism is correct about talk about objects no similar dependence applies to the totality of objects. Reality, understood as the totality of what there is, is independent of us. Reality, in the sense that is the concern of ontology, is independent of us. Reality, in the sense that concerns the propositional, is not independent of us. Thus conceptual idealism combines with ontological realism.

To sum up: whether or not we should accept the ineffability thesis or the effability thesis depends on what we should think we do when we talk about propositions. If externalism is true about such talk, then we should accept the ineffability thesis and the sub-algebra hypothesis. This would be insignificant for most of inquiry and for ordinary life since, on the sub-algebra hypothesis, the ineffable will be there, but it will be hidden from us in a way that makes clear why it is insignificant for these purposes. But the ineffable will be significant for metaphysics, in particular debates about what reality as a whole is like. We should expect that we will naturally be misled to believe that our sub-algebra is all there is. Here ineffability should lead to modesty in ambitious metaphysics.

But if, on the other hand, internalism is true about talk about propositions, then the effability thesis will be true, and we can explain why it seems to us that some aspects of reality should be ineffable for us. No modesty would follow for metaphysics from this, but the metaphysical picture of the propositional that is tied to the internalist view of talk about propositions is itself a substantial consequence. It combines a version of realism, in that reality as the totality of things is independent of us, with a version of idealism, in that what there is to say about reality can all be said by us, as we are right now, not by mere accident, but for a reason. The internalist picture of the propositional makes clear why content cannot be beyond us and thus all there is to say about reality can be said by us. Internalism thus implies conceptual idealism, but it is compatible with ontological realism.

The question of whether internalism or externalism is true about our talk about propositions is a largely empirical question about what we do when we talk about propositions. It is a question about our actual use of certain expressions in natural language, and thus something that we can't settle on the basis of a priori reflection. The crucial question on which this issue depends is thus one about language and its use, which is a largely empirical question. Idealism, properly understood, and the effability thesis should be accepted if things turn out one way, ineffability and modesty should be accepted if they turn out another. If the former, then not only would it support idealism, which might sound bad enough, but furthermore it would support idealism on empirical grounds, which might sound even worse. Nonetheless, since I have argued in the chapters above that internalism is true, I side with idealism and the effability thesis.

11

Objects, Properties, and the Problem of Universals

After considering the significance of internalism about talk about natural numbers for the grander philosophical questions tied to natural numbers, i.e. the philosophy of arithmetic, in chapter 6, and the significance of internalism about talk about propositions for metaphysical debates related to propositions in chapter 10, it is now time to do the same for internalism about properties. When it comes to metaphysical problems connected to properties there is no more important metaphysical problem than the problem of universals, which will be the focus in this chapter. At the end of this chapter we will finally be in a position to state a unified metaphysical view tied to our four cases of ontological questions.

11.1 The Problem of Universals: What It Is and Why It Matters

The problem of universals is widely regarded as a central metaphysical problem. It is a metaphysical problem related to objects and their properties. But what problem there is supposed to be about objects and their properties is not so clear, and in particular which problem deserves the name "problem of universals" is controversial even among those who see a substantial metaphysical problem here. The problem of universals has a long history in philosophy, and which problem was labeled "the problem of universals" certainly differed over the centuries. I take there to be two main classes of problems about objects and their properties, and both are called "the problem of universals" by different philosophers. The first is a problem about the ontology of properties; the second is a problem related to explaining something about having properties. The internalist view about talk about properties is the key to solving both. The next couple of sections are devoted to getting clearer on what the problem of universals is supposed to be and how the internalist conception of talk about properties can solve or dissolve it. We will discuss the problem of universals in its two versions, with the ontological problem coming first, and the explanatory problem coming next.

The problem of universals is a problem related to universals. Universals are generally contrasted with particulars, which in turn are paradigmatically regular ordinary objects: tables, chairs, and the like. Universals, contrary to particulars, are supposed to be repeatable, present in more than one particular.[1] We will not in the following work with this somewhat technical notion of a universal, but rather deal with properties more generally, although we will keep the label "problem of universals" for the problem formulated below. Properties are repeatable in the sense that more than one thing can have them, and the motivations for the problem of universals discussed here all can be stated by simply talking about properties in a non-technical sense. Some philosophers take universals to be properties, others take universals to be something other than properties, and others again take universals to be only special properties.[2] It is overall best to state the problem of "universals" in more neutral terms as a problem about properties.

The problem of universals can first and foremost be seen as an *ontological problem*. It can also be seen differently, as an *explanatory problem*: a problem tied to giving a certain explanation. This is a slightly less common way to think of it, and we will discuss the problem of universals as an explanatory problem in detail below. We will first discuss the problem of universals as it is most commonly taken to be: a problem in ontology. Here we can distinguish two parts of it, with one of them the part where most of the work is actually carried out. The first part is to say whether besides individuals there are also properties, which these individuals somehow have, and which different individuals can share. It is simply the problem of whether or not there are any properties. Assuming an affirmative answer to this question we come to the second, more substantial, and more heavily debated part of the problem of universals. Given that there are such properties in addition to the individuals that have them, how are they related to the individuals? Are they located where the individuals are located, or are they non-spatio temporal? Are they located at more than one place at the same time, or at no place at all? What is the relationship that has to hold between an individual and a property in order for that individual to have that property? And so on and so forth.

The two parts of the problem of universals as an ontological problem are in fact the primary and secondary ontological questions about properties, as discussed at the end of chapter 1. The primary ontological question about properties was the question of whether or not there are any properties; the secondary ontological question was the question of what they are like in various ways, assuming there are any at all. Most of the work on the problem of universals so understood is in the ballpark of trying to answer the secondary ontological question. Most philosophers agree that

---

[1] To mark the difference between particulars and universals in this tradition more precisely is a substantial task. Universals are not simply supposed to be present in more than one particular (why can't a particular thing do that?) but rather "wholly present," truly multi-located. But what "wholly present" is supposed to mean is not at all clear. Some philosophers recently have tried to make sense of multi-location in a more precise way, in particular, Parsons (2007), Gilmore (2006), and Sattig (2006). For a critical discussion, see Hofweber and Velleman (2011).

[2] See Lewis (1983a).

there are properties, though they disagree on what these properties are like, and how they relate to the individuals that have them. Different answers to this question give rise to radically different metaphysical conceptions of what our world of individuals is like. Some, for example, have argued that the best way to understand how individuals relate to their properties is that they have properties as their parts, and they themselves are nothing but bundles of properties, somehow tied together. Such bundle theories then hold that, in the end, it's all properties, and individuals are just a bunch of properties somehow tied together.[3] Others have argued that individuals are merely bare particulars, the anchors of their properties, but by themselves they are featureless blobs. The features these blobs have are somehow anchored in them, but they are separate things. By themselves individuals are featureless.[4] Some philosophers hold that properties are concrete, some that they are abstract. Some hold that properties are located where their instances are, others that properties are located nowhere at all, and so on and so forth. Whichever view one takes in this debate seems to have major repercussions for how one views the world. The problem of universals so understood thus seems central to one's overall metaphysical picture of the world. The problem of universals is also an excellent case study for the relationship between the primary and secondary ontological questions, an issue introduced in section 1.3, but largely sidelined so far. We will be able to draw some general conclusions about it in this chapter, ones that will also apply to our discussion of numbers and propositions, and see what the internalist conception of talk about properties means for the problem of universals.

11.2 Minimalist and Substantial Approaches

There are two main options for solving the problem of universals, understood as a two-part problem, and both are problematic. First, there is the option to deny that there are properties, in particular also the sense that includes to deny that there is anything that Fido and Fifi have in common, even though they are both dogs. Consequently, there is no work to be done in the second part of the problem of universals, and thus nothing to be said about how the properties relate to the individuals that have them. This option was, in effect, taken by Quine in Quine (1980), but it is completely unsatisfactory. Using Quine's example, he proposed that even though there are red houses, red roses, and red sunsets, there is nothing that they have in common, in particular not being red. But this seems incoherent. If they are all red, then they have being red in common. There might not be something deeper that they have in common on top of that, for example, they might all be red in different ways, but certainly they have being red in

---

[3] See Williams (1953) for a classic version and Paul (2002) for a more recent one.

[4] See Sider (2006) for a discussion of how this view can be understood more precisely, and whether it is as absurd as it might sound.

common when they are red. Maintaining that red things have nothing in common is generally the result of over-ontologizing quantification. Maybe a nominalist doesn't want to accept properties since they are a threat to nominalism, and thus they feel the need to deny that there is something red things have in common. But the mistake here is to think that this use of quantization has that impact. We have seen in chapter 3 that quantifiers are sometimes used in their internal, inferential reading, and on this reading it is undeniable that red things have something in common.

A second, and commoner, option is to affirm that there are properties and universals, that red things do have something in common, and furthermore that we need to tell a substantial story about what it is that they have in common and how these properties relate to the individuals that have them. Although this is the common approach in academic metaphysics, it is notoriously difficult to motivate the project of finding out what these properties are like to those outside of the discipline. As everyone knows who has tried to do this, it is easy to get non-philosophers to accept that there is something that red things have in common, but hard to motivate that we need to find out what this thing is, where it is, what it is like in various ways, and so on. There is a real asymmetry in trying to find out about the thing which is red, and its redness. Everyone accepts that we can legitimately ask what the red thing is like in various ways, but there is great hesitation to do the same with the redness of the thing or with the universal of being red. That common reaction might, of course, simply be confused, or it might be on to something. And there certainly is a good argument that it is confused: the red things have something in common, so there is some thing which they all have in common, and so we can ask what that thing is. But that argument might seem dubious in light of what we have seen above, and so might this second way of answering the problem of universals: affirming that there are properties, and putting in a major effort to find out what they are like.

The standard way of approaching the problem of universals is thus to accept our talk about properties as often literally true, and to take it to motivate a substantial philosophical project into finding out what properties are like. We can call this the *substantial approach to the theory of properties*. It should be contrasted with a *minimalist approach to the theory of properties*, which holds that even though what we say about properties is often true, there is no substantial philosophical theory to be given about what properties are like. A defense of a minimalist theory of properties will require an explanation of why attempting such a substantial theory is mistaken. Even though red things have something in common, no substantial theory of properties is legitimate. A defense of minimalism would thus explain why a substantial theory of properties is mistaken, even though our talk about properties is often true. A minimalist approach to the theory of properties would in effect include an answer to the secondary ontological question about properties. As we can now see, internalism about talk about properties gives rise to a minimalist approach to the theory of properties and solves the problem of universals as an ontological problem.

11.3 The Internalist's Solution to the Ontological Problem

The internalist conception of talk about properties defended in chapter 8 holds that property nominalizations are non-referring expressions, and quantification over properties is based on the internal reading of quantifiers. Such talk is nonetheless literally true. In particular, the internalist holds that the reasoning in the trivial argument is valid in the first couple of steps. If Fido is a dog, then Fido has the property of being a dog. If Fifi is a dog, too, then she, too, has that property. Thus there is something they have in common. But just as in our solution to the first puzzle about ontology discussed in section 3.6, to now move on to ask what that thing, the property of being a dog, is confuses the two readings of the quantifiers. It follows, and follows trivially, from Fido and Fifi being dogs that they have something in common. But this follows only trivially using the internal reading of the quantifier. Does it also follow on the external reading? Internalism guarantees that it does not follow on the external reading of the quantifier. Internalism implies a negative answer to the ontological question about properties, and thus on the external reading there are no properties, and thus there is no property Fido and Fifi have in common, on the external reading. This argument was spelled out in section 4.3, among others.

Internalism answers the primary ontological question about properties in the negative, but it nonetheless holds that there are properties, using the internal reading. Internalism does not deny or reject talk about properties as mistaken, it instead is based on an account of what we do when we talk that way. And it makes clear why internalism implies minimalism. Even though there is something the red house, red rose, and red sunset have in common, it is a confused project to now try to find out what that thing is like, where it is, whether it is present everywhere Fido and Fifi are, and so on. Although there are properties, and being a dog is one of them, there is nothing to find out about what these properties are in general. To be sure, there are many truths about properties. One truth is that being a dog is a property that Fido has, another is that anything that has the property of being a dog has the property of being a mammal. But there is no legitimate philosophical project of finding out what properties as such are like, where they are, how they relate to their instances, and so on. Although there are many truths about properties, there is no legitimate philosophical project of finding out what the properties are like. Internalism guarantees minimalism, and it shows why it is a confused project to try to find out about the nature of properties in general.

Minimalism seems incoherent, since it seems hard to hold both that there are properties but that there can't be a legitimate project to find out what they are. But internalism shows how this is coherent, since there are properties on the internal reading, but there aren't any on the external reading. It is hard to see how minimalism is coherent if one holds that properties exist, but nonetheless there is no legitimate project of trying to find out what they are like. There is an epistemically pessimistic way of understanding this, which is perfectly coherent, but overly pessimistic: properties exist, and there are many facts about them, but we are not in a position to find out what these facts are. On this pessimistic line, the project of trying to find out about properties is confused not because there isn't anything to find out about them, but because it is beyond us to find this out, and thus the project of trying to find this out won't lead anywhere. Minimalism on epistemic grounds is simply too pessimistic. Philosophical theorizing can legitimately be speculative in this area if there is anything to speculate about.

A recent attempt to combine the existence of properties with a minimalist approach to the theory of properties is Stephen Schiffer's pleonastic theory of properties in Schiffer (2003). For Schiffer properties exist, but they have no "hidden and substantial natures" Schiffer (2003: 60) and so there is nothing substantial to be found out about them.[5] Schiffer's position is a form of externalism which holds that property nominalizations refer to a special kind of second class entities. Although Schiffer's minimalism is congenial to the internalist position defended here, his view is nonetheless a form of externalism, and thus in conflict with the considerations against externalism given in chapter 8.[6]

Internalism leads to minimalism. Internalism about numbers and propositions, just as internalism about properties, entails that there are numbers and propositions, but the philosophical project of finding out what these things are is based on a confusion. Although there will be many truths topically about natural numbers, and there is lots to be found out about what these truths are, there is no legitimate project of finding out what the numbers themselves are like, whether they are abstract or concrete, and so on. And similarly for internalism about any other domain of discourse. Internalism thus not only answers the primary ontological question in the negative, it also answers the secondary ontological question, or better shows that it was based on a false assumption in these cases. Internalism about Fs implies that even though there are Fs there is no legitimate project of finding out what Fs themselves are. Not because we can't find out what there is to find out, but because there is nothing to find out about them.

The secondary ontological question is often the focus of ontological theorizing, and the case of properties is a good example of this. This is in part so because the primary ontological question is often seen as having a clearly affirmative answer. What is left to find out is what these things are like. I believe this is a mistake. It is undoubtedly true that there are properties, numbers, and so on, on one reading. But this is not enough to get the secondary ontological question off the ground. What is needed for this is that there are properties in the external reading. If there are, then the secondary ontological question is a good question. If there are not, then it is not. The problem

---

[5] See also Johnston (1988), which Schiffer cites as an influence in taking this line.

[6] Schiffer defended a position more congenial to the one defended here in his earlier Schiffer (1987b). I have discussed Schiffer's two views, their relation to each other, and to the kind of view defended here in Hofweber (2016c).

of universals as a secondary ontological question is a good example where a lack of an understanding of the primary ontological question leads to a bad motivation for a project to answer the secondary ontological question. Even though red houses and red sunsets have something in common, namely the property of being red, there is no legitimate project of finding out what that thing is, how it relates to the things that have it, and so on.

11.4 The Explanatory Version of the Problem of Universals

Some philosophers have taken the stance that the problem of universals is not one about whether there are properties, or what they are like and how they relate to the individuals that have them.[7] Instead the problem of universals is an explanatory problem. There are some things about objects and properties that need to be explained, and that group of explanatory tasks is what the problem of universals really comes down to. If the problem of universals is an explanatory problem, it had better not be a problem like the problem of explaining why mice are nocturnal. The explanation for that is given in other parts of inquiry, not metaphysics. What then is it that needs to be explained? And why is it for metaphysics to explain?

11.4.1 Good and Bad Explanatory Tasks

If the problem of universals is an explanatory problem, then there is something related to objects and properties that needs to be explained and, moreover, explained in metaphysics. Not everything can be explained, however, and not everything needs to be explained. Thus we have to be careful to see if what is demanded to be explained is indeed something that can legitimately be asked to be explained. What can legitimately be asked to be explained is, of course, controversial, but there are some limits that we can set without much controversy. Take, for example

(352)   Why is everything red colored?

I take it that if anything is a conceptual truth, then that red things are colored is one. Whether there are any conceptual truths, and what they would be like if there were any, are of course controversial, but suppose for the moment that the notion makes sense, and red things being colored fits the bill. Can we then explain why red things are colored? The answer is actually not completely clear. Simply because something is a conceptual truth might not rule out that it has an explanation. After all, some mathematical truths can be explained, and it might turn out that mathematical truths are conceptual truths. Thus a mathematical explanation could explain a certain mathematical fact by citing certain other, in some sense more basic, mathematical

---

[7] See Rodriguez-Pereyra (2000).

facts. I won't rule out that conceptual truths can be explained, but if they can be explained, then what explains them has to be another conceptual truth, one that in some sense is more basic. It can't be that a conceptual truth is explained by an empirical truth. And this also applies to attempts to give metaphysical explanations of conceptual truths. If such explanations are possible at all, they themselves will have to be conceptual truths. This reflects on what kind of metaphysical theory will be able to give such an explanation. Most contemporary metaphysical theories state a synthetic metaphysical hypothesis. They subscribe to a claim which is not itself a conceptual truth, but rather a synthetic claim about how the world is. This is contrasted with metaphysics as stating an analytic metaphysical hypothesis: a hypothesis that consists of a series of analytic or conceptual truths. An analytic metaphysical hypothesis is of limited philosophical interest, since most interesting theories involve synthetic claims, arguably all the ones that make statements in ontology.

Sometimes it is tempting to confuse a vacuous explanatory task with a deep metaphysical problem. I have argued in Hofweber (2009c) that this sometimes happens with certain versions of the problem of change. Is there a deep explanatory problem about why change happens in time, why some time must pass for some change to happen? This sounds deep, but if it is a conceptual truth that change is having a property at one time and an incompatible one at a different time, then to ask why change happens in time is thus to ask for an explanation of a conceptual truth. It is highly unlikely that this conceptual truth has an explanation, but even if it does, the usual theories of time and objects won't be able to give it, since they are synthetic metaphysical hypotheses. The details of this don't matter for us now, and so we won't go into them. What matters now is that we must stay alert and evaluate not just whether a demand for an explanation is legitimate, but also what kind of an explanation could be expected if it is.

11.4.2 Candidates for an Explanation

Different philosophers have made different proposals about what needs to be explained about objects having properties. And some have argued that to meet some of these explanatory tasks properly we need to assume the existence of something like properties. The explanatory tasks usually come in two, not unrelated forms. One is to explain how it is possible that p. To explain how it is possible that p is to show how p can be the case in light of some other considerations that point to not-p. So, the strength of this explanatory task is closely tied to how strongly we should believe p to be the case, and how strong the other considerations are that suggest not-p. In this sense it would be silly to demand an explanation of how it can be that everything is mental, since we don't have good reason to think that everything is mental. What should be explained here better be something we take to be reasonably uncontroversial.

The second kind of explanatory task the problem of universals can be tied to is to give an explanation of why something is the case. This task is slightly different from the one of explaining how something is possible. In both cases the force of the task is tied to the strength of our belief that what should be explained is indeed the case. But when we ask for an answer to the question of why something is the case, then this might not be tied to reasons that it isn't the case. We might have no reasons to think that not-p, but still wonder why p. This makes this explanatory task distinctly different from that of explaining how it is possible that p.

What then needs to be explained? Here there are some of the more popular candidates:

1. How is it possible that two things have the same property? If there are two things *a* and *b*, then they are different. But if they have the same property, then they are at least partly identical. They are identical in a way. But how can two different things nonetheless be partly identical?
2. Why does *a* have the property of being F? What about *a* is it that makes it have that property? In virtue of what does *a* have that property?
3. Why do *a* and *b* share a property? In virtue of what are they of the same type? What makes it true, or what makes it the case, that they are of the same type?
4. Why do all and only *Fs* have the property of being *F*? What explains this correlation?

These tasks are different, but related. A number of philosophers have argued that the problem of universals properly understood is a problem of this kind. It is not a problem in ontology, but a problem connected to explanation.[8]

11.5 The Internalist Solution to the Explanatory Problem

Although the internalist account of talk about properties is quite clearly related to the problem of universals as an ontological problem, it is also tied to the problem of universals as an explanatory problem. Some answers to this group of problems are independent of internalism, but some rely on it. Internalism is not only tied to ontology, but in particular is an account of what we do when we talk about properties. And this is what is relevant here.

Everyone should reject the first explanatory task as motivated above as confused. There is no tension in two different things being "partly identical." They are not partly identical in the sense that they have the very same part, although that wouldn't be a problem either. They are simply partly identical in the technical sense that they have a property in common. And unless we have some reason for there needing to be an explanation of how it is possible that different things can have a property in common, talk of partial identity and difference does not give us such a reason.

Tasks 2 and 3 in the above list have often been rejected by nominalists as confused, but I think they have not taken the proper line in response to them. Some nominalists

---

[8] See Armstrong (1989), Rodriguez-Pereyra (2000), and many more.

have endorsed "ostrich nominalism" and suggested that there is nothing in virtue of which a is F.[9] But on the natural reading of "in virtue of" this is not correct. On this reading this is simply to ask why a is F. And this has often, but not always, a perfectly good answer. If F is an ordinary property like being a German car, then there is clearly something to be said on why a is a German car, or in virtue of what it is a German car. And the answer is that it was built in Germany, or designed by a German car company, or the like. If F is a property like being an electron, however, then there might well be no answer in virtue of what a is an electron. It just is one. On the other hand, there might be an answer, and it is an empirical question whether there is an answer. In any case, universals as entities have nothing to do with it either way.

What makes it that a and b both have a common property of being F? What makes it that they are of the same type? That they are both F, of course, as is made clear by Michael Devitt in Devitt (1980). There might well be further questions about what makes a an F, and what makes b an F. And the answers might be different in the different cases. But the question what makes a and b of the same type when they are both F has an obvious answer: they are both F.

David Lewis, in Lewis (1983a), accepted this account for the case of a particular, given property being F, but he held it is inadequate for the case of what it is for two things to share some properties or other, possibly an unknown one. Here the believer in properties or universals has a leg up, he holds. And this might well be true when we compare a believer in universals with a nominalist about properties who agrees with the version of Quine discussed above. For Quine there is nothing that red houses and red sunsets have in common. Quine won't be able to make sense of what it is for two things to share a property, in fact he would, implausibly, simply deny it ever happens. Similarly, Devitt could not simply use the "ostrich nominalism" move that worked for a given property being F, since it relied on having that property being given. For this problem, however, internalism provides the answer, since it gives us an account of what we do when we say that two things share some property or other. For an internalist this poses no substantial further difficulty since it in a sense reduced to the case of sharing a particular property, except it is the infinite disjunction of them. We know what it is for a and b to share some property or other: for them to be either both F or both G or both H or . . . , as spelled out by the internalist truth conditions. And given our discussion of inexpressible properties in chapter 9, this holds even for properties that are not in fact expressible by us due to the limitations of expressibility discussed in that chapter. There is thus no substantial explanatory task for metaphysics to say how it is possible, or what it is in general, for two things to share a property.[10]

---

[9] See Devitt (1980).

[10] Armstrong in Armstrong (1978) quite absurdly demands a reductive analysis of what it is to have a property, an account that doesn't rely on having a property, but accounts for it in general. But why that is a legitimate demand remains mysterious. And how this should be possible, although it's besides the point, is another question. It seems no better than demanding a reductive analysis of meaning: giving an account of

Finally, what explains the correlation between *Fs* and things that have the property of being *F*? The account given in chapter 2 of the relationship between "a is F" and "a has the property of being F" gives the answer. On this account the difference between the two is one of focus, not truth conditions. Thus the two statements are truth conditionally equivalent, and our linguistic competence alone is enough for us to realize this. This equivalence can thus be taken to be a conceptual truth in a non-technical and broad sense of the term. Even though conceptual truths might have explanations, as discussed above, it is unlikely that a metaphysical hypothesis can explain a conceptual truth, and in this particular case it is ruled out. What explains the truth, if anything, are the kinds of considerations put forward in chapter 2. No metaphysical story about properties or universals has explanatory work to do here.

11.6 A World of Only Objects

There is a real problem about the ontology of universals or properties, and internalism answers that problem: there are no properties. Nonetheless, objects have properties. The former sentence employs the external reading of the quantifier, the latter the internal one. Consequently, a substantial theory of properties is a mistake and minimalism is true: even though objects have properties there is nothing to be found out about what these things, aside for the moment the properties, are. Leaving issues of the ontology of events, which are left open by what has been said so far, we get the picture of a world that contains objects, objects which have properties, but it doesn't contain properties in addition to the objects which have them.

In this section I would like to consider whether it is a coherent metaphysical picture to hold that all there is are just objects, but no properties, no propositions, and no numbers. I should clarify right away that my goal is not to evaluate whether our world might contain only objects, but not force fields, spacetime regions, events, and various other things that we haven't discussed at all. Our focus instead will be various largely metaphysical considerations that suggest that a world of only objects, but not properties, propositions, or numbers, is metaphysically problematic. We will in particular focus on whether there might be just objects, but no properties. I argued above in chapters 7 and 8 that we have good reason to hold that ordinary objects exist, but properties do not exist. But this was looking at each case separately, not to assess whether the package that there are objects, but no properties is a stable combination. There are a number of arguments in the philosophical literature that we need more than just objects, and we will look at a selection now.[11] I will conclude that a world of only objects is coherent and metaphysically stable.

---

what meaning is without using things that have meaning, like words. An account congenial to the position defended here is in van Cleve (1994).

[11] Some arguments against only objects have already been discussed above, for example, the one that properties are needed as semantic values of predicates. See 8.3 for a discussion of why this is not so.

Bare particulars If there are only objects, but no properties, doesn't it mean that these objects are all bare particulars, from an ontological point of view? Isn't the view defended here similar to a position that consists of bare particulars and transcendent universals that stand in some having relation, expect without the transcendent universals and the having relation? In short, isn't the present view the absurd view that all there is are bare particulars: featureless, naked individuals?

No. Objects do have properties. They are not featureless. Internalism about talk about properties has this consequence, and everyone else should hold this, too. Everyone holds that objects have properties. The question remains whether on some views about what properties are and how they relate to the individuals that have them it makes sense to call the individuals "bare particulars." This becomes a possibility, properly spelled out, if we take the object and the property to be separate entities, only related in some external way.[12] But on an internalist conception of talk about properties there is no such consequence. The object and the properties it has are not separate things, since the properties are not things at all. Still, objects have properties. Simply because properties do not exist does not mean that all objects are featureless bare particulars. To the contrary, to make sense of this option requires one to hold that properties do exist, but are sufficiently separate from the objects that have them.

Objective similarity Some have argued that we need an ontology of sparse universals to account for the obvious fact that some things resemble each other more than others. What, though, is it to resemble some other thing? Not to share a property, since any two things share properties (the property of being a cow or a cup is shared by any cow and cup, despite their being dissimilar), and in fact any two things have infinitely many properties in common, and differ with respect to infinitely other ones. To account for objective resemblance we need to postulate a special group of distinguished properties: the "natural" properties, or the universals. Two things then can be said to resemble each other only if they share natural properties.[13] Such a sparse theory of properties not only accounts for resemblance, but also helps with many other philosophical problems. We thus need to accept the existence of a special group of universals which are expressed by some, but not all, of our or other predicates. Or so the argument goes.

This argument is mistaken on a number of levels. First of all, it tries to solve a real problem with ontology where ontology doesn't help. It is uncontroversial that simply sharing properties, in the abundant sense of "property," doesn't make two things resemble each other. They need to share the right properties, and we make widely shared judgments about this. Fido resembles Fifi, they are both dogs, but not the sun, even though they are both either a dog or a star. These judgments of resemblance are clear, but how could the simple existence of properties or universals corresponding to some predicates, but not others, explain our judgments? That the

---

[12] A nice discussion of some confusions related to bare particulars is in Sider (2006).

[13] A classic example of this line is Lewis (1983a) and Armstrong (1989).

286 OBJECTS, PROPERTIES, UNIVERSALS

property, or universal, of being a dog or a star doesn't exist, while that of being a dog does, doesn't help with this. This non-existence is not transparent to us, it can't be taken recourse to in explaining our judgments of similarity. But that there is a difference of similarity in the above pairs is obvious to us, while the existence and non-existence of the corresponding universals is not.

What accounts for the difference is that not all properties are equal when it comes to leading to similarity and difference. But the difference between those which do and those which don't is not the difference between existing and non-existing. That alone would not help. The difference must be some other difference, and this difference can well be accommodated by those who think that all abundant properties exist equally, or the internalist who accepts property talk, but holds that all abundant properties don't exist equally. All that is needed to account for similarity is a distinction between different kinds of properties. Some properties are relevant for similarity, while others are not. This distinction could be accepted as a primitive, or it might be explained in other terms. Everyone has the option of accepting the distinction as a primitive. The person believing in the existence of properties can hold that some abundant properties correspond to special universals, or say that some properties are natural, while others are not, as Lewis (1983a). The internalist can hold just as well that there is a primitive difference among properties: some are natural and some are not. One doesn't need to believe in an ontology of properties to accept such a distinction, all one has to believe in is that some properties are different from others, and an internalist has no problem holding this. Of course, for an internalist it is not to be understood as the domain of properties being divided into two subdomains. Nonetheless, it can be true that being F is special, while being G is not, and even that this difference can't be spelled out further. Just think of it as a primitive fact that being F is special. If real similarity can be understood by tying it to a primitive difference among properties, then everyone can have an account along those lines. However, such a primitive difference is not what will account for our judgments of similarity. To do this we will need to spell out which properties we take to be relevant for similarity, and which ones we do not. Such an account, once more, is available to everyone. The crucial point then is that holding that some properties exist, or correspond to universals, while others don't, won't be able to play the crucial role in this debate. What difference in properties is tied to our judgments of similarity is a good question. An ontology of universals is not the answer.

Causation and laws of nature Some philosophers have argued that such laws and causal connections can only be understood if we assume the existence of something like universals. Here there are two main routes of an argument. The first is that in order to understand the difference between a law and a mere regularity we need to distinguish natural from unnatural, gerrymandered, properties.[14] The second is that the source of the necessity of a law-like generality comes from a connection among

---

[14] See Lewis (1983a) once more.

properties or universals.[15] Both of these approaches are, of course, controversial, but neither one of them gives us good reason to accept an ontology of something like properties over and above objects. As discussed in the section on similarity above, an internalist can hold that some properties are different from others, and they can either take such a distinction as primitive or spell it out in other terms. Here the internalist is in exactly the same situation as the person holding that we need to accept the existence of properties or something that plays the role properties are supposed to play. When one holds that being an electron is in important ways different from being funny, and that this is for some reason or simply a primitive fact, then all this so far is neutral with respect to whether we are referring to properties as entities in saying this, or using non-referring property terms. On the second argument the situation is only subtly different. An internalist can accept that there is a primitive connection between being F and being G, and because of this it is a law that Fs are Gs. This connection is, of course, not a relationship among entities, but instead is spelled out as something like "being F guarantees being G," where "guarantees" is understood as a primitive predicate that has two argument places for non-referential property nominalizations. This way to do it simply mirrors the externalist proposal that law-like connections derive from primitive relations among universals. To put it differently, a proposal that gives properties a special metaphysical role will be stated in certain terms, and the internalist as such has the option of accepting it as stated, but insisting that the statement be understood the way the internalist takes to be correct. It will require a different and further argument that this account of laws, or causation, or what have you, is not only a good one, but one that requires an externalist kind of a reading, and thus an ontology of properties. Whether the alluded to account is a good account of what a law is is, of course, a completely different question (which I would answer in the negative, but that's a different story). The point now simply is that there is no crucial difference between an internalist and an externalist way of formulating such an account, and that thus there is no good argument for an ontology of something like properties from these debates. Such proposals are generally taken to require an ontology of properties, since externalism is assumed about property talk as it occurs in the proposal. But an internalist reading of the proposal will be just as good, as a proposal about laws, as an externalist one.

Truthmaking Finally, it has been argued that the picture of the world as containing objects, but not properties, is not a metaphysically stable picture.[16] Such a picture does not contain enough ontology, enough existing things, to support what is true. It is in conflict with some version or other of a Truthmaker Principle. Such a principle will precisely capture an inevitable fact about the relationship between what exists and

---

[15] See Dretske (1977) for a representative example.
[16] Versions of this can be found in Armstrong (1989), Armstrong (2004), Rodriguez-Pereyra (2005), as well as many others.

what is true, and on such a principle it can't be that all that exists is just objects. We need more than objects to have truthmaking, or so the argument.

Although many philosophers have in recent years drawn on some version or other of a truthmaking argument to defend their preferred view, I have to confess that I find their arguments generally unconvincing and question-begging, and others, to be cited shortly, have published similar reactions to these arguments. Of course, what is true is not independent of what exists. Some things have to exist for certain propositions to be true. And the existence of some things guarantees the truth of some propositions. But all that does not mean that what is true is determined solely by what exists. What is true is determined instead by what exists and what it is like, but not by what exists alone. The same things can exist, but be different, either in their properties or in their relations among each other, and then different propositions will be true. What there is alone doesn't determine what is true, but only what there is and how it is.[17] Truthmaking so understood is quite unproblematic, but it doesn't give one any argument against the picture of the world as one of objects and only objects, since no one is denying that these objects are a certain way, and thus have properties and relations among each other. Internalism makes clear how that can be while all there is are objects. Any argument for more ontology has to rely on something stronger than that. And many stronger principles have been proposed. Some have demanded that for every truth there must exist something such that the existence of that thing necessitates that truth.[18] But why would one believe that? One reason might be that if one doesn't, then it would be coherent that the same things exist in different worlds, but different propositions are true in them. But the coherence of this is just at issue, and truthmaking in that form motivated that way doesn't give us any argument against this. In general arguments against a world of just objects using truthmaker principles rely on this kind of a move: some plausible sounding principle is asserted ("Truth supervenes on being!") which then is argued to be incompatible with the world containing just objects. But the principle is merely one that is plausible on some reading ("Truth supervenes on what there is and how it is"), and then used in an argument on a different reading ("Truth supervenes on just what there is"), which begs the question in this case. I see no argument against the picture defended here from truthmaking. What exists matters for what is true, but what is true is not merely determined by what exists. It is determined by what there is and what it is like. And since objects are a certain way, even if all there is are just objects, this is perfectly compatible with the metaphysical view defended here.[19]

---

[17] This point was made by Julian Dodd in his Dodd (2001). See also Lewis (2001b) who takes a partly similar line in a different setting.

[18] See for example Armstrong (2004).

[19] Truthmaker arguments are widely discussed in the recent literature. Not congenial to the present view are, for example, Armstrong (2004) and Rodriguez-Pereyra (2005), congenial are Parsons (1999), Dodd (2001), Lewis (2001b), Künne (2003), and Melia (2005). I won't try to repeat the many good points these authors have already made.

11.7   A Defense of a Restricted Nominalism

For the four cases of ontological problems that we have discussed in this book the resulting picture is one of a restricted nominalism: restricting ourselves to these four cases, nominalism is true. That is to say, restricting ourselves to the four cases of natural numbers, ordinary objects, properties, and propositions, we have found that objects exist, but natural numbers, properties, and propositions do not. Nominalism is generally not understood as a restricted view, it instead is the view that all there is are objects in spacetime, or all there is is concrete, or nothing is abstract, or something along those lines. We have seen no argument that nominalism understood as the unrestricted view is true. There are many other cases of ontological problems that we have not considered: events, real numbers, sets, possible worlds, and so on and so forth. For all we have seen so far, it might be that sets exist, and thus nominalism is false. And it might be, for all we have seen so far, that none of these things nor anything else exists, and thus unrestricted nominalism is true.

That this restricted nominalism is coherent in light of the numerous arguments against it is a not insignificant fact. There might be only spatiotemporal objects, big or small, despite the fact that objects have properties, that there is a number of my hands, that arithmetic is true as stated, that some propositions are true, and so on. The discussion in the preceding chapters hopefully made clear that this is a coherent position, and hopefully a case has been made that it might be the correct position. It is a form of nominalism in this restricted sense, that if we consider only objects, properties, propositions, and natural numbers, then the things we generally say and accept about them can all be true, while all there is is just objects. But this restricted nominalism is, of course, not nominalism, the view that all there is is just concrete objects.

Nominalism as such is overrated. Whether nominalism as an unrestricted doctrine is true doesn't really matter, as I hope to make clear shortly, and I never hoped to establish it here or elsewhere. I never had the goal of defending nominalism in this book, and the alleged desirability of nominalism was never a premise in any argument. The defense of the restricted nominalism is an upshot of the results of looking at our four different cases one by one. As it turned out, the results of these four investigations taken together support a restricted nominalism, and show it is coherent. But I never argued that nominalism about arithmetic must be true, since otherwise we couldn't know about numbers, nor that nominalism about properties must be true, since otherwise we couldn't understand how properties relate to objects, and so on. Our considerations above proceeded on a case by case basis. Looking at what we do when we talk about natural numbers we saw that internalism is true for such talk; consequently such talk is not about any objects, and consequently no objects are natural numbers. This led to a form of rationalism about arithmetic, which also is a form of nominalism.

No considerations of this kind should lead us to think that unrestricted nominalism is true. We might look at many other cases that were so far left open and even if we find that for all of them internalism is true, doubtful as this might be, unrestricted nominalism would still not be established. It would leave open that there are many abstract objects that we simply never talk about. Reality could contain mostly abstract things, but creatures like us never talk about them. None of the things we talk about are abstract then, but nominalism would be false, even though internalism is true almost everywhere. I take it here that there is no separate domain of talk about abstract objects and thus that it doesn't make sense to think that one could establish internalism about talk about abstract objects as such. Contrary to talk about natural numbers, where it was clear that if anything is a natural number then the number 1 and the number 2 and so on are natural numbers, there is no similar characterization of the abstract objects as just these ones: abstract object x, abstract object y, etc. For an object to be abstract is a secondary consideration about it. There is no list of terms in our language such that if none of these terms refer, then we can conclude that there are no abstract objects. Thus considerations about internalism are not enough to establish unrestricted nominalism.

What matters primarily is to understand what we do when we talk about various things, and what is required for what we say to be true. That is a question tied to restricted forms of nominalism, restricted to the things we talk about. Whether there are abstract objects in addition to that is a good and interesting question, but it won't affect what is going on with what we do.

My defense of the restricted nominalism given here is different from standard defenses of unrestricted nominalism. A common defense of nominalism is one where abstract objects are seen to be problematic in some form or other, often epistemically. Because of this, the argument goes, we need to reject them and understand our discourse in nominalistic terms instead.[20] But this line is problematic in a number of ways. First, it is not so clear what an abstract object is more precisely, since the characterization as being outside of spacetime is not completely satisfactory. Second, it is not so clear why it would be problematic if there were such things, and why we couldn't know about them.[21] Whoever is right in this respect in the end is not fully clear,[22] but the defense of nominalism in its restricted form given here did not proceed like this. As far as we are concerned here, there is nothing wrong with abstract objects and our knowing about them. Clearly there are puzzles about how we could do this, but none of them played any role here. We simply looked at what we do when we talk a certain way. And we found out that we don't talk about abstract objects, not because there is anything wrong with them, but because that's just not what we do.

In the end we thus get the following overall view about the world, restricted to the four cases of ontological problems that we have considered, and summarized in somewhat general slogans. Objects exist independently of us, and we have empirical

---

[20] See Field (1989b) for a well-known attempt to achieve this.

[21] These points have been elaborated on by John Burgess and Gideon Rosen in Burgess and Rosen (1997) and Burgess (2008), among others.

[22] I have critically discussed John Burgess's work in this regard in Hofweber (2010).

evidence for their existence. What properties objects have and which facts obtain are in general independent of us. But what the range of the propositional is, and correspondingly the range of properties, is not independent of us: there is a range dependence on us of the propositional, and analogously of the properties objects can in principle have. Although the world contains only objects, nonetheless objects have properties and facts obtain. We can in principle say everything that can be truly said about the world, since reality, understood as all that is the case, is not independent of us, even though reality, understood as what there is, is independent of us. Rationalism is true about arithmetic, and the truth of arithmetic is objective, but not about any objects, nor does it require the existence of anything. The only things required to exist for the truth of what we say about the world when we talk about ordinary objects, natural numbers, properties, and propositions, are ordinary objects.

This concludes our discussion of the metaphysical problems tied to our four cases of ontological questions: natural numbers, objects, properties, and propositions. It is now time to return to our questions about ontology and metaphysics as a whole. In particular, we have not yet solved the third puzzle about ontology from chapter 1: the puzzle about how philosophical such ontological questions are, and whether they should be seen to properly belong to metaphysics, understood as a branch of philosophy, in the first place. This will be our final larger topic.

12

The Philosophical Project
of Ontology

We have by now seen why ontological questions in many cases seem trivial, but really are not. We have seen for four cases how important an answer to the corresponding ontological question is for the truth of what we say. That is, we have a solution to the first two puzzles about ontology. We have found answers to various metaphysical questions connected to objects, natural numbers, properties, and propositions, as we set out to do at the beginning. What remains to be addressed is our third and final puzzle about ontology. The third puzzle was the puzzle of how philosophical ontological questions are and thus concerns whether they properly belong to philosophy and metaphysics. To solve this puzzle we will need not only to see what ontological questions are and how they are to be answered in general, but also how we should think about the relationship between philosophy or metaphysics, on the one hand, and other parts of inquiry on the other hand. In this chapter and the next one I will make a proposal about how the third puzzle about ontology should be resolved, and how we should understand ontology and metaphysics as a philosophical project. And the best way to start with that is to make clear why such a project should seem to be problematic.

12.1 Ambitious, Yet Modest, Metaphysics

Whatever one thinks about the cliché of a certain Continental European approach to philosophy, it clearly has one thing going for it.[1] According to this cliché, philosophy is cultural commentary, a comment on the ongoings in society and the world at large. Such a comment is not intended to be merely description of what is going on, but a side remark, something over and above a description of what is going on. It might be witty and insightful, or dull and stupid, but at least it is clear how there can legitimately be such a project of commentary. There is no conflict with or tension between the philosophers commenting on culture while culture goes on with or without them. But many philosophers, including almost all metaphysicians, have different ambitions for their field. They don't want to be just commentators, but inquirers. They hope to solve

---

[1] Parts of this section overlap with parts of Hofweber (2009a).

problems and to answer questions of fact. They want to find out whether something is or is not the case. But this fact-seeking approach to philosophy and in particular metaphysics is problematic. It gives rise to what I take to be the biggest worry about a certain popular conception of metaphysics, and the biggest concern that metaphysics so conceived might be a confused project. This worry is much more serious than the traditional worries about metaphysics: that metaphysics doesn't lead to knowledge, and that metaphysical statements are meaningless according to some general criterion of meaningfulness. These traditional worries usually relied on a much too strict notion of knowledge and meaningfulness, one that would rule out many other parts of inquiry as meaningless or unknown as well. The more serious worry instead is that what metaphysics hopes to do has already been done by the sciences or other authoritative parts of inquiry. The problem isn't that the questions are meaningless or unknowable. Instead the problem is that the questions are meaningful, but the answers are already known from work done somewhere else. For example, the question of whether there are numbers, say, a question we hoped to answer in ontology and metaphysics, is meaningful, but the answer is given in mathematics. And similarly for other problems in ontology. This worry carries over beyond ontology to other traditional metaphysical questions. Take the problem of change as a further example. It is often stated as the problem to say how change is possible. This is presented as one of the oldest and hardest problems in metaphysics. But if that is supposed to be the problem, then it seems that an answer to it is known from work done in various sciences. It is known, for example, why a candle left in the window during a sunny day changes from straight to bent, and how this change was possible, even though no one touched the candle. The answer is not easy to give, and involves various complicated accounts, including the effects of sunlight on solid matter, the particular properties of wax, the effects of gravity, and so on. Considering all these together gives us an account of how that change happened, and how it was possible even though no one touched the candle. But when we know how this change was possible, what else is there to find out about how change in general is possible? The question asked in the problem of change surely wasn't intended to be answered this way, but then how was it intended? As it was stated it seems to be answered in an unintended way, but answered nonetheless. Maybe the question was thus badly stated, and what we really wanted to ask should be stated differently. If that is so, it is not completely clear how it should be stated, but this is one possibility. Or maybe it was stated correctly, it was simply answered in unexpected ways. In that case it would seem that the question we ask in metaphysics is answered in other parts of inquiry.[2]

---

[2] We won't discuss the problem of change in any more detail here, but see Hofweber (2009c) for a discussion of how the problem of change might be stated more precisely and whether it is a metaphysical problem in any of its formulations. We will focus on ontological problems, in particular primary ontological questions, in this chapter instead.

For metaphysics in general, and ontology in particular, to make sense as we commonly think of them, this situation can't obtain. It can't be that we are asking questions in metaphysics that have already been decisively answered. Any defense of metaphysics as a legitimate part of inquiry must make clear that this situation does not in fact obtain. Maybe such a defense is easy, at least for some examples, but there is a prima facie case to be made that such a defense needs to be given. Not because there is anything wrong with metaphysics or philosophy as such, but because a prima facie case can be made that the questions that are in fact asked in metaphysics indeed are long answered, and our ontological questions about numbers, objects, or properties are good examples of this.

One unacceptable way to defend metaphysics is simply to reject the results of the sciences and mathematics. This defense would maintain that the question of whether there are infinitely many prime numbers, and the question of how the candle changed its shape, are not in fact answered by the sciences or mathematics, appearances to the contrary. Instead they are open problems yet to be answered in metaphysics. Sure, there are some good reasons to accept that there are infinitely many prime numbers, like Euclid's proof, but there are also good, maybe even better, reasons to reject it, like arguments for nominalism using Ockham's Razor. But to take this stance would be absurd. To take it to be reasonable to think that those metaphysical considerations simply trump mathematical considerations when it comes to deciding whether or not there are infinitely many prime numbers relies on a somewhat perverse view of their respective strength and with it of the authority of the fields of mathematics and metaphysics. As David Lewis made vivid in Lewis (1991: 58f.), to reject mathematics on philosophical grounds is comically immodest on the philosopher's part. The success of the field of mathematics is so great and the success of philosophy in settling questions of fact is so slim in comparison that it would be absurd to expect that we can outright reject the results of the former because of one's preferred view on the latter.[3] Not that it couldn't happen that philosophy discovered that mathematics got it wrong, but we should not put our hopes of there being a legitimate project of metaphysics on this possibility obtaining. If metaphysics has any work to do, this is not how it should expect to do it.

A different, and subtler, way to save these questions for metaphysics is defended by E. J. Lowe, in particular in his Lowe (1998). The main idea is this: The sciences by themselves do not answer the question of how the candle changed its shape, or whether change is possible, and mathematics by itself does not answer the question of whether there are numbers. Rather they assume or presuppose that change is possible at all / that numbers exist at all. And only under these assumptions do they then establish that there are prime numbers / that and how the candle changed. These assumptions can't be discharged by the sciences themselves, instead they are left for metaphysics

---

[3] Although Lewis's sentiment here is rather popular in contemporary philosophy, not everyone agrees. See Daly and Liggins (2011) for a dissenting voice.

to cash in. The sciences thus need metaphysics to discharge assumptions that they simply made at the outset, or so Lowe. This makes metaphysics into a discipline of the greatest importance. All scientific results depend on the work of metaphysicians for their being established without assumptions. But, of course, this could turn out unexpectedly badly. If metaphysics sides against change, then the sciences were simply wrong. And if metaphysics sides against numbers, then mathematics was based on one big mistake. Not that it is the duty of metaphysics to deliver good news, but the point is that if metaphysics delivers bad news then everything goes down the drain. If there are no numbers or there is no change, then mathematics or science is simply based on a mistake. This situation would be no different from a detective deciding to turn a missing person investigation into a murder case, even though no body was found. The detective might arrest a suspect and accuse them of the murder. But when the missing person turns up there is nothing left to do but apologize and let the suspect go. The arrest and accusation of murder were based on a false assumption: that the missing person is dead. Similarly, mathematics might be based on the false assumption that there are numbers at all, and science might be based on the false assumption that change is possible at all. Mathematics can still be useful, even if it is based on a false assumption. Just as it can be useful to keep someone locked up who is innocent. In either case, though, something has gone badly wrong.

Whatever metaphysics hopes to do, a minimal constraint on it has to be modesty towards the incredibly successful parts of inquiry like mathematics, the physical sciences, and disciplines of similar stature. I take *modest metaphysics* to be metaphysics that doesn't conceive of the work it has to do as relying on the possibility that the mature and generally authoritative parts of inquiry turn out to be mistaken. Modest metaphysics should have its project intact even if, as we have good reason to believe, the answers given in these other parts of inquiry are correct as stated. Thus modest metaphysics can't conceive of itself as aiming to answer questions that have an answer immediately implied by other, authoritative parts of inquiry. Whatever questions modest metaphysics tries to answer, it can't be the mathematical question of whether or not there are infinitely many prime numbers. We have very good reason to take that question as having been answered, and what metaphysics so conceived hopes to do should not be seen as trying to overrule or verify that answer.

Modest metaphysics can try to understand the results of the sciences, it can interpret them, it can unify them, it can add to them, but it can't think of its own project as depending on overruling or rejecting the sciences, nor to give their final vindication. Not that metaphysics can't help the sciences, nor that it might not in fact be able to show that certain scientific theses are false. It might turn out that mathematics gets something badly wrong, unlikely as it is. Maybe some result relied on inconsistent assumptions, maybe a generally accepted proof is incorrect nonetheless, and maybe it took a philosopher or metaphysician to figure that out. Then a metaphysician would be in a position to overrule mathematics. This might well happen, but nonetheless the project that metaphysics hopes to carry out should not depend on this possibility.

Modest metaphysics will try to carry out a project that makes sense even if we grant, as we have good reason to, that mathematics and the mature sciences are correct in their domain.[4]

Modesty in metaphysics is a stance about metaphysics as a whole. It is not blind science worship. A modest metaphysician, just like anyone else, can and should accept the possibility that science might be wrong. But the modest metaphysician will hold that metaphysics has its project intact even if science is right. In addition, a modest metaphysician can and should participate in the project of science in any way they are fit to do. They can point out that some of the concepts used in the sciences are not as precise as one might wish, they can propose new interpretations of scientific theories, or present arguments or data against certain theories or their interpretation. The metaphysician can work in the scientific project. The question is, though, how metaphysics, the discipline, relates to the sciences. Maybe there isn't a legitimate part of inquiry which is anything like metaphysics. Maybe metaphysics is just a field where people get special training that is useful in a certain way in helping out in the sciences. There is no guarantee that there are any problems or questions at all that are properly addressed in metaphysics. The metaphysician does not have to be more modest than the scientist when they try to answer scientific questions. But the discipline of metaphysics has to be modest towards the sciences if there is such a discipline as a part of inquiry at all. Whatever metaphysics so understood hopes to find out, it must be modest metaphysics in doing this.

Metaphysics certainly does not have to be modest towards all other parts of inquiry. Astrology has no standing to demand modesty from metaphysics when it comes to the questions it addresses, but mathematics does. Which discipline has what standing is naturally a substantial question, but we won't need to dwell on it, since only the clear cases will be relevant for our discussion.

Modesty can be taken too far. Overly modest metaphysics can be unambitious. It might simply hold that there are no questions left for metaphysics to address. Metaphysics carried out with this attitude is like popular science journalism. Take some questions that have traditionally been considered part of metaphysics and philosophy: Is space continuous? Is time travel possible? Is all matter composed of atoms? To answer these questions we look at the sciences and spell out what they have said about them. The metaphysician with the unambitious attitude will hold that the field of metaphysics is limited to that. These questions have historically been addressed in philosophy, but all that is left for us to do now is to see how the sciences have answered them. Those that have not yet been answered are ones that we philosophers will have to wait on some more before we can see what the answer is that the sciences will give us. But metaphysics can't hope to answer any questions by itself. It can only look at how they have been answered elsewhere. Just like popular science journalism, metaphysics so conceived looks at the results of the sciences and their consequences without adding to them, and then presents them for a general audience.

---

[4] Thanks to Martin Thomson-Jones for discussions about the proper formulation of modesty.

Like immodest metaphysics, unambitious metaphysics is not worth the name. If there is such a discipline as metaphysics at all, it must be *ambitious metaphysics*: metaphysics that has some questions left that it is supposed to answer. Metaphysics, if there is such a discipline at all, must be both modest as well as ambitious. But ambition and modesty are in tension. Can there be ambitious, yet modest, metaphysics?

We can say quite generally that whatever metaphysics is supposed to do is the *task* of metaphysics. Everyone can agree that there is a task for metaphysics to carry out. It might simply be to make some useful distinctions and to point out to those trying to find out what reality is like that they should not confuse what should be distinguished. Or it might be to report on the history of the discipline, or the results of the sciences. Unambitious metaphysics can have a task, and even anti-metaphysical philosophers can accept a task for metaphysics. Carnap, for example, could hold that metaphysics can contribute to the useful project of coming up with alternative ways to describe the world and to make them available to those who hope to answer questions of fact.[5] And although metaphysics so understood would be a helpful supporter in inquiry, it would not have the standing of itself being a part of inquiry that aims to settle questions of fact. It would be useful for those who answer questions of fact, just as janitors are useful, but janitorial services are not a part of inquiry in this sense, and without a bigger role neither is metaphysics. For that to happen metaphysics must not just have a task, but a *domain*: it must have some questions of fact that it properly aims to answer. Ambitious metaphysics doesn't merely have some task or other, helpful as it might be, it has a domain. So understood its task is not merely to help out others, but to answer the questions in its domain.

Suppose we hold that metaphysics should be ambitious, but also modest. If it is ambitious, then there must be some questions that are properly addressed in metaphysics, that is, metaphysics has a domain: there are some questions that *it* should address. But if metaphysics is also modest, then it not only has to have a domain, it has to have its *own domain*: there have to be questions that are properly addressed in metaphysics and on which the other parts of inquiry towards which it is modest have to be sufficiently silent. There must be some questions that are to be addressed in metaphysics, and only metaphysics. If another authoritative part of inquiry addresses this question as well, then modest metaphysics will not have its question left open if we accept the results of the parts of inquiry it should be modest towards. Furthermore, it can't be that the questions in the domain of metaphysics have an answer *immediately implied* by the results in other parts of inquiry that are authoritative on the topic the question is about. This is the case with our question of whether or not there are numbers. Questions about numbers are addressed in a part of inquiry that is authoritative about them. Of course, you would not hear the question of whether there are numbers in the mathematics department directly, unless there was a philosophical conversation going on. But an answer to it is immediately implied by results that are established in the mathematics department: that there are infinitely many prime

5 See Carnap (1937).

numbers, among many others. I am leaving aside just for the moment the view about polysemous quantifiers defended in chapter 3, to which we will return momentarily. The example, however, illustrates a general point: metaphysics can't be seen as asking questions that have an answer immediately implied by results of the part of inquiry that is authoritative about the subject matter of that question. If there was such an implication, but it was hard to see whether it obtained, then this would be different. There might be subtle and hard to establish implications across disciplines, answers to properly metaphysical questions might be implied in a subtle way by the results of other parts of inquiry. That should not be precluded at the outset. But it can't be a straightforward and trivial implication, as concluding that there are numbers from that there are infinitely many primes. If metaphysics is both ambitious and modest then the questions that are in the domain of metaphysics can't have answers that are immediately implied by answers to the questions that are in the domain of the sciences, in particular the discipline that is authoritative on the subject matter of the question. And it isn't just the answers that are already given by the sciences that matter, but the ones that might be given to the questions in the domain of other authoritative parts of inquiry. Ambitious, yet modest metaphysics, if there is such a discipline at all, has to have its questions left open, or at least an answer to them not immediately implied, by how things might go in these other fields. Ambitious, yet modest, metaphysics has to have its own domain.[6]

This gives rise to two questions we hope to make some progress on here. Metaphysics worth the name has to be ambitious, yet modest. And that requires that it has its own domain. The first question is thus *the question of the domain:*

(D)   What questions are properly addressed in metaphysics?

There should be some questions of fact that metaphysics has to answer and whose answer is not immediately implied by the results of the mature sciences. In other words, there should be some propositions, some contents, for which metaphysics should figure out whether they are true or false, and other parts of inquiry don't immediately settle this. We should note again that one can only reasonably require that the other parts of inquiry don't immediately imply an answer to the question. It would be much too strong to require that they don't imply an answer at all. An implication can be complex and difficult, and cross disciplines in unexpected ways. It might be that there is a sense of implication according to which the physical description of the world implies the psychological description, but any such implication will be so non-transparent that the two fields of physics and psychology can operate as separate fields, side by side. And similarly for the relation of metaphysics to other parts of

---

[6] Talk of modest and ambitious metaphysics, in particular ontology, also appears in Bas van Fraassen's van Fraassen (2002), but with a different meaning. For van Fraassen modest ontology only studies the consequences of the sciences for what there is, while ambitious ontology asks questions that the sciences don't ask, in particular, modest and ambitious ontology exclude each other. See van Fraassen (2002: 11). Thanks to Jason Bowers for this reference.

inquiry. If there is some non-transparent implication between them, then this doesn't make the fields pointless. The fields can both try to answer their questions, side by side, as long as it isn't immediate that the results of one settle the questions of the other. But if the implication is immediate, then metaphysics so understood would be just pointless. Our paradigm case of a pointless project is to ask whether there are numbers even though the answer "yes" is immediately implied by the results of mathematics. If the metaphysical questions are just like that, then there is nothing left to do. Thus if metaphysics has a domain in this sense, then this does not guarantee complete autonomy from other parts of inquiry. But it does guarantee *a limited form of autonomy*: whatever metaphysics is trying to do, it is not immediately settled in other parts of inquiry. Metaphysics must have this form of autonomy if there is to be any discipline of metaphysics at all. One way this could be is if metaphysics has a distinct subject matter from other parts of inquiry, and that the subject matter of metaphysics is sufficiently isolated from the rest. But there might also be other options, ones where although metaphysics has no distinct subject matter, there are nonetheless distinct metaphysical questions of fact. I will defend this option shortly.

Assuming that we can answer the question of the domain, that there thus are propositions that are in the domain of metaphysics, and that we have some idea what these questions are, the next question is *the question of the method:*

(M) How should metaphysics attempt to answer the questions in its domain?

Does metaphysics have a distinct methodology, some special way in which it figures out the answers to its questions? If metaphysics has a distinct subject matter it might well have a distinct method with which questions in its domain are addressed. Mathematics has a distinct subject matter and at least one reasonably distinct method with which questions in its domain are answered: precise deductive proof.[7] Although other parts of inquiry might employ such proofs insofar as they rely on mathematics, no other part of inquiry uses proof as its method to answer questions. Here mathematics is special. Metaphysics might be just like that, with a distinct domain and a distinct method. Or it might just use the general ways in which we find something out that most parts of inquiry also rely on. Or it might use the same methods that some other specific part of inquiry uses, despite having its own domain. To answer the question of the method we will first have to see what the answer is to the question of the domain. And in particular, if the question of the domain has an answer at all.

---

[7] That is not to say that there is only deductive proof in mathematics, nor that all that matters is deductive proof, nor that only mathematics uses deductive proofs. Some insights are achieved in other ways as well, with the help of computers, or with inductive support. And it is not to say that more than a deductive proof isn't desired, for example a revealing or explanatory proof. Nor that deduction isn't employed in other parts of inquiry. Still, deductive proof is a method reasonably distinctly employed in mathematics and so it overall has a special method as its own, besides what it shares. No other part of inquiry relies so much and so heavily on just deductive proof as mathematics.

If metaphysics has a domain in the above sense, then it has a limited sense of autonomy, in that it has questions whose answers are not immediately implied by other authoritative parts of inquiry. This might seem contrary to a dominant view about how different parts of inquiry relate to each other, and to the widely held form of holism about inquiry made popular by Quine.[8] But this is not so. Metaphysics does not have to be separate and independent of other parts of inquiry to have a domain, just as psychology, say, doesn't have to be for it to have a domain. All that is required is that the questions it tries to answer are not immediately answered with authority elsewhere. For that to be the case it can well be that results from other parts of inquiry are central to answering the questions of metaphysics, or psychology, that the different disciplines cross-fertilize and draw on each other's insights, and so on. In fact, we should expect such connections. It might even be that the answer to a metaphysical question is closely tied to a different question in the sciences, although this connection can't be an immediate one, but it could be one that is worked out in a philosophical theory. None of this is denied by demanding that metaphysics worth the name has to have its own domain. It is a standard that metaphysics might or might not live up to, and even if it doesn't live up to it, it can still be useful and have a legitimate task to do. The task of metaphysics might not include to answer questions in its domain, since it might not have a domain, but rather be something else. It could be to explore alternative ways of thinking about or describing the world, and how they might be more or less useful.[9] Or alternatively metaphysics might not establish new facts, but rather construct models of reality as it is described in other parts of inquiry. Such models could focus on some aspects of how reality is taken to be, and different models might in the end be compatible with each other, since their difference might simply be focusing on different aspects of what reality is like.[10] Or metaphysics might contribute to inquiry by training philosophers to have a special eye on errors and omissions scientists might make. All these things are perfectly fine and good, but the question remains whether that is all metaphysics can and should do. If it has a domain in our sense, then it should do more, but whether and how it could do that is the question. Thinking about actual examples of metaphysical projects makes it prima facie problematic that metaphysics can do more. The question remains how it can do more, and what that might be. This is what we hope to find out here, in particular for ontology.

12.2 The Question of the Domain and the Third Puzzle

Our third, and final, puzzle about ontology is closely tied to the question of the domain. The third puzzle was the puzzle of how philosophical ontological questions are. On the one hand, they seem to be paradigmatic philosophical questions; on the other hand,

---

[8] See Quine (1960).   [9] See Carnap (1937) or Carnap (1956).
[10] See Godfrey-Smith (2006) and Paul (2012) for a conditional endorsement of metaphysics as modeling.

THE PHILOSOPHICAL PROJECT OF ONTOLOGY 301

classic examples of ontological questions seem to be answered in mathematics or the sciences. To solve this puzzle is, in effect, to make clear whether ontological questions are ever in the domain of metaphysics, and if so, why. In earlier chapters we have seen everything we need to see to now solve this puzzle, and we are now ready to state the solution.

For a question to be in the domain of metaphysics requires that the answer to the question isn't immediately implied by other authoritative parts of inquiry. For example, the ontological question about natural numbers can only be in the domain of metaphysics if the answer to that question isn't immediately implied by mathematics. As I argued above, the ontological question about natural numbers just is the question of whether there are any natural numbers, on the external reading. This question would have an answer immediately implied by mathematics if externalism about talk about natural numbers were true. But, as we saw, externalism isn't true for talk about natural numbers, internalism is true instead. Furthermore, we have seen that on the internalist view defended in chapters 5 and 6, arithmetic as well as ordinary talk about natural numbers does not imply an answer to the external question about natural numbers. Thus the authoritative discipline about natural numbers does not imply an answer to the ontological question about natural numbers. And thus this question is simply left open by that part of inquiry, and available for other parts to claim as their own. Since it is traditionally debated in metaphysics, it is perfectly legitimate and justified for metaphysics to claim the ontological question so far left open to belong to its domain. And thus there is an ontological question in the domain of metaphysics: the question of whether or not there are any natural numbers. The question belongs to the domain of metaphysics not because it asks about a distinctly metaphysical subject matter; it is a question about numbers, and questions about them are asked elsewhere as well, mostly in mathematics. What puts the question in the domain of metaphysics is simply the fact that it is left open by other parts of inquiry and it is traditionally grouped with metaphysics. Thus putting it into metaphysics is perfectly legitimate whether metaphysics in the end has a special unified mission and a distinct subject matter, or whether it is instead a field with a diverse group of questions in its domain.

On the other hand, the question of whether or not there are any objects is not in the domain of metaphysics. As I have argued in chapter 7, an answer to it is immediately implied by most parts of inquiry, and no matter what one's standards for being authoritative are, at least some of them will meet it. There are two aspects of why this is true. The first is the fact that externalism is true for talk about ordinary objects. The second is that talk about ordinary objects is part of other authoritative parts of inquiry. There certainly are many interesting questions about ordinary objects, and some of them likely can legitimately be put into the domain of metaphysics, as discussed at the end of chapter 7. These were questions that were left open by other parts of inquiry, and that traditionally fell to philosophy. But the question of whether there are any objects at all is not like that. It is immediately answered elsewhere. The answer given elsewhere might be wrong in the end, but the question about ordinary objects has a different status for metaphysics than the question of whether there are any natural numbers, on the external reading. The latter is simply left open by mathematics, while the former is immediately answered in other sciences.

To have the goal for metaphysics to be ambitious, yet modest, is to set a high goal. It requires for metaphysics to have a limited degree of autonomy from other parts of inquiry that address the same subject matter, and it wasn't clear originally whether metaphysics could live up to this requirement. But we have now seen that there are some questions, and in particular some ontological questions, that meet this requirement. And thus metaphysics has some questions in its domain. We have so far identified the ontological questions about natural numbers, properties, and propositions as being in the domain of metaphysics. We also have identified the question of whether there are any objects not to belong in the domain of metaphysics. And for many other questions we have not settled the issue: other mathematical objects, events, possible worlds, etc.

This solves the third puzzle about ontology, at least for our four cases. On the one hand, ontological questions appear to be immediately answered in various parts of inquiry outside of philosophy. But these parts of inquiry only answer them if externalism is true in that domain, which it sometimes is, as in the case of talk about objects. But if internalism is true, then the truth of what we say in these domains does not immediately imply an answer to the external, ontological question for that domain. If internalism is true, then the external question is left open by what we have established in that domain. And since the question is simply left open, and since it is just the question that has traditionally been asked in metaphysics, it is perfectly legitimate to consider the question to belong to metaphysics, without assuming that metaphysics has a distinct subject matter. And for the questions that we have found to be in the domain of metaphysics we have also found the answer. The question in the domain of metaphysics just is the question

(353)   Are there numbers/properties/propositions?

on the external reading of the quantifier. And the answer in all three cases is "no." Internalism about talk about these things guarantees that the answer to the external questions is "no," as discussed in general in section 4.3 and in more detail in the chapters that followed.

12.3 The Philosophical Project

The situation we are now in is somewhat peculiar. After long, hard work we were finally in a position to answer the question of the domain in the affirmative: there are questions that ambitious metaphysics can claim for itself, and they are the questions that we thought they were all along, questions like "Are there natural numbers?" and "Are there properties?" That the former question is not immediately answered in mathematics was a consequence of internalism about talk about natural numbers. If internalism is true, then the true mathematical statements don't imply an answer to the external, ontological question. And thus the authoritative discipline on the subject doesn't answer this question and it is thus left open for metaphysics to claim as its own by historical precedent, within the confines of ambitious, yet modest, metaphysics. Thus at least in this case, metaphysics has something to figure out.

But then it turns out that once all the work was done to show that this question was indeed available for metaphysics to tackle as one of its own questions, that work was sufficient for us to know what the answer to the question must be. What guaranteed that the question can belong to metaphysics also guaranteed a particular answer to it. If externalism were true about talk about numbers, then mathematics would have settled the ontological question about them, and modest metaphysics could not have tied its hopes for a domain on overruling mathematics. But given that internalism is true about talk about natural numbers, metaphysics can claim the external question about natural numbers for itself. The truth of internalism about talk about numbers makes the external question about numbers available for the domain of metaphysics. But if internalism about talk about natural numbers is true, then the external question about natural numbers is guaranteed to have a negative answer. Internalism guarantees that whatever the world contains, none of it is a natural number. Thus if internalism is true, then the ontological question has a negative answer. That this answer is negative should not, however, take away from the status of the question. The ontological question about numbers is not answered in mathematics, and it is a fully factual and non-trivial question. It turns out that the answer to this question is "no." Internalism guarantees that the question can be in the domain of metaphysics, and also that the answer to the question is "no." It is the correct answer to the ontological question about numbers as originally intended. And the same situation obtains also for talk about properties and propositions.

But this situation also generalizes further, to other areas of discourse that we have not considered in this book. For any domain of discourse, talk about events, say, there is the question of whether internalism or externalism is correct about that domain. If internalism is correct, then the external question is not answered by what is implied by the true things we say in that domain. The external question can thus fit into the domain of metaphysics. And then the answer to this external question is guaranteed to be "no." If externalism is correct, then the external question might still be one for metaphysics to consider. Suppose externalism is true about talk about Fs. Metaphysics might still hold that the external question about Fs is in its domain if talk about Fs does not figure in any other area of inquiry that metaphysics should be appropriately modest towards. But maybe talk about Fs is also employed in the results of a part of inquiry that modest metaphysics will be modest towards. We can call this situation an *overlap case*. Fs are an overlap case if finding out whether there are Fs is of interest to metaphysics, and talk about Fs also occurs in another, authoritative part of inquiry. Which cases are overlap cases is, of course, hard to tell, since it depends among others on which part of inquiry has how much authority. But numbers, properties, propositions, events, and ordinary objects clearly are overlap cases, assuming any reasonably modest metaphysics. Possible worlds, souls, god, and many others likely are not overlap cases. Although which cases are overlap cases is hard to tell, the notion of an overlap case is important for understanding the philosophical project of ontology. If externalism is true about talk about Fs, and Fs are an overlap case, then the ontological question about Fs is not in the domain of modest metaphysics. Since Fs are an overlap case, talk about Fs is part of an authoritative science. Since externalism is true about talk about Fs, the results of the authoritative science will imply an affirmative answer to the external question about Fs.[11] Modest metaphysics will then not claim the external question about Fs to be in its domain. So, if we assume modesty, then the question of whether internalism or externalism is true about talk about an overlap case F is the crucial question. If externalism is true, then the ontological question about Fs is not in the domain of metaphysics, and the reasonable answer to hold is "yes." If internalism is true, then the external question can be in the domain of metaphysics, but the answer is guaranteed to be "no." Thus, assuming modesty and overlap, whenever an ontological question is in the domain of metaphysics, the answer is guaranteed to be "no." This is quite significant. It means that, again assuming modesty and overlap, whenever we find that an ontological question belongs to metaphysics we thereby know what the answer must be. There are two sides to this coin. On the one hand, we have seen that ontological questions can indeed belong to metaphysics, in the sense that they are available for metaphysics for the taking, since they are left open by the results of the relevant other parts of inquiry, and metaphysics has a historical claim on them. In this way metaphysics can have a domain and even its own domain. And with having your own domain comes a form of autonomy. On the other hand, whatever questions about overlap cases are in its domain are guaranteed to have a negative answer. It is not that they have a negative answer because metaphysics is a negative discipline. But what guarantees that the question is in the domain of metaphysics also guarantees that the answer to this question has to be negative. In this sense ontology as a metaphysical discipline has no freedom: the answers to its questions are always guaranteed to go a certain way. Combining these two aspects we get the following thesis, again assuming modesty and overlap:

(AF) **Autonomy without Freedom:**[12] There are ontological questions that are in the domain of metaphysics, but whenever that is so, the answer is guaranteed to be negative.

---

[11] To guarantee this to be true one strictly speaking will need to say more about how talk about Fs is supposed to figure in the results of the relevant science. If it only occurs in quotations, or within the scope of negation, or the like, then this won't be enough. But since this isn't really relevant for all the cases of ontological questions that in practice arise I won't attempt to spell out the precise condition further. For all our cases, talk about Fs will imply an affirmative answer to the external question, assuming externalism.

[12] I owe the suggestion of this terminology for that view to Kit Fine.

This does not mean, however, that there is no work to do in metaphysics. On the contrary, what the thesis of Autonomy without Freedom shows doesn't speak against metaphysical projects, but instead points to what work there is to do here. We can distinguish the following two main projects as part of the work that needs to be done.

First, there is the question for non-overlap cases. These are cases where modesty plays no role, since other parts of inquiry do not address this question, or at least not with great authority. Many ontological questions belong in this domain. The question about the existence of a god is a paradigm example. To be more precise, non-overlap by itself is not quite sufficient for the external question to fall in the hands of metaphysics. More accurately we should say that non-conflict is sufficient for that. Non-overlap just means that talk about Fs does not appear in any of the other authoritative domains of inquiry. *Non-conflict* instead means that the existence of Fs is not immediately implied or ruled out by the other authoritative parts of inquiry. To illustrate, consider the case of entities. No other discipline than metaphysics might use the term "entity." Nonetheless, the ontological question of whether there are any entities is not in the domain of metaphysics, since an answer is immediately implied by almost any part of inquiry. Any object or thing is an entity, and thus virtually any discipline immediately implies that there are entities. Even in the case of the existence of God one might argue that the existence of God is ruled out by other parts of inquiry. To illustrate the possible conflict with an extreme example: take God, by definition, to be the being who created earth and all life on it a few thousand years ago. The existence of God so understood is in immediate conflict with other authoritative parts of inquiry.

For non-conflict cases the ontological questions are completely left open by what we have said so far. Our discussion was motivated by long-standing ontological debates about numbers, properties, and propositions, and thus we focused on overlap cases. The conclusions drawn so far apply to overlap cases alone. It should thus be clear that what we say here should not be taken to apply to metaphysics or ontology in general, in the sense that all there is to do for it is just what we discuss here. Rather it should be seen as a partial characterization of metaphysics and ontology: part of the work these fields should do is the work discussed here. Whether there is further and different work to be done in other cases remains open by all this.

Second, within the constraints of overlap and modesty, there is lots of work to do and we can see what it is. We have looked at four cases in this book of whether internalism or externalism is true for a domain of discourse. Settling in favor of internalism in three cases led to a negative answer of the external, ontological question in these cases. But each time it was a substantial task to find this out, one that relied on a number of somewhat involved considerations. And many other cases remain. The question of whether internalism or externalism is true applies to all other overlap cases just as well. And there as well we should expect it to be a substantial question which one it is. This question is crucial for finding the answer to the ontological question about

Fs. If internalism is true about talk about Fs, then the ontological question about Fs has a negative answer. If externalism is true, then the part of inquiry where talk about Fs occurs implies an affirmative answer, and thus makes an affirmative answer the reasonable thing to believe. Since many of our primary ontological questions are still left open, but are overlap cases, this is the project that is legitimately called the core project for ontology:

(CP) **The core project for ontology:** Find out whether internalism or externalism is true for talk about Fs.

To carry out the core project is not trivial. Although we looked at four cases here, many others are left open, and we can't conclude from our cases how others will turn out. As we have seen, each case deserves to be considered on its own merits, and it is not trivial to settle cases one way or another. Carrying out this project is the proper route for finding the answer to the primary ontological question for overlap cases.

To answer some of the classic ontological questions that were left open so far we need to pursue the core project. To do that we need to find out whether internalism or externalism is true about a domain of discourse, and to do that we have to investigate that discourse. We have to find out what we do when we talk a certain way and about certain things. A clear first step here is to look at the semantics of the relevant phrases, and the study of language more generally, but that by itself might not be enough. In the case of talk about natural numbers discussed above in chapter 5, for example, it was crucial to look at issues about cognition as well. Cognitive type coercion, which was a central part of the account of the occurrence of number words in singular term position, goes beyond semantics proper, and brings in issues of cognitive psychology. Similar considerations might be relevant for other domains of discourse as well. But none of the considerations we brought up to investigate whether internalism is true for our main cases involved methods that are distinct to metaphysics. Internalism and externalism are positions about language, communication, representation, speakers' intentions, and so on. Thus to carry out the core project we need to investigate them. And to do that we employ the methods of the study of language and mind. We can thus answer the question of the method for the cases of ontological questions that are captured in the core project, that is, overlap cases:

(MCP) **The method for the core project:** We need to study language and mind, but we won't employ a distinctly metaphysical method.

This answers the question of the method for the cases which were our main focus here. It leaves open what method should be employed in cases that fall outside of our main focus, in particular non-overlap cases. It is thus not an answer to the question of the method for all of ontology, but only for the central subpart of it that we called the core project.

12.4 Objections and Refinements

The defense given above of a domain for metaphysics in the case of ontology invites an important objection. This objection is closely tied to questions about the proper kind of autonomy that metaphysics should hope for, how metaphysics is tied to a distinct subject matter, and how the question of the method relates to the question of the domain. The objection is this: If the method for settling ontological questions is as it was claimed to be above, then this shows that ontological questions do not belong in the domain of metaphysics, but in the domain of linguistics. After all, what we need to do to answer ontological questions is to find out whether internalism or externalism is true. To find that out we need to study language, and so engage in the science that does just that: linguistics, maybe with some psychology thrown in. But then the question is settled in linguistics, not philosophy and not metaphysics. The above arguments don't show, the objection concludes, that metaphysics has a domain, but instead that ontological questions belong to linguistics. All we established was that which science settles the ontological questions is not what we originally thought, but it is a science, and not philosophy, after all. The ontological questions thus don't belong to metaphysics, and for all we have seen metaphysics has no domain. Not only are they not distinctly metaphysical, they are in fact distinctly linguistic.[13]

This objection is clearly quite tempting, but I think it is mistaken. It is true that the method that we now need to employ to make progress on the ontological questions, and possibly to answer them, is one about language, and thus is tied to linguistics, and possibly psychology as well. But that does not mean that the questions we hope to answer that way belong to linguistics. The subject matter of the question is not language. The question is a question about numbers, say, namely whether or not there are any. That question, as it turned out, could be answered in the negative if facts about language turn out to go one way, and in the positive if facts about language go a different way. But this is only in a larger context. What was left in answering this question about numbers was a question about language and about number words in particular. If that question is answered one way, then we should hold that there are numbers, given the authority mathematics has. And if it goes another way, then we should hold that there are no numbers, given all the things I argued for over so many chapters in this book. The question about language is simply the final step on which we could see it all depends. But it is a question about numbers, not about language, and thus it doesn't belong to linguistics, even though linguistics and considerations about language can help us settle the question one way or another.

All this is no different than when a problem in physics, for example about a feature of electrons or matter in general, can be reduced to a mathematical problem, given a particular theory of matter. It might be that if a mathematical equation has one

---

[13] For an objection along the lines that I try to turn metaphysics into linguistics, see Tahko (2012).

solution, then electrons are this way, and if it has another solution, then they are the other way instead. The problem comes down to a mathematical problem, given lots of background assumptions: a particular physical theory of matter, its mathematical formulation, and so on. But that doesn't show that questions about electrons belong in the domain of mathematics. They belong in the domain of physics, even if we can reduce a problem about electrons to a mathematical one, given a certain background. Our case is analogous. The ontological problem about numbers can be seen as reducing to a problem about language, given various things. The problem is one about numbers, not about language, even though the last step in its solution is one about language. Despite the crucial role of considerations about language in settling the problem, the problem doesn't belong in the domain of linguistics.

In a sense, much of this is just bookkeeping. Which questions belong to what discipline is not really all that important. Maybe thinking of disciplines in this regard is an overstatement of the differences between them, although there certainly are clear cases about some questions belonging here, and others there. What was crucial for us to have a domain of metaphysics was not to demarcate metaphysics in a precise way, but to deal with the prima facie good arguments that there is nothing legitimate for it to do. These were concerns about autonomy and authority. It seemed that a prima facie case could be made that the questions metaphysics hopes to answer have an answer immediately implied by the results of other parts of inquiry that are authoritative on the subject matter of the question. Thus it seemed, prima facie, that metaphysics is trying to ask questions that have already been answered. For metaphysics to be legitimate it must therefore have some autonomy from other parts of inquiry. But does it have enough autonomy from linguistics to claim the core project of ontology to be its own? To answer this in the negative is to restate the objection: given the central role of considerations about language in the philosophical project of ontology as spelled out above, metaphysics doesn't have sufficient autonomy to be worth the name, as we put it above. And with no autonomy comes no domain, or so the objection continues.

But this way of objecting is also mistaken. Metaphysics must have some form of autonomy, but it doesn't have that much autonomy. We should distinguish *absolute autonomy* from *relative autonomy* here. Metaphysics would have absolute autonomy if its aim were to answer questions that are simply independent of all other parts of inquiry. Some philosophers might be understood as thinking that metaphysics has that kind of extreme autonomy. But to demand that kind of an autonomy for there to be a discipline of metaphysics worth the name at all is demanding too much. What we need to demand instead is relative autonomy: the questions that metaphysics hopes to answer don't have answers immediately implied by parts of inquiry that are authoritative on the subject matter of the question. If we are asking something about numbers, for example whether there are any, then an answer to this question can't be immediately implied by mathematics, otherwise that question can't be seen as giving metaphysics a domain. Absolute autonomy is not something we can reasonably demand of metaphysics or any other discipline, but relative autonomy in the way specified is what we can demand. Metaphysics and ontology can have relative autonomy even though a crucial part of finding the answer to the questions in the domain of metaphysics involves considerations about language.

Is the connection between the results about language and the answer to the ontological question too trivial to give metaphysics even relative autonomy? One might argue that once we know that number words are like this, and that quantifiers are like that, then it trivially and immediately follows that the answer to the ontological question about numbers is negative. And thus linguistics immediately implies an answer to the ontological questions about numbers. But this would downplay what we had to go through over the main parts of this book to answer the ontological questions that way. It is a discovery that some ontological questions can be answered by reflecting on language, besides their subject matter being something totally different like numbers. There is a crucial difference between how close the connection is between there being infinitely many prime numbers and there being numbers, on the one hand, and between internalism about number talk and there being no numbers, on the other hand. One is a trivial and immediate implication, the other is not. But maybe for some philosophers the connection would still be too close for metaphysics to have the kind of autonomy that they had hoped for. In the end it is a matter of degree, and nothing really hangs on where one wants to draw the line for sufficient autonomy to legitimately call this part of inquiry metaphysics. What matters is that the prima facie arguments for metaphysics being simply confused and trying to answer questions that have long been answered can be resisted in the way we have seen. And seeing how they are to be resisted is part of answering the questions as they were intended in the first place.

All this gives us a partial picture of how to understand metaphysics. I have not proposed an account of what metaphysics is supposed to do in general, only that there is something that metaphysics is supposed to do, and how roughly it should be done. The domain for metaphysics defended so far only includes a couple of ontological questions, and for those I proposed answers. The method suggested for answering further ontological questions only applies to overlap cases so far. We have not seen a method for answering non-overlap and non-conflict cases, nor have we seen a general method for metaphysics. In particular, we can conclude from our discussion above that metaphysics and ontology do not have a distinct subject matter. The ontological question about numbers is no different in its subject matter than mathematical questions about numbers. Both are about numbers. Furthermore, ontological questions are not just asked in metaphysics. Whether there are any white bats is a question about what there is in its external reading, and thus the same general kind of question that we ask when we want to know whether there are numbers. There is no distinct metaphysical terminology in the questions in the domain of metaphysics, nor are these questions about a distinct subject matter. This contrasts the present approach with what I take to be its main alternative, which also affirms a domain of metaphysics with some form of autonomy. According to this alternative, metaphysical questions are different, since they are expressed with terms which are distinctly metaphysical and which are not relied upon in other parts of inquiry. Contrasting the present approach with this alternative, and showing that and why the former is to be preferred, is the topic of our final substantial chapter.

13

Esoteric and Egalitarian Metaphysics

Metaphysics worth the name must be both ambitious as well as modest. For it to be ambitious, yet modest, metaphysics it must have its own domain: some questions that are properly addressed by it, but that are also not immediately answered by other, authoritative parts of inquiry. Metaphysics so conceived must have a limited form of autonomy. Furthermore, we have seen that some ontological questions are in the domain of metaphysics so understood, including the questions "Are there natural numbers/properties/propositions?" The questions we found so far to belong in the domain of metaphysics have no distinct subject matter. They are, for example, about natural numbers, just like the questions in arithmetic, and they are expressed with the same notions as questions that properly belong to other parts of inquiry. Metaphysics, in particular ontology, has a domain and a limited form of autonomy, but no distinct subject matter, nor a distinct methodology with which the questions in its domain are to be addressed. We have seen in chapter 12 why all that is so.

In this chapter we will investigate what I take to be the main alternative for defending a domain for metaphysics. On this alternative account, metaphysics does have a distinct subject matter, and the questions in its domain are expressed with terms that are not shared with other parts of inquiry. Philosophers who hold this position come in different kinds, and they often say things like that metaphysics is concerned with the fundamental structure of reality,[1] or with what grounds what,[2] or with what is true in reality and what is grounded in what is true in reality.[3] To evaluate the success of proposals like these we will have to look at what role various notions like "fundamental," "reality," "ground," and other related ones, like "dependence" or "in virtue of," can and should play in metaphysics. To do this we will approach the issue from the outside: with an extreme case. In the following section we will look at an overly radical, made-up proposal about what metaphysics is supposed to do, and why it quite clearly fails. Then we will look at an actual proposal by Kit Fine, and with the lesson from the radical proposal in mind, we can see that it, too, has to be rejected. After that we will consider the case for giving some notion of metaphysical priority a central role in what metaphysics is supposed to do. Here there is real potential, but

---

[1] Sider (2012).   [2] Schaffer (2009b).   [3] Fine (2001).

as we will see, the candidates proposed in the literature so far either fall short in their ambitions, or are to be rejected outright. However, future work might legitimize giving such notions a greater role, and we will see in outline below what needs to be done for that to occur. In any case, I will conclude that the subject matter of metaphysics is not tied to these notions in a unified way.

13.1 Against Esoteric Metaphysics

To have a domain for metaphysics is for there to be some questions that are properly addressed by metaphysics and that are not immediately answered by other authoritative parts of inquiry. One idea of defending a domain for metaphysics is simply this: the questions that metaphysics is concerned with are not immediately answered by other parts of inquiry, since they involve some vocabulary, some terms or words, that don't appear in other parts of inquiry.[4] No wonder these questions are reserved for metaphysics, no one else even uses the terms to state them. One example of this is the term "entity." We might well assume that no other part of inquiry uses it, and no other part of inquiry asks the question

(354) Are there entities?

However, this won't be enough to put this question into the domain of metaphysics. Although no other part of inquiry might ask that question, the results of almost all parts of inquiry nonetheless immediately imply an answer to it. Although this term is novel, with respect to the rest of inquiry, it is not sufficiently inferentially isolated. Any thing or object is an entity. Thus any part of inquiry that implies that there is some thing or object, which is basically any part of inquiry, immediately implies that there are entities. Although the case of (354) isn't good enough, maybe something in the neighborhood is. If we combine *novelty* and *inferential isolation*, then we might be able to state a question that wouldn't be immediately answered anywhere else, and that thus might give us a domain for metaphysics. Although in principle there are many options of doing this, in practice one phrase stands out as being frequently employed, at least in conversation: "really." The basic idea is here simply the following. On the one hand, it is clear that there are numbers, and mathematics has shown it to be so. On the other hand, this proposal goes, there is the philosophical and metaphysical question of whether there *really* are numbers. This question is not answered by the observation that there are numbers. Whether there really are any numbers is left open by there being numbers. And similarly for basically any other question. This notion of "really" is distinctly metaphysical, and answering questions stated with it is properly what metaphysics is supposed to do, or so the proposal. Although I am not sure who takes this exact position in print, I have certainly heard it many times in conversation.

---

[4] We briefly discussed this idea in chapter 12 in connection with the distinction between non-overlap and non-conflict cases.

For the moment we take it simply to be a radical proposal of a more general strategy in order to see what is wrong with it, and then to look at proposals that are indeed defended in print.

This proposal is a non-starter on a common use of "really." That phrase is widely used not to add anything to the content of a sentence, but to comment on the status of that content as being surprising. In this sense one might say

(355)   I really ate the whole pizza.

or even

(356)   There really are arbitrarily large gaps among the prime numbers.

But none of this is related to metaphysics. Expressions like "really" in this use are usually called *discourse markers*, since they don't have the function of adding something to the content of what is said, but rather to comment on the status of what is said in the discourse as either surprising, or novel, or a change of topic, or something along these lines. Other examples of discourse markers include "anyway," "in fact," and many others. "really" in this sense won't be of help to us.[5]

Of course, "really" in metaphysics generally is not intended to be used in this way. It is intended in a different, metaphysical sense, one that we might not be able to spell out any further. It is meant to be used in the sense in which it is coherent to say that although there are numbers, *really* there are only concrete objects. And although it is wrong to steal, *really* there are no moral facts. To make this metaphysical sense of "really" explicit, let us introduce a new notion which is intended to capture it, and which won't be confused with the discourse marker sense of "really." Let us say that something is *metaphysically the case* just when it is really the case in the metaphysical sense of the term. Thus the metaphysical questions about numbers will include the question of whether it is metaphysically the case that there are numbers. More generally, we can use this notion to define a broad domain for metaphysics. Metaphysics has to try to answer the following question:

(357)   What is metaphysically the case?

This notion of being metaphysically the case is explicitly taken as a primitive in this radical proposal. We might not be able to spell it out in any other way, but nonetheless, it is used in the question that defines the domain of metaphysics. But it isn't that we can't say anything about it. We can say a few things about its inferential behavior. What is the case doesn't have to be metaphysically the case. After all, it was intended to allow us to formulate the position that even though there are numbers it is not metaphysically the case that there are numbers. And what is metaphysically the case doesn't have to be the case. After all, it might be metaphysically the case that there are

---

[5] See Schiffrin (1987) for a discussion of discourse markers.

no numbers, even though, of course, there are numbers. Thus "it is metaphysically the case that p" doesn't imply "p," and the other way round.

This then gives us the two crucial components for how to defend a domain for metaphysics with metaphysical terms. First we have novelty: a new, primitive, expression that can't be spelled out in other ways. Second, we have inferential isolation: no easy way to draw inferences to or from what is expressed with the new metaphysical terms. The radical proposal thus is this: the domain of metaphysics is defined by the question (357). Could this be a legitimate way to defend a domain for metaphysics, and if not, what precisely is wrong with it?

When the domain of metaphysics is defined with primitive, new metaphysical terms that can't be spelled out in a more accessible way, then we can call the resulting discipline *esoteric metaphysics*. It is properly called "esoteric," since it turns metaphysics into a discipline for insiders, with no access to those who are on the outside. You need to be a metaphysician to be able to grasp what questions metaphysics is supposed to answer. If you can't grasp the defining questions, you won't be able to figure out what metaphysics is supposed to do, and so you won't be able to join in. Only insiders are on the in, outsiders must stay out.

It is not enough for metaphysics to count as esoteric that the questions that define its domain contain metaphysical notions, even ones only understood by metaphysicians. What is important is instead that these notions can't be spelled out in more accessible terms. It might well be that metaphysical notions occur in the questions that define the domain of metaphysics, but these notions are perfectly transparent to everyone or can be explained in accessible terms. Everyone can then be an insider. Esoteric metaphysics instead has its domain defined by notions that are not generally accessible. Using new, primitive metaphysical notions is one clear way of doing this.

Esoteric metaphysics should be contrasted with an approach to metaphysics that holds that the questions that define the domain of metaphysics are universally access-ible. Everyone can in principle grasp these questions, since the notions that occur in the question are simply ones that either are part of our shared repertoire of notions, or they can be spelled out in terms of them, and thus are accessible from our shared position. Metaphysics so understood will be called *egalitarian metaphysics*, since it stresses the requirement of universal and equal accessibility of the defining questions. Egalitarian and esoteric metaphysics, as understood here, are intended to be exclusive, but they are not exhaustive. It might be that metaphysics has no domain at all, and then neither approach would be correct. Both egalitarian and esoteric metaphysics hold that metaphysics has a domain, but they disagree on how the questions that define the domain are to be expressed.[6]

---

[6] The literal opposite of "esoteric" is "exoteric," but since the latter notion is much less well-known than the former, and since "egalitarian" captures the spirit of the alternative proposal nicely, I prefer the terminology proposed in the main text.

Esoteric metaphysics should be rejected. We should reject it in the same sense that we rejected immodest metaphysics in chapter 12. Not because it can't be done, but because it is not worth doing. It can be done in the sense that someone might propose theories about what is metaphysically the case, and others might propose different theories. They could debate them at conferences, start a journal, and so on. But esoteric metaphysics is a pointless project. It has at least two problems.

First there is *the problem of insufficient value*. Why would we care about finding out what is metaphysically the case? What value could we attach to finding out that it is metaphysically the case that there are no numbers? Would it solve any of our puzzles about ontology? Would it help us understand mathematics better in any way? It would not, since being metaphysically the case is inferentially isolated, and primitive. It would simply be a result on its own, with no significance for other issues. It wouldn't affect other parts of inquiry, since they are not concerned with what is metaphysically the case. It wouldn't have an impact for other parts of philosophy as far as anyone could tell. It would only be relevant for what else is metaphysically the case. For example, it would follow, I would assume, that if it is metaphysically the case that there are no numbers, then it is metaphysically the case that there are no prime numbers. But there would be no value to finding that out as well. If we could spell out what "being metaphysically the case" means in more egalitarian terms, then real significance of the results of what is metaphysically the case might follow. Depending how this notion is spelled out, such discoveries might have significant consequences. But if it is not spelled out at all, we have no reason to see any value in investigating what is metaphysically the case.

The second problem is *the problem of insufficient content*. If some of the terms in the question that define the domain of metaphysics are novel and primitive, then what do they mean and how do they get that meaning? In our example of being metaphysically the case, it can't get its meaning from what metaphysics is supposed to do. It can't just mean or be explicated by: according to metaphysics, or according to metaphysics at the end of inquiry. The dependence is the other way round. We wanted to know what metaphysics is supposed to do, and we tried to specify that with the notion of it being metaphysically the case. If the latter notion is just a primitive and novel term then no meaning has been given to it. Things would be different if the notion could be spelled out in more accessible and egalitarian terms. But without such an account of what being metaphysically the case is supposed to be we have insufficient content, besides insufficient value.

Esoteric metaphysics in this extreme version is to be rejected. But this extreme version is unfortunately not just a straw man version of metaphysics. Some metaphysicians have toyed with an approach to metaphysics that is just as extreme, others have endorsed it, although it is usually better disguised. We will discuss a prominent example in the next section. After that we will need to look more closely at the difference between esoteric and egalitarian metaphysics, in particular in what sense the relevant notions have to be accessible from our shared repertoire, and which notions fall where.

13.2 Fine on Reality

In a series of recent papers Kit Fine has proposed an intriguing approach to the debate about realism and anti-realism, and with what method we should attempt to make progress on particularly controversial cases.[7] Fine wants to clarify the notion of a proposition being factual, as that notion would naturally be used in the expression of a realist position. For example, a realist about morality might be inclined to express their view that it is a fact that murder is wrong, but this notion of a fact can't just be the minimal notion according to which the equivalence between p and it is a fact that p holds. A different, more substantial notion of fact must be in play to characterize realism, and one of Fine's goals is to make one available for this purpose and propose a method for settling what is factual.

We will not focus on Fine's account of realism in general here, but instead the proposal for a "realist metaphysics" Fine (2001: 28) outlined in the essay on realism and developed further in some of his other essays.[8] Fine's proposal essentially relies on two notions: that of what is real, and that of ground. We will discuss these two notions separately, however, since one is arguably esoteric, while the other one is arguably not. In this section we will focus on Fine's notion of what is real and the role it is supposed to play in metaphysics. In later sections below we will discuss the notion of ground along with other related notions of metaphysical priority.

Fine holds that in order to properly understand the debate about realism, and in order to see what metaphysics has to do more generally, we need to accept a notion of reality which does not simply coincide with, nor is it easily explicated by, what is true, what exists, or what is a fact. Instead it is supposed to capture something like "how reality is in itself."[9] With such a notion we can then ask what is true in reality, or how are things in reality, so understood. But how should we think of this notion of reality? It is uncontroversial that it is not to be understood in the way in which "real" or "reality" is often used in situations outside of metaphysics. In this metaphysically innocent use it either contrasts with what is false or non-existent, or else with fiction or make-belief. So, when someone says that the Loch Ness Monster isn't real, then this is no different from saying that the Loch Ness Monster doesn't exist. To contrast make-belief with non-make-belief the word "reality" is often used as in the following:

(358)   In his mind he is a superhero, in reality he is an accountant.

These are not the uses of "real" and "reality" that Fine relies on. Can the notion of "reality" or "reality as it is in itself" be spelled out? As Fine discusses in Fine (2001),

---

[7] See in particular Fine (2001), but also Fine (2005b), Fine (2009), and Fine (2012).

[8] See in particular Fine (2005b), Fine (2009), and Fine (2012).

[9] Fine distinguishes between two notions of reality: that which is fundamentally real, and that which is factually real. Our concern is with the former. The latter is to be spelled out in terms of ground and the former. The factually real is what is fundamentally real or grounded in the fundamentally real. See Fine (2001).

there are a series of attempts to do this, but they are generally considered to be unsuccessful. Fine holds that it might well be correct that we cannot spell out this metaphysical notion of reality in any other terms. The metaphysical notion of reality is different from the ordinary notion which is at work in the above innocent examples. But even if we can't spell it out in any other terms, Fine holds we nonetheless should accept it as a legitimate notion to play a major role in metaphysics.[10] If we have it available, then we can build metaphysics around it. Since this metaphysical notion of reality is different from the innocent notion of reality that contrasts with make-belief we should capitalize the metaphysical one as "Reality," and leave the other one as "reality." Having the notion of Reality at hand we can then ask

(359)   How are things in Reality?

This question can then, in part, define the domain of metaphysics. But is the notion of Reality an esoteric notion, and is metaphysics understood as the project of finding the answer to (359) esoteric metaphysics? We need to answer two questions to determine this: first, is the notion of Reality accessible from our shared repertoire, and second, is it inferentially isolated?

It might well be that there is a notion of Reality which is as intended by Fine and which we can spell out in some sufficiently clear way. As I see it, this is a major open problem in metaphysics. What is striking about Fine's use of the notion is that he takes it to be good enough to be given a central place in metaphysics even if we can't spell it out any further at all. Thus even if we have to accept it as a primitive notion, distinct from the ordinary notion of reality, we nonetheless should use it, and give it a major role in metaphysics. If the notion can be spelled out in egalitarian terms, then there certainly is nothing objectionable in its use. The controversial case is just the one where we have to take it as a primitive notion that we can't spell out any further. Fine holds that it is acceptable even then, and thus for the following we should evaluate that option. Can we use the notion of Reality to define the domain of metaphysics even if we can't spell it out any further?

Simply because we can't spell out this notion doesn't mean that it is esoteric. Many egalitarian notions can't be spelled out further, and in fact those of reality (lower case "r"!) or fact or truth might well be among them. What is crucial is to see whether the notion of Reality is part of our shared repertoire even if it can't be spelled out any further. And here the question will be what evidence there is for or against it.

In general I see two sources of evidence for a notion being a primitive, but egalitarian, notion. First there are examples of the use of this notion in contexts that can be seen as largely shared situations of use. The latter is to be contrasted, for our purposes here, with uses that are restricted to metaphysical debates. If all the examples where the notion is supposed to play a role are from discussion among professional metaphysicians, then I would take this as at best weak evidence for the notion being

---

[10] See Fine (2001: 11ff.).

egalitarian. A second source of evidence for a primitive notion being egalitarian is an account of what role this notion is supposed to play in reasoning, communication, or the like. Here, too, the evidence would be greater if the account would give it a wide role, say in reasoning that is of significance to all, as opposed to a narrow, specialized role, say in reasoning in metaphysical debates alone.[11]

Evidence against a primitive notion being egalitarian is most naturally the doubt by otherwise competent thinkers and speakers that they have access to that notion. This can only be prima facie evidence, of course, but is still not to be discounted.

On all these counts it does not look good for the notion of Reality to be an egalitarian notion. No account of its role in reasoning is forthcoming, other than it playing a role in the metaphysical debate about realism, tense, ontology, and to define the domain of metaphysics. Examples of its use are restricted to specialized metaphysical debates. And in these debates it is highly controversial even among the metaphysicians engaging in them whether we have access to this notion. In fact, Fine's proposal to accept such a notion in the debate about realism even when we can't spell it out any further is generally seen as a radical methodological proposal.

What is uncontroversial, however, is that there is an egalitarian notion of reality. But Reality, if there is such a notion at all, is different from it. The notion of reality can not play the role in the debate about realism that Reality is supposed to play, and it can't be used to define the domain of metaphysics as Fine hopes to. It is important to keep these two notions in mind, and the plausibility of one being egalitarian should not be carried over to the other simply because they are pronounced the same way.

Fine uses the example of Democritus' atomism to motivate that we have access to such a notion of Reality. On this version of atomism, the ordinary description of reality in terms of tables and chairs is correct, but still, in the end, all there is is atoms in the void. This, Fine maintains, is coherent, and the notion of how things are in Reality shows how it is so: even though there are tables, and tables are not atoms, how things are in Reality is that all there is are atoms. The intuitive coherence of atomism is evidence, Fine holds, that we do have access to such a notion of Reality. But is this position coherent? Fine says:

Of course it is always open to the sceptic to doubt the coherence of Democritus' position. It simply follows from the existence of chairs he might say, there *is* more to the world than atoms in the void since there are also chairs. But I hope that I am not alone in thinking that such a philosopher is either guilty of a crass form of metaphysical obtuseness or else is too sophisticated for his own good. Fine (2009: 175) (emphasis in the original)

I don't want to claim sophistication, but I do confess to obtuseness here. Whether a position like Democritus' position can be spelled out coherently is a good question, and, if so, how is an even better question. But if it is coherent, it must be shown how

---

[11] We will discuss these issues in more detail again below, when we consider the harder to settle cases of whether grounding or other notions of priority are esoteric notions.

it can be coherent, how we are to spell out that on the one hand there are tables, but on the other hand everything is simple: just atoms in the void. It should be seen as a serious contender in the debate that this is not coherent at all. On the face of it, this position does not seem to be coherent. Whether Democritus' position is coherent, and how to coherently formulate it if it is, should be seen as open questions. To give it a coherent formulation should be understood as a project for egalitarian, not esoteric, metaphysics. The notion of Reality thus either has to be spelled out in more egalitarian terms or it should be taken as an esoteric notion. If we are willing, with Fine, to accept it as a novel primitive, then we are accepting an esoteric notion. And if we use that notion in the account of what metaphysics is supposed to do, then we accept esoteric metaphysics.

A notion can help define the domain of metaphysics, and give metaphysics so conceived some autonomy, if it is both novel and isolated, as discussed above. Leaving aside the issue whether the notions of Reality, or how things are in Reality, or what is true in Reality, are esoteric, are these notions sufficiently inferentially isolated? Are there any immediate implications from some premises to how things are in Reality, or the other way round?

The most natural interpretation of this notion in Fine's written work suggests that it is inferentially isolated. What is true doesn't have to be true in Reality. That is part of the point of this notion, in particular when it is supposed to be used in the debate about realism. We can all agree that it is true that murder is wrong, but whether morality has any place in Reality is thereby supposed to be left open, in particular whether there are any moral truths in Reality. Similarly, it is true that there are tables, but that leaves open whether there are tables in Reality. What is true doesn't have to be true in Reality. However, it is natural to understand the notions such that the inference in the other direction doesn't hold either. What is true in Reality doesn't have to be true. The failure of this implication plays a role on a natural reading of Fine's discussion of atomism as a metaphysical hypothesis in Fine (2009). Here Fine holds that it is coherent to have the metaphysical hypothesis that all there is are atoms in the void, while at the same time ordinary objects like tables and chairs, which are not atoms, exist as well. For this to be coherent we need the notion of Reality: ordinary objects exist, but it is true in Reality that everything is an atom in the void. If these two claims together are supposed to be coherent, then the notion of truth in Reality can't be factive: the inference from it is true in Reality that p to p or to it is true that p, can't be valid. If it were, then it being true in Reality that everything is just atoms in the void would imply that everything is just atoms in the void. But the former is supposed to be consistent with there being tables and chairs, which are not atoms in the void. And thus the natural understanding of truth in Reality and the metaphysical work it is supposed to do makes it inferentially isolated.

In personal communication Fine informed me that he intended "how things are in Reality" or "it's true in Reality" to be factive. Atomism, he holds, is not to be understood as the view that it's true in Reality that everything is an atom. Instead it is understood as the view that everything is such that it exists in Reality only if it is an atom. In particular, the quantifier has wide scope over the "Reality" operator. This proposal about the notion of Reality is more palatable than the proposal that we get with inferential isolation. But it is more problematic as a proposal aimed to capture the spirit of atomism as a metaphysical view. On this formulation of atomism it is true that

(360)   Everything is such that if it exists in Reality then it is an atom.

But it can't be true that

(361)   In Reality everything is an atom.

The fact that there are some things which are not atoms, say chairs, guarantees that it can't be true in Reality that everything is an atom, assuming the operator is factive. There being chairs guarantees that it isn't true in Reality that everything is an atom, even though these chairs don't exist in Reality. But shouldn't only the things that exist in Reality matter for whether everything in Reality is an atom? It is a perfectly coherent and well-formed question whether in Reality everything is an atom. As long as there are chairs the answer, on the factive construal of the "in Reality" operator, must be "no," even if everything which exists in Reality is an atom, and the things which are not atoms don't exist in Reality. Unless there other things, which do exist in Reality and are not atoms, it should be true in Reality that everything is an atom. But it can't be if the operator is factive.

Having the in Reality operator as factive makes it less inferentially isolated, but doesn't make it coherent as it is intended. To allow for a coherent formulation of atomism the operator must be inferentially isolated in the above sense, in particular it can't be factive.

Fine's use of the notion of Reality in his account of what metaphysics is supposed to do is the clearest example of an actual proposal of esoteric metaphysics that I know of, although several others are close seconds. The notion of how things are in Reality is essential for the question that defines the domain of metaphysics. On the version discussed above, which Fine is comfortable with, it can't be spelled out in other, more egalitarian terms. It is distinctly metaphysical, with no known role outside of metaphysics. It is as good an example of an esoteric metaphysical notion as any. Furthermore, as intended it is best seen as inferentially isolated as well. In the end, how things are in Reality is no different from what is metaphysically the case, although its esotericness is better disguised, since the notion of reality is egalitarian, although that of Reality is esoteric.

Fine's metaphysical project is based on two notions: that of Reality, and that of ground. The former has to be abandoned, at pain of leading to esoteric metaphysics. The situation with the latter is somewhat more complicated, and we will discuss it, and related notions, in the second half of this chapter. Before we take a closer look at notions of metaphysical priority, of which ground is an example, and whether reliance on them leads to esoteric metaphysics as well, we should clarify a few points about the above rejection of esoteric metaphysics.[12]

13.3 Compare and Contrast

Esoteric metaphysics is metaphysics whose domain is defined by questions that contain novel metaphysical terms not generally accessible. Egalitarian metaphysics has the questions in its domain defined in generally accessible terms. It is easy to misunderstand the difference and to take esoteric metaphysics to be something that is clearly unproblematic. For example, to reject esoteric metaphysics is not to demand that questions in metaphysics do not involve primitive unanalyzable terms, or to demand that every notion can be spelled out in other terms. This would be absurd; many terms can't be explained in any other way and are rightly taken as primitive. But if they are, they should be taken to belong to our shared repertoire, and not be a novel term introduced in metaphysics. It is also not required in egalitarian metaphysics that the notions are accessible to absolutely everyone. But any exceptions should have some explanation of why they are exceptions: Alzheimer's, short attention span, etc. are perfectly good exceptions why some people won't be able to grasp the relevant notions, although in general they are accessible to all. And it is of course not fully clear what "accessible" is supposed to mean more precisely. There are clear cases of success, like an explicit definition, and clear cases of failure. For example, telling those who don't get a certain metaphysical notion that they need to do more metaphysics to get it, and if they still don't get it, to tell them metaphysics probably just isn't their thing. But there are other subtler ways in which esoteric metaphysics needs to be contrasted with things that are unobjectionable. We will focus on three of them now.

13.3.1 Esoteric Questions and Esoteric Answers

It is important to make clear that what is objectionable is to rely on esoteric notions in the questions that define the field. It is not objectionable, as far as I can tell, to rely on whatever notions one pleases in one's attempts to answer well-defined questions. If the questions are expressed in terms accessible to all, then the project of trying to answer these questions is a well-defined project. Once this project has been established it might well prove useful to introduce new and esoteric terms in one's attempt to answer these questions. These might be theoretical terms, undefined except for their role in the proposed answer to a well-defined question. Of course, such a term could not be the answer by itself, just a part of the answer. Here there are clearly better and worse

---

[12] Our discussion in this chapter focuses on the two most central notions that might define a unified domain for metaphysics: reality and priority. There are also other candidates in the literature that go by various names, including metaphysical structure, fundamental quantification, ontologese, ultimate truth, etc. Although I won't be able to discuss these cases in detail here, I maintain that the same worries brought out in the cases I do discuss carry over to them as well, *mutatis mutandis*.

ways of having such a novel term in the proposed answer to a well-defined question. But whatever might go wrong with the use of esoteric notions in the answers, it at most affects the quality of the proposed answer to a well-defined question. It doesn't reflect badly on the question that defined the field, and thus it doesn't reflect badly on the field itself. Esoteric notions in the answers might point to a bad state of progress in the field, esoteric notions in the defining questions point to a bad field.[13]

13.3.2   Mathematics as Egalitarian Inquiry

What applies to metaphysics also applies to other parts of inquiry. Esoteric inquiry is just as problematic as esoteric metaphysics. What matters for a field of inquiry to be esoteric is that the defining questions of the field are not accessible. It is unproblematic for a field to be esoteric in the sense that it is hard to access because the questions are difficult, or only of interest to a few, or require years of study. This doesn't make a field esoteric in our sense. Are other parts of inquiry esoteric in our sense, in particular, are other perfectly acceptable parts of inquiry esoteric?

A good example to look at, and of the contrast between the two senses of being esoteric, is mathematics. Mathematics, the academic discipline, is hard, requires years of training, and isn't for everyone. But is it esoteric inquiry in our sense? On a first pass it might seem like a paradigm case of an esoteric discipline. All the questions that mathematics tries to answer are full of mathematical notions. To take one example: What is recursion theory supposed to find out? What the mathematical properties of recursive functions are. This seems to be a perfect example of esoteric inquiry. The central notion of recursion theory, namely the notion of being recursive, or recursion, appears in the question that defines the field of recursion theory. And similarly for

---

[13]   A good example of the difference is the use of "natural" in David Lewis, on the one hand, and in Ted Sider on the other. The role of the natural non-natural distinction in Lewis is certainly debatable, but I find the following account to be a plausible and friendly reading of Lewis: In metaphysics we ask a number of ordinary, egalitarian questions like: which statements of the form "if X were the case then Y would be the case" are true (i.e. which counterfactuals are true)? What is the relationship of causation? And so on. These perfectly egalitarian questions are clear enough as questions, but hard to answer. Lewis proposes, in an attempt to answer these questions, that we accept that there are degrees of naturalness among properties, something which is not spelled out further. That is, he proposes that there is a primitive difference of naturalness among properties which is then relied upon in an attempt to answer the egalitarian questions. Naturalness is thus analogous to a theoretical term in a theory that is the proposed answer to a perfectly meaningful question. I am no fan of Lewis's proposed answer to these questions, in particular when it comes to his reliance on natural properties, but I don't think he engages in esoteric metaphysics when he gives his account of causation, laws, counterfactuals, and so on, at least not as I understand him here. He engages in a perfectly clear project when he tries to answer a well-defined question. I am not a fan of his particular answer, but I am a fan of his trying to answer these questions. See Lewis (1983a).

Sider's use of naturalness goes further and does turn his approach to metaphysics into esoteric metaphysics. In his Sider (2009) and, in particular, Sider (2012) his project is quite clearly esoteric. For Sider, the subject matter of metaphysics is to find out what the fundamental structure of the world is like. And this is tied to finding out which sentences in a language where all expressions express perfectly natural things are true. What metaphysics is supposed to do is defined using the primitive metaphysical notion of naturalness or structure. Here the problem is not with the answer given to a perfectly good question, but it is the question itself that is problematic. As I understand Sider, he is explicitly, and unapologetically, esoteric.

many other branches of mathematics: what is topology supposed to do? Investigate the mathematical features of topological spaces, etc. However, a discipline is not esoteric simply because terms from that discipline occur in the questions that define the field. It would be esoteric if terms in the questions that define the field were not generally accessible. And this is exactly not true in mathematics. The notions of a recursive function and topological space are not primitive mathematical notions, not to be spelled out in other terms. To the contrary, they are defined in accessible terms. In these examples, as in many other examples in mathematics, there was an intuitive, but imprecise notion at the start, one which was good enough for a while, but later was made precise and explicitly defined in other, more accessible terms. All this speaks for the discipline to be egalitarian, and not esoteric. New notions can arise from egalitarian notions, and they can be made more precise and given a special mathematical meaning via a definition in egalitarian terms. The commitment to precisification and explicit definition is a sure sign of an egalitarian discipline.

Mathematics appears to be perfectly esoteric, but in fact it is egalitarian through and through. Even the most basic notions like that of a collection or set, a function, or a number, are egalitarian. Although the notion of a set in the mathematical sense by itself has no clear role in our shared repertoire, it is quite easily accessible from it. This is what is commonly done in introductory courses as follows:

(362)   A set is a collection of objects where it doesn't matter in what order they are collected, just which ones are collected.

This is a perfectly egalitarian account of what a set is. It doesn't answer all questions about set. And the notion of a "collection" used above could be spelled out further, too, for example to make clear whether sets are different from pluralities or mereological sums, or whether to leave that open. This doesn't mean that such things as the empty set or a singleton set are intuitive or easily acceptable to those newly familiarized with the notion of a set so understood. But it also doesn't mean that the notion of a set is esoteric, just that there are supposed to be some somewhat strange sets.[14] Similar remarks carry over to the notion of a function, a number, and so on. They are egalitarian notions that get precisified in egalitarian ways. For the notion of a number, for example, there is the paradigmatic instance of it in the natural numbers and there are options for how a more general concept of a number should be developed from there: should we count the complex "numbers" as numbers, even though they do not

---

[14] In Raven (2011a) Michael Raven objects to a similar argument in Hofweber (2009a) that the mathematical notion of a set is esoteric, and the ordinary notion of a collection is incoherent and to be rejected outright. Since, Raven takes it, the ordinary notion is governed by the incoherent naive comprehension schema, it must be rejected, and we can't see the mathematical notion of a set as being a precisification derived from it. This strikes me as mistaken. The ordinary concept of a collection would only be incoherent if it were a conceptual truth that naive comprehension held for it, in the style of Eklund (2002). But this isn't the case, certainly not on the strict reading, as opposed to a generic one, as discussed in Hofweber (2008). Instead naive comprehension is an initially plausible, but mistaken, conception of what collections there are, which was improved upon in the development of axiomatic set theory.

come with a natural linear ordering? Precisifications of the concept of a number could go either way here, but however it will go, it doesn't turn the notion of a number into an esoteric one. It's just an egalitarian precisification of an egalitarian notion.

That mathematics is committed to egalitarian inquiry is made vivid by the role that definitions have in it. Although basic notions are often characterized axiomatically, making clear what mathematical features an egalitarian notion like a set or natural number is explicitly taken to have, most mathematical concepts or notions are explicitly defined in terms of already understood ones. New concepts are generally introduced with an explicit definition, and in cases where such definitions are not given there is an implicit commitment to providing them as soon as it is feasible to do so. Contrary to first appearances, mathematics is in its practice and its methodology thoroughly egalitarian. This commitment is one that should be adopted in metaphysics as well.

13.3.3 Esoteric Stepping Stones

The arguments presented against esoteric metaphysics should not be taken to object to there being any role for esoteric questions or esoteric notions in metaphysics. Sometimes esoteric notions appear in a question that can't yet be stated in any better way. "reality in itself" might be a good example of this, "ultimately" is another. Does reality in itself contain value? Is mathematics ultimately our product? It is not completely clear what these questions are supposed to be more precisely, but they are suggestive, and possibly in a fruitful way. It shouldn't be objected that one starts a project this way, with a suggestive, although so far not sufficiently clear, question. What is objectionable is to leave the question in this state. An esoteric question must be seen as a stepping stone and only a first step towards an egalitarian alternative. The project of answering these questions must include a commitment to getting clearer what the question is more precisely. As stepping stones esoteric notions can certainly be useful and important. Thus we shouldn't object to their use in principle, in particular their temporary use until we can do better. But we must object to their use in questions when we also hold that this is the best we should do. A number of metaphysical debates rely, at present, on such a stepping stone, be they esoteric notions or metaphors. Debates about realism and idealism, about the philosophy of time and about persistence are good examples of this situation. These debates should not be rejected, even if competing positions are stated in terms like an object being wholly present at every moment, or our values being projected onto the world, or the structure of reality being grounded in the structure of our minds. But the project must be to overcome these stepping stones, and to spell out the opposing views in egalitarian terms. In the end the alternative views should be stated in perfectly accessible terms, but how to do that is not easy.[15] Sometimes part of the progress is to state the question

---

[15] A wonderful case of this is Jonathan Bennett's discussion in Bennett (1963) of dualism and monism in Descartes and Spinoza. Since Descartes is a dualist, and Spinoza by contrast is a monist, there should be

better, and sometimes it is to see that the question was ill conceived. In any case, we should not leave it standing in esoteric terms and be comfortable with it.

13.4 Metaphysical Priority

The commonest contemporary way to hold that metaphysics has a distinct subject matter is to hold that this subject matter is tied to some form of metaphysical priority. Metaphysics is supposed to find out, possibly among other things, what is more basic than what, in some sense of being more basic. In the second half of this chapter we will mostly critically evaluate this proposal, in particular with an eye on whether it leads to esoteric metaphysics, since the notion of being more basic has to be taken as an unexplained metaphysical primitive. This will lead us to the hard question of how to determine whether a particular notion is egalitarian and to be accepted, or esoteric and to be rejected. We will need to proceed by looking at such a proposal for a domain of metaphysics in more detail.

The basic, and somewhat simplified, idea of tying the domain of metaphysics to metaphysical priority is to hold that while the sciences and other parts of inquiry try to find out which facts obtain, metaphysics tries to find out which facts are more basic than which other facts. Metaphysics thus concerns the hierarchy of facts, while the sciences concern the obtaining of facts. Both can live happily side by side, each with their own domains. Metaphysics is thus centrally, although possibly not exclusively, concerned with metaphysical priority. There are a variety of notions which have been proposed to serve this role, and since they are notions of a priority ordering which is supposed to play a central role of metaphysics, it is common to use the term "metaphysical priority" as an umbrella term for such notions.[16] Particular candidate relations of metaphysical priority are *being more fundamental than, holding in virtue of, being more basic than, being more natural than, dependence, being grounded in,* and so on.

It is uncontroversial that we have notions of priority in various senses, and that several of these notions can play a role in metaphysics. We will look at some uncontroversial cases shortly. However, it is also fairly clear that these uncontroversial cases aren't good enough to play anything like this domain defining role outlined just above. The interesting and controversial issue is whether a much broader notion of priority can play a more central role in metaphysics. Here neo-Aristotelians[17] hold,

---

a question stated in accessible terms to which Descartes would answer "two," while Spinoza would answer "one." It turns out, it is not so clear what that question is. A more recent version of this issue arises from the problem on how to state presentism in the philosophy of time. See Crisp (2004) and Ludlow (2004).

[16]   For a discussion of the relationship of such general concepts of priority to more specific metaphysical priority relationships like material constitution, functional realization, etc., see Bennett (2011) and Wilson (2014).

[17]   The term "Neo-Aristotelian" is also used for a number of other philosophical movements. In ethics it generally stands for forms of virtue ethics. In metaphysics it is also associated with those who hold that objects are in some sense compounds of matter and form. See Koslicki (2007) and Fine (1999).

with Aristotle, that priority is a central focus of metaphysical investigation.[18] To assess the neo-Aristotelian project we need to see whether the egalitarian notions of priority go far enough, and whether the notions that go far enough are esoteric.

13.4.1 Some Egalitarian Notions of Priority

There are many notions of priority that are perfectly egalitarian, although many have taken them to be distinctly, and even problematically, metaphysical. Among these unproblematic notions are modal notions, counterfactuals, causal notions, temporal notions, and many more. These are unproblematic for our purposes in that they are straightforwardly egalitarian. The might well be problematic in other ways in that they give rise to intriguing philosophical problems, but such questions, when stated using these notions, are perfectly accessible to all. Although given the recent history of philosophy one might think that necessity and modal notions more generally are especially dubious metaphysical notions, ones reserved for idle speculation, this is mistaken. Modal notions play an important role in ordinary, everyday reasoning. Counterfactuals, for example, are important for reasoning about the past and to plan for the future. If I want to learn from my mistakes, it is important to think about what would have happened if I had done things differently. If I want to think about my options, it is important to keep in mind what has to happen given that something else happened. Modal notions in general are perfectly egalitarian. And the fact that they are egalitarian is made vivid by the role they play in everyday reasoning, outside of metaphysics. Their having a role in reasoning that everyone engages in is clear evidence for their status as egalitarian notions. That they are made problematic in metaphysics doesn't take away from that in the least. And refinements of modal notions, distinguishing different kinds of modality, say, are as egalitarian as the terms with which they are refined. Thus even isolating a notion of metaphysical modality can be perfectly egalitarian if it is characterized in accessible ways.

Given that modal notions are egalitarian, anything that can be defined from them is just as egalitarian. In general, if a notion can be defined in terms of egalitarian notions, then the result is also egalitarian. Explicit definition is the paradigm of extending our repertoire without increasing its expressive power. Less than explicit definition might well be enough for something to count as egalitarian, but explicit definition certainly is enough. Many notions of metaphysical priority are definable in terms of modal notions. Paradigmatically among them are the various notions of supervenience.[19] Some notions of supervenience won't be good enough as notions of metaphysical priority, since they will not give us any asymmetry. But some will satisfy some minimal constraints that a notion of priority has to exhibit. A second paradigmatic case of an egalitarian notion of priority, or dependence, is counterfactual

---

[18] See Aristotle's well-known passage Aristotle (1984: Vol. 2, 1688).   [19] See Kim (1993).

dependence: *a* wouldn't be the case if *b* weren't the case. In fact, this is maybe the most natural way of understanding that something depends on something else: if the latter hadn't happened the former wouldn't have either.

These notions of dependence, or priority, are unproblematic, and metaphysics should use them as much as it pleases. So far all we get is egalitarian metaphysics. But it is also well known that such notions of priority and dependence are of limited power when it comes to carving out a domain for metaphysics, or to stating traditional metaphysical problems in terms of them. The use of supervenience in philosophical theorizing of the last couple of decades or so is a good case in point. Although the notions used there are unproblematic, the results achieved are limited. In particular modal notions can't distinguish between things that necessarily go together. And some of the intended uses of neo-Aristotelian priority are to do just that. If we ask whether the truths about numbers are grounded in the truths about sets we are asking about things that necessarily both are the way they are (or so we can assume for now). And similarly when one wonders whether being water is grounded in being H_2O (also assuming they necessarily go together, of course). The notions of metaphysical priority that are supposed to do the heavy lifting are intended to be hyperintensional: they can distinguish between what is necessarily equivalent.

It is important to make perfectly clear from the start that there is nothing wrong with hyperintensionality. Many egalitarian notions are hyperintensional: explanation, intentional notions like believing, and many more. One can be surprised by one mathematical fact, but not be equally surprised by a necessarily equivalent one, but that doesn't make surprise a dubious notion. What is potentially problematic with metaphysical priority as it is often employed is not its hyperintensionality, but its esotericness. Metaphysics can't be based on esoteric notions, whether they are extensional, intensional, or hyperintensional. Metaphysics has to be based on egalitarian notions, whether they are extensional, intensional, or hyperintensional.

The question in the following will thus be whether there is more than the uncontroversial egalitarian notions of priority, in particular something more that might play a central role in defining the domain of metaphysics. The issue is not whether we have an egalitarian notion of priority. The answer here is clearly "yes," and the above cases are uncontroversial examples. The issue also isn't whether we have an egalitarian and metaphysical notion of priority. The answer is again "yes," at least assuming an innocent enough understanding of what counts as metaphysical, which would include supervenience, as well as causal and counterfactual dependence. What is at issue is whether we have, as we could call it, a *substantial notion of priority*: one which is egalitarian and can play a central role in defining the domain of metaphysics. A notion of priority is substantial in the present sense only if it goes beyond the uncontroversial modal and counterfactual ones, but nonetheless plays a central role in the questions that define the domain of metaphysics. I take this to rule out notions of priority as substantial which are primarily about conceptual connections. Metaphysics deals with facts or things, not primarily with how we represent the facts or things. If we do have such a notion, then the neo-Aristotelian project can get off the ground. If not, it is doomed to esoteric metaphysics.

Do we have a substantial, egalitarian, notion of priority available? As it turns out, there is good reason to think that we do. In the following we will have a critical look at some of the best arguments that we have such a notion available. We will start with one widely used notion in the present neo-Aristotelian context: fundamentality. After that we will consider various considerations which suggest that we have an egalitarian notion of ground that goes beyond the uncontroversial cases.

13.4.2 Fundamentality and Explanation

One of the most widely used, and probably initially the most promising notion for a neo-Aristotelian metaphysics is that of fundamentality. This notion is very promising in that it seems clear enough in many cases. Physics is more fundamental than chemistry. The physical features of molecules are more fundamental than their chemical features. And this notion of being more fundamental is easily shown to be egalitarian, not only by these and other examples, but it can also be spelled out as follows: all things being equal, the chemical features are explained in terms of the physical ones. Being more fundamental corresponds to an explanatory ordering of the facts. Some are more basic than others in that they explain them. And this, in part, is the proposed subject matter of neo-Aristotelian metaphysics.

Relying on a notion of being more fundamental, or just being fundamental, gives rise to a possible way of stating ontological questions. In the case of numbers the ontological question is not whether there are numbers. It is rather the question of whether numbers are fundamental, or whether fundamentally there are numbers, or whether fundamental reality contains numbers, or whether "fundamentally speaking" there are numbers.[20]

Although the notion of fundamentality is clearly egalitarian as understood above, the question remains whether it can have the role it is supposed to have in metaphysics. In particular, can we connect what is more fundamental, or what explains, with metaphysical priority as it is intended. One way of doing this is to hold that, in a sense, being more fundamental is being more real. And being fundamental is being ultimately real, what fundamental reality is like. But if being more fundamental is understood as an explanatory connection, why would we connect it with what is or isn't more real? Why shouldn't we think that all facts are equally real, but some nonetheless explain others?[21] It might well be that there is a more metaphysical notion

---

[20] The latter phrase is from Dorr (2008), who proposes to state ontological questions in this way.

[21] Fine mentions the problem of why an explanatory connection should be seen as corresponding to something being more or less real in Fine (2001: 26). He thinks of it mostly as a challenge that a quietist about debates about realism would pose, and hopes to get around it by accepting a notion of Reality as discussed, and rejected, above. As we will see shortly, there are some non-quietist reasons to think such a connection should not be made.

of Fundamentality, one where it is clear that such a connection with what is real or more real obtains. But we are working with a notion of being more fundamental that is tied to explanation. Why should we think that such a connection obtains in this case?

Although it might be tempting to make such a connection, and to hold, for example, that the physical is more real than the chemical, and that ultimately, it's all physics, or the like, I don't think drawing such a conclusion is justified from an explanatory connection. That this is so is nicely illustrated by a case different from physics and chemistry, but where explanatory connections and fundamentality nonetheless play a crucial role.

There is a clear sense in which in arithmetic the prime numbers are more fundamental than the even numbers, and in which the prime numbers are more fundamental than the composite, i.e. non-prime, numbers. Every composite number is just the product of some prime numbers. Every truth about numbers can be understood as a truth about prime numbers, since all composite numbers can be understood as simply the result of the product of some prime numbers. It isn't for nothing that the result that all numbers uniquely decompose into a product of prime numbers is called the "Fundamental Theorem of Arithmetic." In arithmetic, work on prime numbers is central and results about all numbers are often derived from results about prime numbers. Results about composite numbers can sometimes be explained by results about prime numbers. In particular, no other subset of all the numbers is as central as the prime numbers. Prime numbers are fundamental, but does this show that fundamentally, it is all prime numbers? In analogy with the relationship between chemistry and physics someone might hold that the non-fundamental description of the numbers includes all the usual numbers, the whole number sequence

$$0, 1, 2, 3, 4, 5, 6, 7, 8, 9, 10, 11 \ldots$$

whereas the fundamental description of the natural numbers will only include the prime numbers, and leave out all the composite numbers, which are merely products of prime numbers:

$$2, 3, 5, 7, 11 \ldots$$

Even though prime numbers are more fundamental in arithmetic than composite numbers, it would be a mistake to think that fundamentally there are only prime numbers, or that fundamentally the number after 5 is 7, or that fundamentally there is no number between 5 and 7, or that fundamental arithmetical reality contains only prime numbers. Instead one should hold that all numbers are equally real, but some of them are special among the numbers. The way they are special is not that of being especially real, but rather being special in explaining, proving, and so on. Fundamentality does not lead to greater reality. And this applies to physics and chemistry just as well as to prime and composite numbers. That the prime numbers are more fundamental than the even numbers is an important fact about arithmetic.

And that physics is more fundamental than chemistry is an important fact about the natural sciences. The question "what is more fundamental than what?" is a good, and often difficult, question. It just doesn't have the metaphysical implications that some take it to have, in particular it shouldn't lead one to think that the subject matter of metaphysics is "fundamental reality."

This, of course, only holds on a notion of fundamentality that we can access from our shared conceptual repertoire. On a primitive notion of Fundamentality, one where by stipulation a more substantial form of metaphysical priority is built in, this won't be the case. But then we won't have any reason to think that in that sense physics is more Fundamental than chemistry, or the prime numbers are more Fundamental than the even numbers. The notion of fundamentality is a great example to illustrate the crucial dilemma for neo-Aristotelian metaphysics: on an ordinary reading of a notion of priority it won't give the intended results, and on a metaphysical reading it turns into esoteric metaphysics. Reliance on a notion of being more fundamental is one example of this dilemma.

13.4.3 Are There Egalitarian, but Primitive, Notions of Metaphysical Priority?

Some notions of priority are perfectly egalitarian, but too weak to base much metaphysics on them. Other notions of priority, like being more fundamental than, are clear enough via their connection to explanation, but in the sense in which they are clear there is no reason for, and some reason against, taking them to have a connection to metaphysical priority as intended. But still, a number of philosophers hold that metaphysical priority is a, if not the, central notion for metaphysics.[22] Some of these neo-Aristotelians are surely happy to endorse priority as an esoteric notion, citing its use in asking metaphysical questions and formulating metaphysical theses as one of the reasons for accepting such a notion. For example, Jonathan Schaffer, in Schaffer (2009b), holds the view that metaphysics should be concerned not with issues of existence, but with issues of priority. Existence claims, he holds, are almost always true. What should be controversial is not what exists, but what is prior to what. Schaffer considers attempts by various philosophers to spell out the notion of metaphysical priority in ways we also considered above, but finds such egalitarian accounts insufficient for metaphysical purposes. He thus proposes that we take such a notion of priority, or grounding, as primitive:

Grounding should rather be taken as *primitive* [. . .] Grounding is an unanalyzable but needed notion—it is *the primitive structuring conception of metaphysics.* Schaffer (2009b: 364) (emphasis in the original)

---

[22] A partial list includes Jonathan Schaffer in Schaffer (2009b), discussed further below, Kit Fine in Fine (2001) and Fine (2012), Ted Sider on being more natural in Sider (2009) and Sider (2012), Kathrin Koslicki in Koslicki (2012), Benjamin Schnieder in Schnieder (2006b), Gideon Rosen in Rosen (2010), Fabrice Correia in Correia (2005), Ross Cameron in Cameron (2018), and many more.

Grounding and metaphysical priority, Schaffer argues, need to be accepted as legitimate notions, since without them metaphysical debates and positions are hard, if not impossible, to articulate. But this is not a good argument for the acceptance of a notion of ground as a primitive metaphysical notion. It might well be correct that metaphysical debates can't be articulated without such a notion. It does not follow, though, that metaphysical debates can be articulated with such a notion. It might be that there is nothing to articulate in these debates. Even if egalitarian metaphysics is impossible as a legitimate part of inquiry, esoteric metaphysics doesn't thereby become a legitimate part of inquiry. That metaphysics is a pointless project is one of the options on the table, one that must be taken seriously. If metaphysics is to have a domain we must articulate it, and adding primitive esoteric notions to do so won't help us to establish the field. If egalitarian metaphysics has to go, esoteric metaphysics won't be able to take its place. If priority has a role in metaphysics it can't be as a primitive esoteric notion.

But there is another option. One might hold that notions like ground or priority are neither definable in modal, counterfactual, or other terms, nor are they to be taken as primitive esoteric concepts introduced by metaphysics. Instead one might try hold on to the ideal of an egalitarian metaphysics and at the same time endorse the neo-Aristotelian project: by claiming that a substantial notion of priority is part of our shared conceptual repertoire. Maybe the notion of "ground" is just as ordinary as "would" and other modal notions. Maybe priority can be relied upon as an unexplained concept in metaphysics since it is a shared one, and just happens to be one of the shared concepts that can't be explained in other terms. Maybe there is just one such substantial notion of priority among our shared notions, or maybe there are in fact several that might each play some central role in metaphysics, inviting a large-scale project of spelling out how these notions relate to each other.

To be sure, all this has some plausibility to it. After all, notions like "ground," "prior," "more fundamental," and "in virtue of" are commonly used. We often have clear judgments of priority or ground in various cases that seem to be philosophically relevant. In this section we will consider whether our judgments of priority or ground should be seen as evidence that we have a notion of priority among our shared repertoire that can be used to define the domain of metaphysics. We should thus look at whether we indeed have an egalitarian primitive notion of metaphysical priority available. I will distinguish two kinds of arguments for this conclusion: arguments that illustrate the availability of such a notion by cases, and arguments that aim to show that we have such a notion by its role in our mental lives.

BOTTOM UP: THE METHOD OF CASES

By far the commonest way to argue for our having a notion of metaphysical priority readily available is by giving an example where we allegedly employ it. Prima facie promising examples of a substantial, but egalitarian, notion of priority are examples like these:[23]

(363) a. ϕing is wrong in virtue of it causing suffering.
        b. The wrongness of ϕing is grounded in its causing suffering.
        c. ϕing is wrong because it causes suffering.

However, examples like these simply point to a counterfactual dependence of the wrongness of ϕing on the suffering that it causes. In the simplest case: if ϕing wouldn't cause suffering, then it wouldn't be wrong. This is not to say that counterfactual dependence and "in virtue of" come down to the same thing. But in these above cases the judgments of priority are simply judgments of counterfactual dependence. Although with counterfactual dependence, as with counterfactuals in general, the precise situation of an utterance has a great effect on how we should understand a counterfactual, these cases are naturally cases of counterfactual dependence on an ordinary situation where they would be uttered. Such judgments of priority do not express a substantial notion of priority as discussed above since they are simply employing the uncontroversial notion of counterfactual dependence.[24] There are, however, examples that are not to be understood in these terms. Such examples can't be understood in terms of modal asymmetries or counterfactual dependence, and they are widely used by neo-Aristotelians like Kit Fine:

- The truth of a true conjunction is grounded in the true conjuncts.[25] A conjunction is equivalent to the conjuncts, taken together. But there is an asymmetry: the truth of the conjunction is grounded in the truth of the conjuncts. The truth of the conjuncts is not grounded in the truth of their conjunction. The conjunction is true *because* the conjuncts are true, and not the other way round.

- Singleton Socrates, the set containing only Socrates, i.e. {Socrates}, is grounded in Socrates.[26] The set exists exactly when Socrates exists, but there is an asymmetrical dependence relationship between them: the existence of the set is grounded in the existence of Socrates, and not the other way round. The singleton exists because Socrates exists, and not the other way round.

---

[23] See, for example, Correia and Schnieder (2012) or Audi (2012).

[24] Whether the perceived asymmetry in the above cases can be understood simply by the counterfactual dependence going one way, but not the other, or in some other way is an open question by this. The inverted conditional—if it wouldn't be wrong, then it wouldn't cause suffering—is much harder to parse than the original one, and this is of some importance even if it has a true reading. There are also acceptable readings of the inverted "because" statements, for example, but we won't be able to go into the details of the question of asymmetry and its connection to conditionals now, since it would require a much more detailed discussion of conditionals than is possible here. However, see the discussion of counterpossible conditionals below in footnote 35 for a bit more.

[25] See Fine (2012) and Schnieder (2011).

[26] This is a well-known example of Fine's, originally introduced in the debate about essence. See Fine (1994).

- Having a certain density is grounded in having a certain mass and a certain volume. Even though density, mass, and volume are such that any two determine the third, there is a special grounding relation among them. Volume and mass ground density, and no other pair grounds the third. A thing has a certain density because it has a certain volume and mass, and not the other way round.
- What it is to be water is to be H_2O. "water" is not defined as "H_2O," so this is not a verbal definition. But it is a *real definition*, it makes explicit what it is to be water, what the nature of water is. Real definitions give us some metaphysical structure or grounding or priority, maybe among properties, or kinds of stuff, or something else. It is not a definition of our concept of water, but of what it is to be water itself.

All these examples have the following features: they give rise to judgments of priority, ground, or explanation. The latter simply since they involve "because" judgments. These judgments are quite universal and robust. And they in general can't be accounted for using modal or causal notions of priority. The conjunction and the conjuncts together are necessarily equivalent. The singleton and its member necessarily (we can assume) exist together. Water, we shall also assume, is necessarily H_2O. Density is necessarily mass divided by volume. Since there are these necessary connections between the two sides, and since we universally nonetheless make asymmetric priority judgments, it can be taken to show that we, universally, have access to a hyperintensional notion of metaphysical priority. This would be just the substantial notion of priority we were hoping for to start an egalitarian neo-Aristotelian metaphysics.

Although I take this to be a promising attempt to bring an undefined notion of priority into metaphysics, I do not think that examples like the above make this case.[27] To understand these examples we not only need to make sense of the fact that in these areas we make judgments of priority, but more importantly of the fact that we share these judgments: we all judge the same. We will have to understand why these judgments are so uniform in these cases. One possibility is, of course, that these judgments are judgments about metaphysical priority, and somehow we agree on what is prior to what. But how can this account explain that we are so uniformly, and easily, in agreement here? How would it explain that these judgments are so robust? It can't be assumed, of course, that metaphysical priority is a subject matter that should be especially controversial, or inaccessible. But nonetheless, it remains unclear why there is such great uniformity in judgment in these cases. What remains in all these cases is to meet *the uniformity challenge*: the challenge to explain why our judgments of priority are so uniform in these cases.

---

[27] Here I disagree with Chris Daly in Daly (2012), who argues that trying to establish a notion of grounding for metaphysics by giving examples of it is simply question-begging: those who reject grounding will just reject the examples as pointless or meaningless. But that strikes me as too stubborn on the side of the skeptic about grounding. The examples certainly have some force, and shouldn't be rejected outright as Daly suggests. The question is, though, what they show and what notion of priority is at work in them.

In the following we will look at what the explanation is of why we judge uniformly in these cases. We shouldn't assume, and I will argue it isn't the case, that the explanation is the same in every case. And we can't assume that the explanation given for the above cases will carry over to other cases. However, there is a common theme in the explanation of the uniformity of these cases. And this theme is in conflict with understanding the subject matter of the judgments as being metaphysical priority. The fact that we make judgments of priority that are hyperintensional speaks in favor of us having access to a notion of metaphysical priority. But the fact that we judge so uniformly and robustly speaks in favor of a different explanation of what we judge here, one that speaks against a substantial notion of priority.

The uniformity in judgment might seem puzzling at first, but the explanation is in several of these cases fairly clear. It isn't enough simply to insist that we, somehow, have access to the priority facts. How we access them, and why we seem to agree on them, and thus apparently we are either all wrong or all correct, and why we feel strongly about our judgments need to be explained. And for several of our cases an explanation is forthcoming. Let's consider the case of density, mass, and volume. We uniformly judge that a has density d because it has a certain ratio of mass and volume. And that is to say, in the terminology presently under discussion, that the fact that a has density d is grounded in the fact that it has a certain ratio between mass and volume. The two sides of the "because" or "ground" statements are necessarily equivalent. Furthermore, it is a conceptual truth that they are equivalent. The concept of density is conceptually tied to the ratio of mass to volume. That density is the ratio of mass to volume is not a substantial discovery, but a conceptual truth. This might make it even more mysterious why we have asymmetrical judgments of priority given that our conceptual competence would allow us to conclude that they are necessarily equivalent. But conceptual competence does give us an asymmetry between the two sides. This asymmetry explains our judgments of ground. There is an asymmetrical conceptual relationship between our concept of density on the one hand and our concept of mass and volume on the other.[28] Our concept of density is, somehow, derivative on, or less basic than, our concept of volume and mass.[29] To say that a concept is less basic than another is simply to say that the less basic concept is understood in terms of the more basic one, leaving open the substantial empirical details of how concepts depend on each other. All that is needed for our account here is that there is some hierarchy among some concepts, and that our case is one that fits into this hierarchy. How the hierarchy itself is to be understood is a substantial empirical question, one I presume is largely open. But that there

---

[28] It is arguable that our ordinary concept of density is in fact weight by volume, not mass by volume, but I don't think this essentially changes the story of our judgments of ground given here. We might distinguish weight density from mass density, and tell a comparable story in each case.

[29] Marc Lange pointed out to me that Newton apparently thought that density was physically more basic than mass. If so then for Newton physical priority differs from conceptual priority in this case.

is a dependence of some concepts on others is beyond question, with the case of definition an uncontroversial example, although not the example we are dealing with here. In particular, this hierarchy is among the concepts themselves, and not what the concepts express. Suppose then that, somehow, there is a hierarchy among some of our concepts, that there is a sense in which some concepts depend on others, and that for some cases like density, mass, and volume, there are conceptual connections between them, in particular asymmetrical ones. How does it explain our judgments of ground?

An asymmetrical dependence of concepts will give rise to an asymmetrical procedure for determining whether a concept applies. This procedure should not be seen as the all things considered best procedure to apply the concept, but instead as a default procedure, one that conceptual competence suggests: if you want to find out about density, find out about volume and mass. It might well be true that in the end it is easier to find out about density than about mass or volume. But by default our competence with the concepts of density, mass, and volume suggests to look for mass and volume to find out about density, and not the other way round. An asymmetrical conceptual relationship gives rise to an asymmetrical default epistemic relationship. And this fact is not lost on those competent with the concepts. It is this relationship, I propose, that explains our asymmetric judgment of priority in this case. This explanation does not carry over to all other cases, as we will see shortly, but it is an account of the relationship that explains the uniformity and strength of our judgments. Our competence with our concepts is the source of our asymmetric judgment, not a metaphysical relationship between facts. Our judgments of priority arise from an asymmetrical relationship among the representations we use to think about certain facts, not an asymmetrical relationship among the facts themselves. And since the asymmetrical relationships among the representations are shared among us we uniformly share our judgments of priority, or ground, in these cases. The uniformity challenge can thus be met quite directly on this account.

This account also applies to several, but not all, of our other cases. Take the case of the conjunction being true because the conjuncts are true. Here, too, there is an asymmetrical conceptual relationship between the conjunction and the conjuncts. It is a conceptual truth that a conjunction is true just in case the conjuncts are true. But at the level of the representations of the conjunctive fact and the two facts corresponding to the conjunctions there is an asymmetry. When we try to assess the truth of a conjunction there is a default strategy to do so: figure out whether the conjuncts are true. This default strategy might not be the best one in a particular case all things considered, but it is the default strategy that we have in connection with our competence with the concept of conjunction. It is because the default strategy goes this way, and not the other way round, that we uniformly judge priority or ground in these cases the way we do. It is an asymmetrical relation at the level of the representations that explains our judgments of ground, not an asymmetrical relationship among the facts expressed with these representations. There might well be an asymmetrical relation among the facts, but its holding isn't responsible for our judgments of ground, and isn't what our judgments are about, at least not in the cases we have discussed here. Other cases might be different, of course, and I do not want to claim that they all fall into a single camp. To the contrary, I hold that they are a diverse bunch, and the examples discussed in the following are partly similar and partly different to the above ones.

A similar story applies to our judgments of priority in Fine's well-known example of the relationship between Socrates and the singleton of Socrates, i.e. the set that contains just Socrates as its member. It is natural to think that singleton Socrates exists because Socrates exists, and not the other way round. And it is natural to think that both necessarily exist together: necessarily, one exists if, and only if, the other one exists. But if they necessarily exist together, then how can we explain our shared judgment of an asymmetrical relationship between them? Our above strategy of explanation was to exploit an asymmetrical conceptual connection. But this strategy might seem problematic in this case. Is there a conceptual connection between the concept of Socrates and the concept of singleton Socrates? In particular, is it a conceptual truth that one exists just in case the other one exists? It is widely held, however, that there are no conceptual truths about what exists. Existence is not to be settled by conceptual investigation, but is a synthetic matter: that certain objects exist is conceptually coherent, and that they don't exist is equally coherent. Thus matters of existence are not settled conceptually, and so a conceptual connection can't explain our judgments of priority in the singleton Socrates case.[30]

It is, of course, not a conceptual truth that

(364)   Socrates exists.

However, it is a conceptual truth that

(365)   If the set containing only Socrates exists then Socrates exists.

Let's distinguish an *absolute existential claim* from a *conditional existential claim*. An absolute existential claim is simply made by a sentence of the form "a exists." We can grant here that absolute existential claims can't express conceptual truths, or what we can take to be equivalent for present purposes, aren't analytically true. A conditional existential claim is expressed by a sentence of the form "if a exists then b exists." We can also grant that a conditional existential claim can't be analytic, or equivalently express a conceptual truth, when "a" and "b" are replaced with proper names. There are no conceptual ties between proper names that could turn a conditional existential claim into a conceptual truth. However, when "a" and "b" are conceptually richer singular

---

[30] The classic debate about whether or not matters of existence can be conceptual matters is the debate about the ontological argument for the existence of god, and in particular Kant's critique of it in Kant (1781). A more recent version of this debate arose from Frege's argument in Frege (1884) for the existence of numbers.

terms then the conditional existential claim can be a conceptual truth. Richer singular terms can have conceptual ties between them, and these ties can be strong enough to make a conditional existential claim come out true. Examples of such conditional existential claims are:

(366)   If the married couple Bill and Hilary exists, then each of Bill and Hilary exists.

(367)   If a set exists, then each of its members exist.

It is a conceptual truth about sets that they depend on their members for their existence, in the sense that if one of the members doesn't exist, then the set doesn't exist. This is not a substantial result about sets, but a conceptual truth about sets. And a special case of it is just the conceptual truth (365). The set {Socrates} exists only when Socrates exists, and this is a conceptual truth.[31]

However, the opposite direction is not a conceptual truth. It is not a conceptual truth that

(368)   If Socrates exists, then the set containing only Socrates exists.

It is not a conceptual truth that sets exist at all, and it is conceptually coherent that Socrates exists, but no sets exist at all. Thus contrary to (367), which is a conceptual truth,

(369)   If some things exist, then a set containing just those things exists.

is not a conceptual truth and most likely not even a truth since it implies the existence of a universal set.

Thus there is an asymmetrical conceptual relationship between "Socrates exists" and "Singleton Socrates exists." It is a conceptual truth that the latter guarantees the former, but it isn't a conceptual truth that the former guarantees the latter. And this asymmetrical relationship explains our judgments of ground. The asymmetrical relationship is tied to our concept of set, and thus universally accessible to those who share this concept. This asymmetry holds even if it turns out that sets exist necessarily, and Socrates and singleton Socrates necessarily exist together.

The situation is somewhat different with plausible examples of real definitions. Real definitions are accounts of what it is to be an F. The general form of a real definition is: to be an F is to be a G. Philosophers sympathetic to real definitions give them a central role in philosophy. They hold that philosophical investigation is often tied to finding real definitions: what is it to know something? What is it to be good? and so on. Real definitions are supposed to reveal the nature of something, and are to be contrasted with verbal definitions, which illustrate the meaning of a term. Real definitions are

---

[31] See Hofweber and Velleman (2011), which applies this to give an account of how objects relate to their spatial and temporal parts.

asymmetric: if being G is what it is to be F, then this is not supposed to hold the other way around as well. Real definitions are supposed to go from the less revealing to the more revealing.[32]

But there is also a rather different and to me more compelling way to think about statements like "to be an F is to be a G," which I'll call *to-be statements* for short. To-be statements do not reveal the natures of things and uncover a metaphysical hierarchy of some kind. Instead, in the paradigm cases of alleged examples of real definitions, they are a kind of identification, broadly an identity statement, in particular an identity statement on a generic reading. These identifications are asymmetric for a reason having nothing to do with metaphysical priority, and that we will discuss shortly. First we should see that even on the paradigm examples of alleged real definitions they are neither strict property identities, nor are the involved properties guaranteed to be necessarily co-extensional, nor even actually co-extensional. It is hard to see how any account of what a real definition is supposed to be can live with this failure of co-extensionality, and so this would speak against real definitions and strict identity of properties. First, consider

(370)   To be British is to be either English or Scottish or Welsh.

Subtleties aside, this is true, but the property of being British and the property of being either English or Scottish or Welsh do not necessarily have the same extension. The Scots almost seceded from Britain and they certainly could have done so. Nonetheless, (370) is true, informative, asymmetric in the sense to be discussed shortly, and just as good an example of a real definition as the examples usually mentioned. But even worse for the neo-Aristotelian, the most paradigmatic examples of alleged real definitions involve properties that are not even actually co-extensional. Consider one example:

(371)   To be a bachelor is to be an unmarried man.

But not every unmarried man is a bachelor, not the pope, not the man in a long-term relationship without legal marriage, and so on. And not every bachelor is unmarried. A man who legally married a stranger just to get a visa, and who hasn't seen his legal spouse since the documents were signed, can well be a bachelor. Nonetheless, I take it to be true that to be a bachelor is to be an unmarried man. Instead of a being a strict identity claim about properties, or a real definition, it should be seen as a generic identification: in general, all things being equal, bachelors are unmarried men. This generic identification is not unrelated to generic readings of sentences like birds fly. They allow for exceptions, but require an important

---

[32] Not all agree that such statements are asymmetric in an important sense. One exception is Agustín Rayo in Rayo (2013).

connection. Birds fly, and to be a bird is to fly, despite penguins, baby chickens, and so on.[33]

What remains to be explained is how the perceived asymmetry of alleged real definitions arises. Identifications are symmetric, but to-be statements have a clear sense of asymmetry attached to them, at least in the paradigmatic cases that motivate the neo-Aristotelian take on them. But even though identity statements are logically symmetric, there are many other reasons why an asymmetry is attached to them. This is not only true for identity statements about objects,[34] but also for identifications which are to-be statements. First, an asymmetry can arise in an identification when there is a conceptual asymmetry between the terms involved. There can be an asymmetrical conceptual relationship between the concepts of an F and a G. This explains the asymmetry in cases like "to be a vixen is to be a female fox," and also "to be a bachelor is being an unmarried man." However, asymmetries in to-be statements also occur in cases where there are no such conceptual connections. These are explained differently. The main, second, case is exemplified by the example of water and H_2O: to be water is to be H_2O. Here there is no conceptual connection between being water and being H_2O. Nonetheless, "to be water is to be H_2O" is a true identification. The asymmetry of this identification is to be explained by its informativeness. To say that to be water is to be H_2O is to make an informative identity statement. Identity statements commonly exhibit asymmetric behavior with respect to which part is the familiar and which part is the new information. Although logically they are symmetric, in the way they communicate information and augment the information one already has, they are not. Which part of the identity statement stands for the familiar thing and which part stands for the new thing matters for the order of the terms in the identity statement. The same is true for regular identity statements like "George Orwell is Eric Arthur Blair." This, in outline, explains the perceived asymmetry in the example of the identification of water with H_2O.

To sum up, it is problematic to argue for there being a substantial and egalitarian notion of metaphysical priority by example. The examples given in the literature are generally examples of a different asymmetric relationship than metaphysical priority or perfectly egalitarian notions of metaphysical priority, like counterfactual dependence, which are insufficient to define the domain of metaphysics. The explanation of why our judgments in these cases are uniform illustrates this point. Other examples could be and, of course, have been presented, but I have little hope that the issue will

---

[33] Other examples of to-be statements have an interestingly different use based on a somewhat different reading. They involve what could be seen as a stereotype reading or a more normative reading, as in

(372)   To be an Englishman is to love the Queen.

This difference is not just one about this example being a partial identification, as opposed to an all-out identification, but seems to be based on a different reading of the to-be statement more generally.

[34] Such asymmetries in identity statements were also discussed in more detail above, in the appendix to chapter 2.

be resolved this way.[35] Examples and the method of cases isn't the right approach to settle this dispute. What needs to be done is to adopt a top-down approach instead.

TOP DOWN: THE ROLE OF THE NOTION

If it were true that we had a primitive notion of metaphysical priority available as part of our shared conceptual repertoire, then we should be able to explain, at least in outline, why we have such a notion. What work is it doing for us? Of course, if we do have such a notion it could do work for us in metaphysics. But it is hard to swallow that we have a notion in our shared repertoire that is solely used in metaphysics, and thus likely only drawn upon by a tiny fraction of humanity. Any such primitive notion should have some role outside of metaphysics as well. It should have some other use in reasoning or thinking, likely one upon which metaphysics will draw and which would explain why this notion is so central for it. Contrary to the attempts to argue by example for our having an egalitarian primitive notion of priority, this much more promising route for showing that we have such a notion is, to my knowledge, never taken. But this strikes me as the crucial challenge to meet for anyone who wants to pursue this line. They should aim to meet this challenge:

(373)   **The cognitive function challenge:** why do we have a primitive notion of metaphysical priority as part of our shared repertoire? What function does it play in our ordinary cognitive lives, outside of academic metaphysics?

No complete account of this cognitive role could, of course, be demanded. That would be asking way too much. For many other quite clearly egalitarian, but likely

---

[35]   Among the examples I did not discuss in any detail is the Euthyphro contrast, but its proper treatment would involve too many further issues I can't discuss here. Although in simple cases the Euthyphro contrast seems to be quite clearly merely counterfactual dependence, this is more controversial in other cases and leads into issues about counterfactuals I won't be able to deal with here. To illustrate the issue, take the following example. Are handmade Swiss watches expensive because lots of people want to buy them, or do lots of people want to buy them because they are expensive? This seems like a straightforward difference in counterfactual dependence, one about what would happen if fewer people wanted to buy them vs. if they cost less. In the original Euthyphro contrast, i.e. being pious vs. being loved by the gods, the thought is that the gods necessarily love the pious, and so the counterfactuals come out the same either way. But this is only the case if we hold fixed what is necessarily the case, in this case the connection between the gods' love and what is pious. But counterfactuals don't always do this, and what to make of that is controversial. Counterfactual conditionals that concern not just what isn't the case, but furthermore what couldn't be the case are generally called "counterpossible conditionals." Whether such conditionals are all vacuous is controversial. Nonetheless, it seems perfectly coherent to me to sometimes entertain what would be the case if something wasn't the case, even though that has to be the case. For example, it is true that if there were only odd numbers, but no even numbers, then the number right after 3 would be 5. And it is false that under these conditions the number right after 3 would be 9. Nonetheless others think that all such conditionals would have the same truth value. How such counterfactual conditionals can be understood more precisely is controversial, and I won't be able to try to make progress on this here. For the connection of counterpossible conditionals to issues about metaphysical grounding, as well as further references on counterpossible conditions in general, see Krakauer (2012) and Wilson (2015).

primitive, notions we don't know the details of the cognitive function, but we have a clear enough idea of the outlines of their role. Given this standard, many notions used in metaphysics meet this challenge. We know, roughly, what role thinking about causation, counterfactuals, modality, truth, properties, etc., play in our ordinary mental lives. To establish grounding or some other candidate notion of metaphysical priority among these, a similar story will need to be told, at least in rough outline. But on the upside, if such a story could be told successfully, then we should accept that we have such a notion available after all. The crucial methodology for establishing a notion of metaphysical priority as a legitimate notion that can be used to define a domain of metaphysics, even if it can't be spelled out in other ways, is thus not to give examples where this alleged notion occurs, but to meet the cognitive function challenge. Nothing like it has been done so far, again to the best of my knowledge, but this is what the issue hangs on. I have not given any arguments here that this challenge can't be met, but I do have my doubts. I won't, and don't have to, put my hopes for a domain for metaphysics on this working out. To show that metaphysics has a domain even without such notions was one of the main conclusions in chapter 12 and this book in general. That would stand no matter how the present issue turns out. Any notion of metaphysical priority that could be used to add to the domain of metaphysics will have to be egalitarian, not esoteric.

To sum up, it should be uncontroversial that we have many egalitarian notions of priority: counterfactual dependence, conceptual priority, logical priority, and so on and so forth. Some of these notions are primitive, others can be spelled out. What is controversial is whether we have a substantial notion of priority, one that can play a central role in defining the domain of metaphysics, and possibly give metaphysics a unified subject matter as envisioned by some neo-Aristotelians. If such a notion of priority is taken to be a novel primitive notion, then this leads to esoteric metaphysics and nowhere. If it is an egalitarian notion, then it needs either to be spelled out in egalitarian terms, or to be shown to be a primitive, but nonetheless egalitarian, notion. Reliance on cases to do the latter is deficient, since the cases generally rely on trading a substantial notion of priority for one of several insubstantial ones, which is made vivid by the explanation of why we uniformly judge priority in the same way in these cases. To pursue the approach to a domain for metaphysics via a substantial notion of priority one thus needs to do one of two things. One has to either spell out the notion in egalitarian terms and show how it, so spelled out, can help define the domain of metaphysics, or else meet the cognitive function challenge and show how this notion, with that function, can define the domain. None of this has been done to the best of my knowledge, and I would judge the prospects of it working out to be slim at best. If it would work out, then the domain of metaphysics would be larger than I defended here, but the questions I claimed are in it would still be in it. In particular, that there is a domain for metaphysics at all is independent of this issue, but what it is like is not.

13.5 Resisting the Temptation of Esoteric Metaphysics

There is no question that esoteric metaphysics is extremely tempting. If we just accept a primitive notion of ground, metaphysical priority, or metaphysical reality, then we can formulate all kinds of questions and projects for ourselves that are otherwise very hard to state and very hard to justify as metaphysical questions. And these questions can be pursued in isolation of the somewhat tricky empirical facts. Are there just atoms in the void? Answering yes seems to conflict with most of the sciences, since chemistry, biology, sociology, etc., all talk about things other than atoms in the void. But is it true in Reality that there are just atoms in the void? Whatever you say here doesn't conflict with any science. This gives great freedom to state questions and to explore answers without being bogged down by other parts of inquiry. Esoteric metaphysics gives us the freedom to carry out projects independently of the facts. It only has to be constrained by the Facts, i.e what is metaphysically the case, or what is true in Reality, and so on, but not the facts, or what is true, which are in the domain of other parts of inquiry. It could be inspired by other parts of inquiry, but it doesn't have to take this inspiration on board. In this case great freedom doesn't come with great responsibility.

Using esoteric notions allows one to reformulate many metaphysical positions that were long considered either incoherent or clearly false. Take for example Jonathan Schaffer's recent defense of monism:[36] simply put, the doctrine that there is only one thing, according to him: the whole cosmos. This on the face of it seems absurd, since there are clearly many people, electrons, and numbers. But that confuses simple monism, the clearly false view that there is only one thing, with priority monism, the view Schaffer holds, which is that the whole cosmos is ultimately Prior: it is the one most Basic thing. Simple monism is in conflict with almost every science. But priority monism, and its negation, are compatible with everything else, since the notion of "priority" used is distinctly metaphysical, primitive, and inferentially isolated. Using the notion of metaphysical priority we can state a coherent version of monism, one that doesn't conflict with the results of other parts of inquiry.

The same is true with many other metaphysical doctrines that apparently have long been refuted. Thales held that everything was water, which seems quite absurd and clearly refuted these days. But maybe we shouldn't confuse Thales' view with simple *aquaism*: the view that everything is water. Rather we should take it to be *priority aquaism*: the view that ultimately everything is water. Water is ultimately Prior, in a primitive, unexplained sense of metaphysical priority. Aquaism is clearly refuted. But priority aquaism is still alive and well. After all, none of the results of the sciences refute it. It can consistently be added to all that is known in any part of inquiry outside of metaphysics. Whether priority aquaism is true is an open metaphysical question, for those who are tempted by esoteric metaphysics. And the same is true for *priority aeroism*: the view that air is ultimately Prior. And so on and so forth.

---

[36] Schaffer (2010).

Esoteric terms allow for the statement, or restatement, of many positions that otherwise might be considered absurd. All these positions can be in the domain of metaphysics. It gives hope to overcome one of the biggest obstacles of egalitarian metaphysics. For an egalitarian metaphysician there must be a question stated in ordinary terms for any question that defines the domain of metaphysics. But it is well known from debates in the philosophy of time, the debate about persistence, and others, that it is not clear what ordinary statements mark the difference between intuitively different positions. An esoteric metaphysician has the option of holding that although both parties agree on all ordinary statements, they disagree about statements formulated in terms of esoteric notions: what is true in Reality, what is Prior to what, and so on. The egalitarian metaphysician will reject this option. If there is a real debate here, then there must be a question expressible in egalitarian terms such that the two parties in the debate disagree on the answer to this question. It might well be a substantial task to find that question, and in particular to show that an answer to it is not immediately implied by other parts of inquiry. This might be hard to do, but this is what needs to be done to show that metaphysics has a domain.

Esoteric metaphysics might well have a temporary resurgence in the near future in metaphysics, but we should try to resist it. It used to be completely shunned, but it has become more acceptable in practice. There is some explanation for its acceptance. It has become more and more clear that it is not obvious what work metaphysics can do, and how it can have anything like the autonomy from other parts of inquiry that many have taken it to have. One reaction to this predicament is to pursue questions in other parts of philosophy where this problem can be avoided. Another reaction is to formulate the questions in esoteric terms. Those who have the former attitude won't find their way into metaphysics, whereas those who have the latter will. Metaphysics is in this sense self-selecting, and such selection favors the esoteric, in particular once it is more acceptable. The more esoteric it in practice becomes, the more it will select for those who are comfortable with the esoteric. But the more esoteric metaphysics becomes, the less accessible and relevant it will be for the rest of inquiry. Esoteric metaphysics is not a way to save metaphysics, but a way to give up on it. It turns metaphysics from a discipline that had the hope of making a real contribution to inquiry into one that talks only to itself in a language that not even it can understand.

One might think that the legitimacy of notions like ground, priority, in virtue of, dependence, and so on, is indeed questionable in their intended, substantial sense, but we should err on the side of tolerance. If it is not clear whether we can take recourse to them, why don't we just give it a try and see what happens? This is a great idea, as long as the use of these notions is confined to the answers we propose to clearly stated questions, and they don't occur in the questions themselves. Erring on the side of tolerance is a good idea to further a well-defined project that otherwise got stuck. Erring on the side of caution is a good idea to motivate a project in the first place.

Neo-Aristotelians who hold that ground and dependence have a central role in philosophy often hope to have history on their side. I have heard frequently in conversation, and it is spelled out well in Rosen (2010), that the status of notions like ground and dependence in philosophy now is just like that of modal notions in the 1950s and 1960s. Back then philosophers were deeply suspicious of modal notions, trying to cast them out from philosophy. But such philosophers just poorly understood the notions, and once various work on modality and modal logic was done, it was clear that they made perfect sense and were notions well suited to be used in philosophy. Ground, this line continues, will see the same acceptance with time, in particular once we have spelled out the logical principles governing it. It is merely overly anti-metaphysical philosophers who reject it, just as they rejected modality. Ground and dependence should thus be accepted as legitimate notions in metaphysics.

This strikes me as mistaken. I completely agree that modal notions are perfectly legitimate in metaphysics as they are everywhere else. Although modal notions certainly give rise to various difficulties and puzzles, this doesn't distinguish them from many other utterly uncontroversial notions like that of an object or an event. Notions like "would," "has to," and counterfactual conditionals are perfectly egalitarian: they are universally shared, accessible to all, and they have a clear role in ordinary rea-soning and communication. Although the details of what role, say, counterfactuals have in ordinary reasoning are complicated, in part because ordinary reasoning is complicated, that they have a crucial role in it is beyond question. And this role is not one tied to metaphysics, it is a role in ordinary reasoning in general: reasoning about what would have happened if one took the other path, for example. Such reasoning is important and informative for dealing with similar situations differently in the future, in learning from past mistakes, and the like. Ground and dependence are different in these respects. Although these notions have egalitarian readings, the substantial reading intended for metaphysics, one that can define a domain for metaphysics, is not available. And it does not, as far as anyone has been able to make clear, have any role in ordinary reasoning. The two cases are thus fundamentally different. Philosophers should never have doubted the legitimacy of modal notions, but they should doubt the legitimacy of ground and dependence in the substantial sense, as used in the questions that are supposed to define the domain of metaphysics.

What needs to be done is thus this: those who hold that ground and dependence can play a role in metaphysics, and that they have a central place in the questions that define the domain of metaphysics, need to motivate that these notions are egalitarian notions. To do this we need to either spell them out or understand at least in outline what role this notion might have in thought and reasoning, and this role had better not be restricted to a few examples from metaphysics. We need to see an answer, in outline, to the cognitive function challenge. I have my doubts that it can be done, but that might, of course, be mistaken. There is always the possibility that there is a notion like that of ground which is accessible to all. Maybe such a notion does have a role in ordinary thinking and reasoning, and this would be evidence for its status as an egalitarian notion. We can't rule out this possibility, but we have good reason to think that it isn't true. All this doesn't mean that we shouldn't use "ground" and related notions in philosophy. But when we do, we must take the methodological stance that such notions are merely temporary placeholders for something that is to be filled in later, when we have a better grip on what we are after when we are tempted to describe the situation in these terms.

Overall, then, any reasonable defense of a domain for metaphysics should specify such a domain in egalitarian terms. Whatever the questions are that are to be addressed in metaphysics, they should be stated in terms that are accessible to all. The defense of a domain for metaphysics in the case of ontology given in chapter 12 did just that. The ontological questions that were shown to be addressed in metaphysics were questions like "Are there numbers?" or "Are there properties?" These questions do not rely on an esoteric sense of any of the notions involved. They are perfectly egalitarian questions. We saw in chapter 12 that metaphysics should be ambitious, yet modest, metaphysics. We have seen in this chapter that the domain of metaphysics should be defined in egalitarian terms. The questions that we have seen so far to belong in the domain of metaphysics were our ontological questions of whether there are any numbers, properties, or propositions. These questions meet all three requirements: be ambitious, be egalitarian, be modest. They are stated in egalitarian terms, and the resulting project of answering them is part of ambitious, yet modest, metaphysics. Which other questions belong to metaphysics as well is left open by all this, but we know that whatever they are, they, too, must meet these three requirements.[37]

[37] My thanks to Kit Fine, Marc Lange, as well as Shamik Dasgupta, Boris Kment, and the participants of their 2011 graduate seminar for helpful comments on an earlier version of this chapter.

14

Conclusion

My goal was to make progress on a selection of metaphysical problems that are closely tied to ontological questions. My strategy was to think closely about these ontological questions, what they are, how they should be stated, and whether they are properly connected to the metaphysical problems we hoped to solve. I noted that there was good reason to think that ontological questions are not fully understood and that their role in metaphysics is unclear. Furthermore, there are some good prima facie reasons to think that the project of metaphysics understood a certain way is a confused project. Those reasons were in particular the following three, which were in essence our three puzzles about ontology: the puzzles about how hard, how important, and how philosophical ontological questions are. First, it seemed the ontological questions can be trivially answered for certain cases even though they were supposed to be substantial questions. Second, it seemed that the ontological questions are of the greatest significance for all of inquiry even though it didn't seem unreasonable at first to hold that they are largely theoretical afterthoughts. And, third, these questions seem to be settled not in philosophy, but in other parts of inquiry. In particular, several of the classic ontological questions appear to be already answered in parts of inquiry that are nothing like what metaphysics was supposed to be. Metaphysics, in particular when tied to ontology, prima facie doesn't seem to make much sense, and thus is in need of some defense. To give such a defense we needed to see why these prima facie reasons for metaphysics being a confused project are only prima facie good, but in the end mistaken. And to do this we needed to solve the three puzzles about ontology. With these three puzzles properly solved this should help us make progress on the metaphysical problems we were concerned with in the first place. In the body of this book I hoped to make the case that this indeed is a promising strategy for solving the metaphysical problems that we started with. Once we understand ontology better and how it relates to the metaphysical problems that are tied to ontological questions we are able to solve the metaphysical problems, or so I have argued.

The key ingredients to the proposed account of ontological questions and their role in metaphysics can be briefly reviewed as follows.

Quantifiers have two different functions in ordinary communication which correspond to two different contributions that they can make to the truth conditions. Quantifiers are polysemous expressions with at least two different readings, and they have those readings not for metaphysical purposes as such, but because of two functions they have in ordinary communication. This was essential for our resolution of our first puzzle. Since quantifiers are polysemous, questions like "are there numbers?" have more than one reading. One of them is indeed trivial, but the other one is not, and that one is the one employed when we ask questions in ontology or metaphysics. Both of them are equally factual, and even though the question is trivial on one reading, this does not speak against the status of the question.

We saw that sometimes singular terms are not used to pick out objects, but they have a different function in communication instead. Our main early examples were singular terms as part of a focus construction, but that just opened the door to the possibility that this might occur much more frequently for all kinds of reasons.

This then gave rise to the possibility that whole areas of discourse do not engage in reference and use quantifiers internally, not externally. We distinguished internalism from externalism for a domain of discourse, and saw that it will be crucial to find out which one of them is true for a particular domain to answer the ontological question that is most closely associated with that domain. In particular, we needed to find out whether the possibility of internalism obtains in the cases of domains of discourse tied to our metaphysical problems.

To settle this we had to take a detailed look at what we do when we talk about natural numbers, objects, properties, and propositions. We found out that internalism was true for talk about natural numbers, properties, and propositions, while externalism was true for talk about ordinary objects. This conclusion was drawn from considerations about our talk about such things, and was not motivated by a desired outcome of the ontological questions in these areas. We were able to conclude that the ontological question about natural numbers, properties, and propositions has a negative answer, while the ontological questions about ordinary objects has a positive answer. The former is a consequence of internalism alone, the latter a consequence of externalism and the empirical reasons we have for there being objects. This answers the ontological questions, which were just the questions we originally took them to be. The answers might not be the expected ones in some of these cases, and how these questions were answered might not have been as expected, but nonetheless, it is the answer to the fully factual ontological questions that we wanted to ask.

The truth of internalism in our three domains had significant further consequences besides an answer to the ontological questions themselves. It was the key to answering the metaphysical questions tied to the ontological ones. These consequences were not always the result of internalism as such, but followed from the particular defense of internalism given, that is, from the reasons why internalism is true in these cases. Our three internalist domains of discourse corresponded to three of our most substantial conclusions. First, a version of internalism about talk about natural numbers gave rise to a rationalist conception of arithmetic. Second, our internalist conception of talk about properties gave rise to a nominalist position on the problem of universals. Third, internalism about talk about propositions gave rise to a defense of the effability thesis and conceptual idealism. On the other hand, externalism about talk about ordinary objects was tied to the view that we have empirical support for the existence of ordinary objects. Overall these positions provided answers to several of the metaphysical problems we hoped to answer.

This left us to ask about metaphysics itself. We saw that metaphysics worth the name must have a limited form of autonomy from certain other parts of inquiry, and that it indeed does have that form of autonomy for some of our ontological questions. Metaphysics has a domain, questions that it itself needs to answer, and we saw three examples of such questions: the ontological questions about natural numbers, properties, and propositions. We also saw the answers to these three questions, and that this situation generalizes for a larger group of cases: for a certain range of ontological questions, if the question is in the domain of metaphysics, then the answer is guaranteed to be "no." Metaphysics in this case has autonomy, but not freedom.

Overall, this was a defense of metaphysics globally, but a rejection of various local metaphysical problems as well as a rejection of esoteric metaphysics. As for a global defense, one of my main goals was to investigate whether some traditional metaphysical problems are confused in asking questions that appear to have long been answered elsewhere. I concluded that metaphysics indeed has a domain, and this domain comes with a limited form of autonomy. Furthermore, I proposed answers to the questions that were found to be within the domain of metaphysics. On the other hand, the view defended showed that some traditional metaphysical problems are confused. The secondary ontological question is ill motivated for any domain where internalism is true. Even though our talk about properties is true, there are no legitimate questions about what kinds of things these properties are, where they are, etc. Thus the traditional way of thinking about the problem of universals is based on a mistake. And similarly for other domains for which internalism is true. In addition, I argued that metaphysics can't have its domain tied to what is metaphysically fundamental, ultimately true, nor to similar esoteric ways of understanding it.

It would be unfair to criticize the position defended here as anti-metaphysical because of the answers it gives to various metaphysical questions. To be sure, whenever we found that an ontological question is in the domain of metaphysics, we got a negative answer to that question. Thus the position defended here maintains that there are no such things as numbers, properties, or propositions. But this is simply one of the possible answers to the question whether or not there are such things. The issue first and foremost is whether there is a legitimate metaphysical question here. This determines a defense or a rejection of metaphysics for that particular case. If there is such a question, then, no matter what the answer, this is a defense of metaphysics. To call the view defended here anti-metaphysical would be just as unfair as calling materialism anti-philosophy-of-mind. True enough, if materialism is true, then there isn't really a *mind*, but that is just one of the positions in the philosophy of mind that aim to answer the philosophical questions that need answers. Similarly with the view defended here. True enough, I didn't propose theories about ultimate reality, tied to essences, metaphysical priority, and abstract objects, nor did I defend the existence of any entities other than ordinary objects. But it would be unfair to hold that this then isn't really *metaphysics*. To the contrary, it is one of the views that would answer the metaphysical questions. In particular, I hoped to show that there really are distinctly metaphysical questions that need to be answered. Metaphysics can be ambitious, it has real questions of fact in its domain, and for some I proposed an answer. That the answer goes a certain way doesn't take away from the status of the question nor of the project of trying to answer the question.

Of course, the fact is not lost on me that the position defended here will seem anti-metaphysical to many metaphysicians. That is only fair: depending on what one takes metaphysics to be like, this should seem anti-metaphysical to many. It is clearly contrary to the neo-Aristotelian revival in metaphysics, which many see as the future, but I see as the end. It is also aimed to be contrary to the neo-Carnapian revival, which takes metaphysical questions to be either flawed or trivial. Some are flawed, I agree, but many are not, and they are not trivial either. Some non-flawed ones I hoped to propose a non-trivial answer to. And although I accept with Carnap a distinction between internal and external questions about what there is, this distinction was used above to serve a defense of metaphysics, not to undermine or reject it. If the present approach has to be labeled neo-something at all, the closest might be neo-Kantian. With Kant, the present view defends some parts of metaphysics, while it rejects others, and both stress the importance of thinking about our representation of reality in making progress in metaphysics, although, of course, for very different reasons.

This is what I tried to make a case for in this book. There are lots of further questions left unanswered. First, we only looked at some of the consequences of internalism for the three cases for which internalism was defended. I did not, for example, discuss issues related to the ascription of content, a topic widely discussed in philosophy and one for which internalism about propositions will quite likely be very relevant. I did not spell out in any detail how the internalist picture of arithmetic relates to other parts of mathematics, nor what we should think of these other parts. I also did not discuss the consequences of the internalist account of properties for areas where properties have traditionally played a larger role, for example supervenience, laws of nature, causation, and so on. And I left most questions about the metaphysics of ordinary objects unanswered. Second, I only discussed four ontological questions in this book, thus neglecting most of them. But we saw above that a number of further ontological questions can be approached with a method that carries over from the cases that we looked at. For overlap cases the method is to find out whether internalism or externalism is true for the relevant domain of discourse. The Core Project for ontology is to settle these other cases in this way, but it is not at all clear where the individual cases will fall. This will have to be carefully looked at on a case by case basis. Depending on how the cases turn out, this will potentially have significant consequences for other metaphysical debates. Third, I only focused on four cases of groups of metaphysical questions closely tied to ontological questions about overlap cases. I don't want to suggest in the least that all of metaphysics is like that. It certainly is not. Some central metaphysical problems are like the ones we looked at, but many others are not. The view defended here only addresses a small range of metaphysical problems.

But besides all the questions that are left open, progress has hopefully been made. If anything, it seems fair to accuse me of having tried to answer too many questions in just one book, long as it may be. But I hope by the end it has become clear why it makes sense to look at these four cases of ontological and metaphysical problems together, how they are similar and different, how they all fit into one general picture, and how the resulting position leads to a coherent, although only partial, metaphysical picture of the world. In a nutshell, it is a combination of realism about things with idealism about facts, and rationalism about numbers with empiricism about objects. My strategy was to solve the puzzles about ontology and thereby to make progress on various metaphysical questions tied to ontological ones, and with that to see whether metaphysics can indeed have the ambition to make a real contribution to inquiry by answering questions of fact that are properly its own questions. Thinking about why these problems are philosophical ones in the first place is the key to their answer. I hope to have made a case that this strategy is indeed fruitful.

Bibliography

Adams, R. (2007). Idealism vindicated. In van Inwagen, P. and Zimmerman, D., editors, *Persons Human and Divine*, pages 35–54. Clarendon Press.

Aristotle (1984). *The Complete Works of Aristotle*. Jonathan Barnes edition. 2 volumes. Princeton University Press.

Armstrong, D. M. (1978). *Universals and Scientific Realism*. Cambridge University Press.

Armstrong, D. M. (1989). *Universals: An Opinionated Introduction*. Westview Press.

Armstrong, D. M. (2004). *Truth and Truthmakers*. Cambridge University Press.

Audi, P. (2012). A clarification and defense of the notion of grounding. In Correia, F. and Schnieder, B., editors, *Metaphysical Grounding: Understanding the Structure of Reality*, pages 122–38. Cambridge University Press.

Azzouni, J. (2004). *Deflating Existential Consequence: A Case for Nominalism*. Oxford University Press.

Azzouni, J. (2007). Ontological commitment in the vernacular. *Nous*, 41(2):204–26.

Azzouni, J. (2010a). Ontology and the word "exists." *Philosophia Mathematica*, 18(3):74–101.

Azzouni, J. (2010b). *Talking About Nothing*. Oxford University Press.

Bach, E. (1986). Natural language metaphysics. In Marcus, R. B., Dorn, G., and Weingartner, P., editors, *Logic, Methodology, and Philosophy of Science, VII*, pages 573–95. North-Holland.

Bach, K. (1997). Do belief reports report beliefs? *Pacific Philosophical Quarterly*, 78:215–41.

Bach, K. (2005). Context ex machina. In Szabo, Z. G., editor, *Semantics versus Pragmatics*, pages 15–44. Oxford University Press.

Balaguer, M. (2013). Fictionalism in the philosophy of mathematics. In Zalta, E. N., editor, *The Stanford Encyclopedia of Philosophy*. Fall 2013 edition. <http://plato.stanford.edu/archives/sum2015/entries/fictionalism-mathematics/>.

Balcerak Jackson, B. (2013). Defusing easy arguments for numbers. *Linguistics and Philosophy*, 36(6):447–61.

Balcerak Jackson, B. (2014). What does displacement explain, and what do congruence effects show? A response to Hofweber (2014). *Linguistics and Philosophy*, 37(3):269–74.

Barlew, J. (2015). Focus on numbers. Unpublished manuscript.

Barwise, J. (1975). *Admissible Sets and Structures*. Springer Verlag.

Barwise, J. and Cooper, R. (1981). Generalized quantifiers and natural language. *Linguistics and Philosophy*, 4:159–219.

Båve, A. (2009). A deflationary theory of reference. *Synthese*, 169:51–73.

Beaver, D. and Clark, B. (2008). *Sense and Sensitivity*. Wiley-Blackwell.

Becker, J. P. and Selter, C. (1996). Elementary school practices. In Bishop, A. J., Clements, K., Keitel, C., Kilpatrick, J., and Laboide, C., editors, *International Handbook of Mathematics Education*, pages 511–64. Kluwer.

Benacerraf, P. (1965). What numbers could not be. *Philosophical Review*, 74:47–73.

Bennett, J. (1963). A note on Descartes and Spinoza. *Philosophical Review*, 74:379–80.

Bennett, K. (2009). Composition, coincidence, and metaontology. In Chalmers, D., Manley, D., and Wasserman, R., editors, *Metametaphysics: New Essays on the Foundations of Ontology*, pages 38–76. Oxford University Press.

Bennett, K. (2011). Construction area (no hard hat required). *Philosophical Studies*, 154:79–104.

Blackburn, S. (1994). Enchanting views. In Clark, P. and Hale, B., editors, *Reading Putnam*, pages 12–30. Blackwell.

Boër, S. E. and Lycan, W. G. (1986). *Knowing Who*. MIT Press.

Bonevac, D. (1985). Quantity and quantification. *Noûs*, 19(2):229–47.

Boolos, G. (1984). To be is to be the value of a variable, (or some values of some variables). *Journal of Philosophy*, 81:439–49.

Brogaard, B. (2007). Number words and ontological commitment. *Philosophical Quarterly*, 57(226):1–20.

Bueno, O. (2013). Nominalism in the philosophy of mathematics. In Zalta, E. N., editor, *The Stanford Encyclopedia of Philosophy*. Fall 2013 edition. <http://plato.stanford.edu/archives/spr2014/entries/nominalism-mathematics/>.

Burge, T. (1973). Reference and proper names. *Journal of Philosophy*, 70:425–39.

Burgess, J. (2004). Review of Jody Azzouni's Deflating Existential Consquence. *Bulletin of Symbolic Logic*, 10(4):573–7.

Burgess, J. (2008). *Mathematics, Models, and Modality: Selected Philosophical Essays*. Cambridge University Press.

Burgess, J. and Rosen, G. (1997). *A Subject with No Object*. Oxford University Press.

Büring, D. (1997). *The Meaning of Topic and Focus: The 59th Street Bridge Accent*. Routledge.

Byrne, A. (2005). Perception and perceptual content. In Sosa, E. and Steup, M., editors, *Contemporary Debates in Epistemology*, pages 231–50. Blackwell.

Cameron, R. (2018). Truthmakers. In Glanzberg, M., editor, *Oxford Handbook of Truth*. Oxford University Press.

Carey, S. (2009). *The Origin of Concepts*. Oxford University Press.

Carlson, G. N. and Pelletier, F. J., editors (1995). *The Generic Book*. University of Chicago Press.

Carnap, R. (1928). *Der logische Aufbau der Welt*. Weltkreis Verlag.

Carnap, R. (1937). *The Logical Syntax of Language*. Kegan Paul.

Carnap, R. (1956). Empiricism, semantics, and ontology. In *Meaning and Necessity*, pages 205–21. University of Chicago Press, 2nd edition.

Chalmers, D. (2009). Ontological anti-realism. In Chalmers, D., Manley, D., and Wasserman, R., editors, *Metametaphysics*, pages 77–129. Oxford University Press.

Chomsky, N. (1975). *Reflections on Language*. Pantheon Books.

Collins, J. (2002). The very idea of a science forming faculty. *dialectica*, 56(2):125–51.

Correia, F. (2005). *Existential Dependence and Cognate Notions*. Philosophia Verlag.

Correia, F. and Schnieder, B. (2012). Grounding: an opinionated introduction. In Correia, F. and Schnieder, B., editors, *Metaphysical Grounding: Understanding the Structure of Reality*, pages 1–36. Cambridge University Press.

Crimmins, M. (1998). Hesperus and phosperus: sense, pretense, and reference. *Philosophical Review*, 107(1):1–47.

Crisp, T. (2004). On presentism and triviality. *Oxford Studies in Metaphysics*, 1:15–20.

Cumming, S. (2008). Variablism. *Philosophical Review*, 117(4):525–54.

Dalrymple, M., Kanazawa, M., Kim, Y., Mchombo, S., and Peters, S. (1998). Reciprocal expressions and the concept of reciprocity. *Linguistics and Philosophy*, 21:159–210.

Daly, C. (2012). Skepticism about grounding. In Correia, F. and Schnieder, B., editors, *Metaphysical Grounding: Understanding the Structure of Reality*, pages 81–100. Cambridge University Press.

Daly, C. and Liggins, D. (2011). Deferentialism. *Philosophical Studies*, 156(3):321–37.

Davidson, D. (1968). On saying that. *Synthese*, 19:130–46.

Davidson, D. (1984). The very idea of a conceptual scheme. In *Inquiries into Truth and Interpretation*, pages 183–98. Oxford University Press.

Dehaene, S. (1997). *The Number Sense*. Oxford University Press.

Dever, J. (1998). *Variables*. PhD thesis, University of California, Berkeley.

Devitt, M. (1980). "ostrich nominalism" or "mirage realism." *Pacific Philosophical Quarterly*, 61:433–9.

Dodd, J. (2001). Is truth supervenient on being? *Proceedings of the Aristotelian Society*, 102(2):69–86.

Dorr, C. (2005). What we disagree about when we disagree about ontology. In Kalderon, M., editor, *Fictionalism in Metaphysics*, pages 234–86. Oxford University Press.

Dorr, C. (2008). There are no abstract objects. In Sider, T., Hawthorne, J., and Zimmerman, D. W., editors, *Contemporary Debates in Metaphysics*, pages 32–64. Blackwell.

Dretske, F. (1977). Laws of nature. *Philosophy of Science*, 44:248–68.

Dretske, F. (1981). *Knowledge and the Flow of Information*. MIT Press.

Dummett, M. (1973). *Frege: Philosophy of Language*. Duckworth.

Dummett, M. (1991). *Frege: Philosophy of Mathematics*. Harvard University Press.

Eddington, A. (1929). *The Nature of the Physical World*. Macmillan.

Einheuser, I. (2012). Is there a (meta-) problem of change? *Analytic Philosophy*, 53(4):344–51.

Eklund, M. (2002). Inconsistent languages. *Philosophy and Phenomenological Research*, 64(2):251–76.

Eklund, M. (2009). Carnap and ontological pluralism. In Chalmers, D. M. and Wasserman, R., editors, *Metametaphysics: New Essays on the Foundations of Ontology*, pages 130–56. Oxford University Press.

Eklund, M. (2011). Fictionalism. In Zalta, E. N., editor, *The Stanford Encyclopedia of Philosophy*. Fall 2011 edition. <http://plato.stanford.edu/archives/win2015/entries/fictionalism/>.

Eklund, M. (2016). Carnap's legacy for the contemporary metaontological debate. In Blatti, S. and Lapointe, S., editors, *Ontology after Carnap*, pages 165–89. Oxford University Press.

Elder, C. (2011). *Familiar Objects and Their Shadows*. Cambridge University Press.

Fara, D. G. (2001). Descriptions as predicates. *Philosophical Studies*, 102(1):1–42.

Fara, D. G. (2011). Call me "stupid," just don't call me stupid. *Analysis*, 71(3):492–501.

Fara, D. G. (2015). Names are predicates. *Philosophical Review*, 124(1):59–117.

Feferman, S. (1998). *In the Light of Logic*. Oxford University Press.

Felka, K. (2014). Number words and reference to numbers. *Philosophical Studies*, 168:261–82.

Field, H. (1989a). Platonism for cheap? Crispin Wright on Frege's context principle. In *Realism, Mathematics and Modality*, pages 147–70. Blackwell.

Field, H. (1989b). *Realism, Mathematics, and Modality*. Blackwell, Oxford.

Field, H. (2004). The consistency of the naive theory of properties. *Philosophical Quarterly*, 54(214):78–104.

Field, H. (2008). *Saving Truth From Paradox*. Oxford University Press.

Fine, K. (1994). Essence and modality. *Philosophical Perspectives*, 8:1–16.

Fine, K. (1999). Things and their parts. *Midwest Studies in Philosophy*, 23:61–74.

Fine, K. (2001). The question of realism. *Philosophers' Imprint*, 1(1):1–30.

Fine, K. (2005a). Our knowledge of mathematical objects. *Oxford Studies in Epistemology*, 1:89–110.

Fine, K. (2005b). Tense and reality. In *Modality and Tense*, pages 261–320. Oxford University Press.

Fine, K. (2009). The question of ontology. In Chalmers, D., Manley, D., and Wasserman, R., editors, *Metametaphysics: New Essays on the Foundations of Ontology*, pages 157–77. Oxford University Press.

Fine, K. (2012). Guide to ground. In Correia, F. and Schnieder, B., editors, *Metaphysical Grounding*, pages 37–80. Cambridge University Press.

Fodor, J. (1983). *The Modularity of Mind*. MIT Press.

Fodor, J. (1987). *Psychosemantics: The Problem of Meaning in the Philosophy of Mind*. MIT Press.

Fodor, J. (1990). Fodor's guide to mental representation. In *A Theory of Content and Other Essays*, pages 3–30. MIT Press.

Frank, M. C., Everett, D. L., Fedorenko, E., and Gibson, E. (2008). Number as a cognitive technology: evidence from Pirahã language and cognition. *Cognition*, 108:819–24.

Frege, G. (1884). *Die Grundlagen der Arithmetik: eine logisch mathematische Untersuchung über den Begriff der Zahl*. W. Koebner.

Frege, G. (1950). *The Foundations of Arithmetic*. Translation of Frege (1884). Blackwell.

Gamut, L. (1991). *Logic, Language, and Meaning*, volume II. University of Chicago Press.

Geurts, B. (2006). Take 'five': the meaning and use of number words. In Vogeleer, S. and Tasmowski, L., editors, *Non-Definiteness and Plurality*, pages 311–29. Benjamins.

Gilmore, C. (2006). Where in the relativistic world are we? *Philosophical Perspectives*, 20(1):199–236.

Godfrey-Smith, P. (2006). Theories and models in metaphysics. *Harvard Review of Philosophy*, 14:4–19.

Goodman, C. (2002). *Ancient Dharmas, Modern Debates: Towards an Analytic Philosophy of Buddhism*. PhD thesis, University of Michigan, Ann Arbor.

Goodman, C. (2005). Vaibhāṣika metaphoricalism. *Philosophy East and West*, 55(3):377–93.

Gottlieb, D. (1980). *Ontological Economy: Substitutional Quantification and Mathematics*. Oxford University Press.

Grice, P. (1989). *Studies in the Way of Words*. Harvard University Press.

Hale, B. (1987). *Abstract Objects*. Blackwell.

Hale, B. and Wright, C. (2001). *The Reason's Proper Study*. Oxford University Press.

Hardie, W. F. R. (1968). *Aristotle's Ethical Theory*. Clarendon Press.

Haslanger, S. (2003). Persistence through time. In Loux, M. and Zimmerman, D., editors, *Oxford Handbook of Metaphysics*, pages 314–54. Oxford University Press.

Hawley, K. (2001). *How Things Persist*. Oxford University Press.

Hawthorne, J. and Cortens, A. (1995). Towards ontological nihilism. *Philosophical Studies*, 79:143–65.

Hawthorne, J. and Manley, D. (2012). *The Reference Book*. Oxford University Press.

Heck, R. (2000). Cardinality, counting, and equinumerosity. *Notre Dame Journal of Formal Logic*, 41(3):187–209.

Hellman, G. (1989). *Mathematics without Numbers*. Oxford University Press.

Herburger, E. (2000). *Focus and Quantification*. MIT Press.

Hilbert, D. (1925). Über das Unendliche. *Mathematische Annalen*, 95:161–90.

Hirsch, E. (2011). *Quantifier Variance and Realism: Essays in Metaontology*. Oxford University Press.

Hodes, H. (1984). Logicism and the ontological commitments of arithmetic. *Journal of Philosophy*, 81:123–49.

Hodes, H. (1990). Where do the natural numbers come from? *Synthese*, 84:347–407.

Hofweber, T. (1999). *Ontology and Objectivity*. PhD thesis, Stanford University.

Hofweber, T. (2000a). Proof-theoretic reduction as a philosopher's tool. *Erkenntnis*, 53:127–46.

Hofweber, T. (2000b). Quantification and non-existent objects. In Everett, A. and Hofweber, T., editors, *Empty Names, Fiction, and the Puzzles of Non-Existence*, pages 249–73. CSLI Publications.

Hofweber, T. (2005a). Number determiners, numbers, and arithmetic. *The Philosophical Review*, 114(2):179–225.

Hofweber, T. (2005b). A puzzle about ontology. *Noûs*, 39:256–83.

Hofweber, T. (2006a). Inexpressible properties and propositions. In Zimmerman, D., editor, *Oxford Studies in Metaphysics*, volume 2, pages 155–206. Oxford University Press.

Hofweber, T. (2006b). Schiffer's new theory of propositions. *Philosophy and Phenomenological Research*, LXXIII(1):211–17.

Hofweber, T. (2007a). Innocent statements and their metaphysically loaded counterparts. *Philosophers' Imprint*, 7(1):1–33.

Hofweber, T. (2007b). Review of Jody Azzouni's Deflating Existential Consquence. *The Philosophical Review*, 116(3):465–7.

Hofweber, T. (2008). Validity, paradox, and the ideal of deductive logic. In Beall, J., editor, *Revenge of the Liar: New Essays on the Paradox*, pages 145–58. Oxford University Press.

Hofweber, T. (2009a). Ambitious, yet modest, metaphysics. In Chalmers, D., Manley, D., and Wasserman, R., editors, *Metametaphysics: New Essays on the Foundations of Ontology*, pages 269–89. Oxford University Press.

Hofweber, T. (2009b). Formal tools and the philosophy of mathematics. In Bueno, O. and Linnebo, O., editors, *New Waves in the Philosophy of Mathematics*, pages 197–219. Palgrave Macmillan.

Hofweber, T. (2009c). The meta-problem of change. *Noûs*, 43(2):286–314.

Hofweber, T. (2010). Review of John Burgess' Mathematics, Models, and Modality. *Notre Dame Philosophical Reviews*.

Hofweber, T. (2014a). Extraction, displacement and focus: a reply to Balcerak Jackson. *Linguistics and Philosophy*, 37(3):263–7.

Hofweber, T. (2014b). Rayo's The construction of logical space. *Inquiry*, 57(4):442–54.

Hofweber, T. (2015). The place of subjects in the metaphysics of material objects. *dialectica*, 69(4):473–90.

Hofweber, T. (2016a). Are there ineffable aspects of reality? In Bennett, K. and Zimmerman, D., editors, *Oxford Studies in Metaphysics*, volume 10. Oxford University Press.

Hofweber, T. (2016b). Carnap's big idea. In Blatti, S. and Lapointe, S., editors, *Ontology after Carnap*, pages 13–30. Oxford University Press.

Hofweber, T. (2016c). From remnants to things, and back again. In Ostertag, G., editor, *Meanings and Other Things: Essays in Honor of Stephen Schiffer*. Oxford University Press.

Hofweber, T. and Pelletier, J. (2005). Encuneral noun phrases. Unpublished manuscript, available at www.thomashofweber.com.

Hofweber, T. and Velleman, J. D. (2011). How to endure. *Philosophical Quarterly*, 61(242):37–57.

Hrbacek, K. and Jech, T. (1999). *Introduction to Set Theory*. CRC Press, 3rd edition.

Jackendoff, R. (1972). *Semantic Interpretation in Generative Grammar*. MIT Press.

Johnston, M. (1988). The end of the theory of meaning. *Mind and Language*, 3(1):28–42.

Kant, I. (1781). *Kritik der reinen Vernunft*. Johan Friedrich Hartnoch.

Katz, J. (1997). *Realistic Rationalism*. MIT Press.

Keenan, E. L. and Westerstahl, D. (1997). Generalized quantifiers in linguistics and logic. In van Benthem, J. and ter Meulen, A., editors, *Handbook of Logic and Language*, pages 837–93. MIT Press.

Keisler, H. J. (1971). *Model Theory for Infinitary Logic*. North-Holland.

Kemmerling, A. (2016). *Glauben: Essay über einen Begriff*. Klostermann Verlag, Frankfurt am Main.

Kim, J. (1993). Concepts of supervenience. In *Supervenience and Mind*, pages 53–78. Cambridge University Press.

King, J. (2002). Designating propositions. *Philosophical Review*, 111(3):341–71.

Koch, A. F. (1990). *Subjektivität in Raum und Zeit*. Klostermann.

Koch, A. F. (2010). Persons as mirroring the world. In O'Shea, J. and Rubinstein, E., editors, *Self, Language, and World*, pages 232–48. Ridgeview Publishing.

Korman, D. (2014). Debunking perceptual beliefs about ordinary objects. *Philosophers' Imprint*, 14(13):1–21.

Koslicki, K. (2007). Towards a neo-Aristotelian mereology. *dialectica*, 61(1):127–59.

Koslicki, K. (2012). Varieties of ontological dependence. In Correia, F. and Schnieder, B., editors, *Metaphysical Grounding: Understanding the Structure of Reality*, pages 186–213. Cambridge University Press.

Krakauer, B. (2012). *Counterpossibles*. PhD thesis, University of Massachusetts at Amherst.

Kraut, R. (2016). Three Carnaps on ontology. In Blatti, S. and Lapointe, S., editors, *Ontology after Carnap*, pages 31–58. Oxford University Press.

Krifka, M. (2004). The semantics of questions and the focusation of answers. In Lee, C., Gordon, M., and Büring, D., editors, *Topic and Focus: A Crosslinguistic Perspective*, pages 139–51. Kluwer Academic Publishers.

Kripke, S. (1976). Is there a problem about substitutional quantification? In Evans, G. and McDowell, J., editors, *Truth and Meaning*, pages 324–419. Clarendon Press.

Künne, W. (2003). *Conceptions of Truth*. Oxford University Press.

Leng, M. (2010). *Mathematics and Reality*. Oxford University Press.

Lewis, D. (1983a). New work for a theory of universals. *Australasian Journal of Philosophy*, 61:343–77.

Lewis, D. (1983b). Truth in fiction. In *Philosophical Papers*, volume 1, pages 261–80. Oxford University Press.

Lewis, D. (1991). *Parts of Classes*. Blackwell.

Lewis, D. (2001a). *Counterfactuals*. Wiley-Blackwell, 2nd edition.

Lewis, D. (2001b). Truthmaking and difference-making. *Nous*, 35(4):602–15.

Lewis, D. (2004). Tensed quantifiers. *Oxford Studies in Metaphysics*, 1:3–14.

Link, G. (1983). The logical analysis of plural and mass terms: a lattice-theoretic approach.

In Rainer Bäuerle, C. S. and von Stechow, A., editors, *Meaning, Use and the Interpretation of Language*, pages 303–23. Walter de Gruyter.

Link, G. (1998). *Algebraic Semantics in Language and Philosophy*. CSLI Publications.

Linnebo, O. and Nicholas, D. (2008). Superplurals in English. *Analysis*, 68(3):186–97.

Linsky, L. (1972). Two concepts of quantification. *Nous*, 6(3):224–39.

Lowe, E. J. (1998). *The Possibility of Metaphysics*. Oxford University Press.

Ludlow, P. (2004). Presentism, triviality, and the varieties of tensism. *Oxford Studies in Metaphysics*, 1:21–36.

Lycan, W. G. (1979). Semantic competence and funny functors. *Monist*, 62:209–22.

Marcus, R. B. (1962). Interpreting quantifiers. *Inquiry*, 5:252–9.

Marcus, R. B. (1993). Quantification and ontology. In *Modalities*, pages 75–88. Oxford University Press.

Matushansky, O. (2008). On the linguistic complexity of proper names. *Linguistics and Philosophy*, 21:573–627.

McDowell, J. (1994). *Mind and World*. Harvard University Press.

McNally, L. (1997). *A Semantics for the English Existential Construction*. Garland Press.

Melia, J. (2005). Truthmaking without truthmakers. In Beebee, H. and Dodd, J., editors, *Truthmakers: The Contemporary Debate*, pages 67–84. Oxford University Press.

Menninger, K. (1969). *Number Words and Number Symbols: A Cultural History of Numbers*. MIT Press.

Merricks, T. (2001). *Objects and Persons*. Oxford University Press.

Merricks, T. (2003). Replies. *Philosophy and Phenomenological Research*, LXVII(3):727–44.

Moltmann, F. (2003a). Nominalizing quantifiers. *Journal of Philosophical Logic*, 32(5):445–81.

Moltmann, F. (2003b). Propositional attitudes without propositions. *Synthese*, 35(1):77–118.

Moltmann, F. (2013a). *Abstract Objects and the Semantics of Natural Language*. Oxford University Press.

Moltmann, F. (2013b). Reference to numbers in natural language. *Philosophical Studies*, 162(3):499–536.

Montague, R. (1974). The proper treatment of quantification in ordinary English. In Thomason, R. H., editor, *Formal Philosophy*, pages 17–34. Yale University Press.

Mostowski, A. (1957). On a generalization of quantifiers. *Fundamenta Mathematicae*, 44:12–36.

Nagel, T. (1986). *The View from Nowhere*. Oxford University Press.

Neale, S. (1993). Term limits. *Philosophical Perspectives*, 7:89–124.

Nolan, D. (1997). Impossible worlds: a modest approach. *Notre Dame Journal of Formal Logic*, 38(4):535–72.

Nolt, J. (2014). Free logic. In Zalta, E. N., editor, *The Stanford Encyclopedia of Philosophy*. Summer 2014 edition. <http://plato.stanford.edu/archives/win2014/entries/logic-free/>.

Norton, J. D. (2008). Must evidence underdetermine theory? In Carrier, M., Howard, D., and Kourany, J., editors, *The Challenge of the Social and the Pressure of Practice: Science and Values Revisited*, pages 17–44. University of Pittsburgh Press.

Parsons, C. (1971). A plea for substitutional quantification. *Journal of Philosophy*, 68(8): 231–7.

Parsons, C. (1980a). Mathematical intuition. *Proceedings of the Aristotelian Society*, 80:145–68.

Parsons, J. (1999). There is no "truthmaker" argument against nominalism. *Australasian Journal of Philosophy*, 77(3):325–34.

Parsons, J. (2007). Theories of location. In Zimmerman, D., editor, *Oxford Studies in Metaphysics*, volume 3, pages 201–32. Oxford University Press.

Parsons, T. (1980b). *Nonexistent Objects*. Yale University Press.

Partee, B. (1987). Noun phrase interpretation and type-shifting principles. In Groenendjik, J., de Jongh, D., and Stokhof, M., editors, *Studies in Discourse Representation Theory and the Theory of Generalized Quantifiers*, pages 115–43. Foris Publications.

Partee, B. and Rooth, M. (1983). Generalized conjunction and type ambiguity. In Bäuerle, R., Schwarze, C., and von Stechow, A., editors, *Meaning, Use and the Interpretation of Language*, pages 361–93. Walter de Gruyter.

Paul, L. (2002). Logical parts. *Noûs*, 36(4):578–96.

Paul, L. (2012). Metaphysics as modeling: the handmaiden's tale. *Philosophical Studies*, 160(1):1–29.

Pelletier, J. (2011). Descriptive metaphysics, natural language metaphysics, Sapir-Whorf, and all that stuff: evidence from the mass-count distinction. *The Baltic International Yearbook of Cognition, Logic and Communication*, 6:1–46.

Piaget, J. (1952). *The Child's Conception of Number*. Psychology Press.

Pietroski, P. (2004). *Events and Semantic Architecture*. Oxford University Press.

Pietroski, P. (2007). Systematicity via monadicity. *Croatian Journal of Philosophy*, 21:343–74.

Pollock, J. (1974). *Knowledge and Justification*. Princeton University Press.

Pollock, J. (1986). *Contemporary Theories of Knowledge*. Rowman and Littlefield.

Priest, G. (2007). *Towards Non-Being: The Logic and Metaphysics of Intentionality*. Oxford University Press.

Prior, A. (1971). *Objects of Thought*. Clarendon Press.

Pryor, J. (2000). The skeptic and the dogmatist. *Noûs*, 34(4):517–49.

Pryor, J. (2007). Reasons and that-clauses. *Philosophical Issues*, 17(1):217–44.

Putnam, H. (1981). *Reason, Truth, and History*. Cambridge University Press.

Quine, W. V. O. (1960). *Word and Object*. MIT Press.

Quine, W. V. O. (1966). On Carnap's views on ontology. In *The Ways of Paradox and Other Essays*, pages 203–11. Harvard University Press.

Quine, W. V. O. (1969). Ontological relativity. In *Ontological Relativity and Other Essays*, pages 26–68. Columbia University Press.

Quine, W. V. O. (1980). On what there is. In *From a Logical Point of View*, pages 1–19. Harvard University Press.

Raven, M. (2011a). In defense of ground. *Australasian Journal of Philosophy*, 90(4):687–701.

Raven, M. (2011b). There is a problem of change. *Philosophical Studies*, 155(1):23–35.

Rayo, A. (2002). Frege's unofficial arithmetic. *Journal of Symbolic Logic*, 67(4):1623–38.

Rayo, A. (2013). *The Construction of Logical Space*. Oxford University Press.

Rayo, A. and Yablo, S. (2001). Nominalism through de-nominalization. *Noûs*, 35(1):74–92.

Recanati, F. (2001). What is said. *Synthese*, 128(1–2):75–91.

Resnik, M. D. (1997). *Mathematics as a Science of Patterns*. Oxford University Press.

Rochemont, M. S. and Culicover, P. W. (1990). *English Focus Constructions and the Theory of Grammar*. Cambridge University Press.

Rodriguez-Pereyra, G. (2000). What is the problem of universals? *Mind*, 109:255–73.

Rodriguez-Pereyra, G. (2005). Why truthmakers. In Beebee, H. and Dodd, J., editors, *Truthmakers: The Contemporary Debate*, pages 17–32. Oxford University Press.

Romero, M. (2005). Concealed questions and specificational subjects. *Linguistics and Philosophy*, 28(6):687–737.

Rooth, M. (1985). *Association with Focus*. PhD thesis, University of Massachusetts at Amherst.

Rosefeldt, T. (2008). That-clauses and non-nominal quantification. *Philosophical Studies*, 137:301–33.

Rosen, G. (1994). Objectivity and modern idealism: what is the question? In O'Leary-Hawthorne, J. and Michael, M., editors, *Metaphysics in Mind*, pages 277–319. Kluwer.

Rosen, G. (2010). Metaphysical dependence: grounding and reduction. In Hale, B. and Hoffmann, A., editors, *Modality*, pages 109–36. Oxford University Press.

Rosen, G. and Dorr, C. (2002). Composition as a fiction. In Gale, R., editor, *The Blackwell Guide to Metaphysics*, pages 151–74. Blackwell.

Rosenkranz, S. (2007). Agnosticism as a third stance. *Mind*, 116:55–104.

Russell, B. (1905). On denoting. *Mind*, 14:479–93.

Russell, B. (1913). *Theory of Knowledge*. Routledge.

Rychter, P. (2009). There is no puzzle about change. *dialectica*, 63(1):7–22.

Salmon, N. (1986). *Frege's Puzzle*. MIT Press.

Sattig, T. (2006). *The Language and Reality of Time*. Oxford University Press.

Saul, J. (1997). Substitution and simple sentences. *Analysis*, 57(2):102–8.

Saul, J. (2007). *Simple Sentences, Substitution, and Intuitions*. Oxford University Press.

Schaffer, J. (2009a). The deflationary metaontology of Thomasson's Ordinary Objects. *Philosophical Books*, 50(3):142–57.

Schaffer, J. (2009b). On what grounds what. In Chalmers, D., Manley, D., and Wasserman, R., editors, *Metametaphysics*, pages 347–83. Oxford University Press.

Schaffer, J. (2010). Monism: the priority of the whole. *Philosophical Review*, 119(1):31–76.

Schiffer, S. (1987a). The "Fido"-Fido theory of belief. *Philosophical Perspectives*, 1:455–80.

Schiffer, S. (1987b). *Remnants of Meaning*. MIT Press.

Schiffer, S. (1992). Belief ascription. *Journal of Philosophy*, 89(10):499–521.

Schiffer, S. (2003). *The Things We Mean*. Oxford University Press.

Schiffrin, D. (1987). *Discourse Markers*. Cambridge University Press.

Schlenker, P. (2003). Clausal equations (a note on the connectivity problem). *Natural Language and Linguistic Theory*, 21:157–214.

Schnieder, B. (2006a). Canonical property designators. *American Philosophical Quarterly*, 43(2):119–32.

Schnieder, B. (2006b). A certain kind of trinity: dependence, substance, explanation. *Philosophical Studies*, 129:396–419.

Schnieder, B. (2011). A logic for "because." *Review of Symbolic Logic*, 4:445–65.

Schoubye, A. (2017). Type-ambiguous names. *Mind*, 126 (503):715–67.

Schwartzkopff, R. (2015). *The Numbers of the Marketplace*. PhD thesis, Oxford University.

Shapiro, S. (1997). *Philosophy of Mathematics: Structure and Ontology*. Oxford University Press.

Sidelle, A. (2002). Is there a true metaphysics of material objects? *Philosophical Issues*, 12(1):118–45.

Sider, T. (2006). Bare particulars. *Philosophical Perspectives*, 20:387–97.

Sider, T. (2009). Ontological realism. In Chalmers, D., Manley, D., and Wasserman, R., editors, *Metametaphysics*, pages 384–423. Oxford University Press.

Sider, T. (2012). *Writing the Book of the World*. Oxford University Press.

Sider, T. (2013). Against parthood. In Bennett, K. and Zimmerman, D., editors, *Oxford Studies in Metaphysics*, volume 8, pages 237–93. Oxford University Press.

Stanley, J. (2000). Context and logical form. *Linguistics and Philosophy*, 23(4):391–434.

Strawson, P. (1964). A problem about truth: a reply to Mr. Warnock. In Pitcher, G., editor, *Truth*, pages 68–84. Prentice-Hall.

Tahko, T. E. (2012). In defence of Aristotelian metaphysics. In Tahko, T. E., editor, *Contemporary Aristotelian metaphysics*, pages 26–43. Cambridge University Press.

Thomasson, A. L. (1999). *Fiction and Metaphysics*. Cambridge University Press.

Thomasson, A. L. (2007). *Ordinary Objects*. Oxford University Press.

Thomasson, A. L. (2008). Existence questions. *Philosophical Studies*, 141:63–78.

Thomasson, A. L. (2015). *Ontology Made Easy*. Oxford University Press.

Turner, J. (2011). Ontological nihilism. *Oxford Studies in Metaphysics*, 6:3–54.

Unger, P. (1979). There are no ordinary things. *Synthese*, 41(2):117–54.

van Benthem, J. (1991). *Language in Action: Categories, Lambdas, and Dynamic Logic*. Elsevier Science.

van Cleve, J. (1994). Predication without universals? A fling with ostrich nominalism. *Philosophy and Phenomenological Research*, 54(3):577–90.

van der Does, J. (1995). Sums and quantifiers. *Linguistics and Philosophy*, 16:509–50.

van Fraassen, B. (2002). *The Empirical Stance*. Yale University Press.

van Inwagen, P. (1977). Creatures of fiction. *American Philosophical Quarterly*, 14(4):299–308.

van Inwagen, P. (1981). Why I don't understand substitutional quantification. *Philosophical Studies*, 39:281–5.

van Inwagen, P. (1990). *Material Beings*. Cornell University Press.

Varley, J. (1977). *The Ophiuchi Hotline*. The Dial Press.

Verschaffel, L. and Corte, E. D. (1996). Number and arithmetic. In Bishop, A. J., Clements, K., Keitel, C., Kilpatrick, J., and Laboide, C., editors, *International Handbook of Mathematics Education*, pages 99–137. Kluwer.

von Fintel, K. and Matthewson, L. (2008). Universals in semantics. *Linguistic Review*, 25:139–201.

von Stechow, A. (1991). Focusing and backgrounding operators. In Abraham, W., editor, *Discourse Particles*, pages 37–84. John Benjamins.

Wallace, J. (1971). Convention T and substitutional quantification. *Nous*, 5(2):199–211.

Walton, K. (1990). *Mimesis as Make-Believe*. Harvard University Press.

Walton, K. (1993). Metaphor and prop oriented make-believe. *European Journal of Philosophy*, 1(1):39–56.

White, R. (2006). Problems for dogmatism. *Philosophical Studies*, 131:525–57.

Williams, D. (1953). On the elements of being: I. *The Review of Metaphysics*, 7(1):3–18.

Williamson, T. (2008). *The Philosophy of Philosophy*. Blackwell.

Wilson, A. (2015). Grounding entails counterpossible non-triviality. Unpublished manuscript.

Wilson, J. M. (2014). No work for a theory of grounding. *Inquiry*, 57(5–6):535–79.

Wright, C. (1983). *Frege's Conception of Numbers as Objects*. Aberdeen University Press.

Wright, C. (2000). Neo-Fregean foundations for real analysis: some reflections on Frege's constraint. *Notre Dame Journal of Formal Logic*, 41(4):317–34.

Wynn, K. (1992). Addition and subtraction by human infants. *Nature*, 358:749–50.

Yablo, S. (1998). Does ontology rest on a mistake? *Proceedings of the Aristotelian Society*, Supp. Vol. 72:229–61.

Yablo, S. (2000). A paradox of existence. In Everett, A. and Hofweber, T., editors, *Empty Names, Fiction and the Puzzles of Non-Existence*, pages 275–312. CSLI Publications.

Yablo, S. (2005). The myth of the seven. In Kalderon, M., editor, *Fictionalism in Metaphysics*, pages 88–115. Oxford University Press.

Yablo, S. (2006). Non-catastrophic presupposition failure. In Thomson, J. and Byrne, A., editors, *Content and Modality: Themes from the Philosophy of Robert Stalnaker*, pages 164–90. Oxford University Press.

Yablo, S. (2014). *Aboutness*. Princeton University Press.

Zalta, E. N. (1999). Principia metaphysica. Available at <http://mally.stanford.edu/principia.pdf>.

Zalta, E. N. (2000). Neo-logicism? An ontological reduction of mathematics to metaphysics. *Erkenntnis*, 53(1–2):219–65.

Index

Printed and bound by CPI Group (UK) Ltd, Croydon, CR0 4YY